de Gruyter Series in Logic and Its Applications 2

1749
Walter de Gruyter
250
Berlin · New York
1999

Recursion Theory and Complexity

Proceedings of the Kazan '97 Workshop
Kazan, Russia, July 14–19, 1997

Editors

Marat M. Arslanov
Steffen Lempp

Walter de Gruyter
Berlin · New York 1999

Editors

Marat M. Arslanov
Department of Mathematics
Kazan State University
420008 Kazan
Russia

Steffen Lempp
Department of Mathematics
University of Wisconsin
480 Lincoln Drive
Madison, WI 53706-1388
USA

1991 Mathematics Subject Classification: 03-06; 03Dxx

Keywords: Computability, recursion, complexity

♾ Printed on acid-free paper which falls within the guidelines of the ANSI to ensure permanence and durability

Library of Congress – Cataloging-in-Publication Data

Kazan '97 Workshop (1997 : Kazan , Russia)
Recursion theory and complexity : proceedings of the Kazan '97 Workshop, Kazan, Russia, July 14–19, 1997 / editors Marat M. Arslanov, Steffen Lempp.
p. cm. – (De Gruyter series in logic and its applications, ISSN 1438–1893 ; v. 2)
"This volume contains the proceedings of the International Workshop 'Recursion Theory and Complexity Theory' (WORCT '97), which was held from July 14 to July 19, 1997 at Kazan State University, Kazan, Tatarstan, Russia."
ISBN 3-11-016587-2 (alk. paper)
1. Recursion theory Congresses. 2. Computational complexity Congresses. I. Arslanov, M. M. (Marat Mirzaevich) II. Lempp, Steffen, 1959– . III. Title. IV. Series: De Gruyter series in logic and its applications ; [2].
QA9.6.K39 1997
511.3'5–dc21 99-32783
CIP

Die Deutsche Bibliothek – Cataloging-in-Publication Data

Recursion theory and complexity : proceedings of the Kazan '97 workshop, Kazan, Russia, July 14–19, 1997 / ed. Marat M. Arslanov ; Steffen Lempp. – Berlin ; New York : de Gruyter, 1999
(De Gruyter series in logic and its applications ; 2)
ISBN 3-11-016587-2

ISSN 1438–1893

Printed in Germany.
Typesetting using the Authors' TEX files: I. Zimmermann, Freiburg – Printing and binding: WB-Druck GmbH & Co., Rieden/Allgäu – Cover design: Rainer Engel, Berlin.

Preface

This volume contains the proceedings of the International Workshop "Recursion Theory and Complexity Theory" (WORCT'97), which was held from July 14 to July 19, 1997 at Kazan State University, Kazan, Tatarstan, Russia.

The meeting was attended by more than 40 participants in the Recursion Theory part and about 20 participants in the Complexity Theory part from 9 different countries. The Workshop was organized by Kazan State University, the V. A. Steklov Institute of Mathematics (Moscow) and the N. G. Chebotarev Research Institute of Mathematics and Mechanics (NIIMM, Kazan). Financial support was provided by the Russia Foundation for Basic Research (RFBR) and the Tatar Academy of Sciences. The participants from the United States were supported by the National Science Foundation.

The preparation of the two main branches of the Workshop was coordinated by M. M. Arslanov and S. Lempp (Recursion Theory), and F. M. Ablayev and A. A. Razborov (Complexity Theory).

This volume reflects a part of the Recursion Theory branch of the actual program of the Workshop. The volume contains also the list of open problems in recursion theory mentioned at the open problem sessions.

M. Arslanov
S. Lempp

Table of contents

Priority method in generalized computability

Igor V. Ashaev

1. Introduction

This paper contains some research results on recursively enumerable Turing degrees in generalized computability (computability over arbitrary algebraic structures). The main result is a proof of two analogues of the Sacks Splitting Theorem for some classes of algebraic structures, including fields of the real, complex and p-adic numbers. The proof proceeds by the finite injury priority method based on using formulas of a language $L_{\omega_1\omega}$. It shows that such methods can be used in generalized computability in spite of dealing with nonconstructive structures. Our result applied to the structure $\langle \mathbb{N}, +, \times, 0, 1\rangle$ gives the Sacks Splitting Theorem for arithmetic, so it can be viewed as another variant of the proof of the classical Splitting Theorem.

We use a formalization of generalized computability from [ABM]. Briefly, it consists of three steps. Let $\mathcal{A} = \langle A, \sigma\rangle$ be a structure in a finite signature σ. In the first step, we introduce a machine over a structure $\mathcal{A}$. It is a modification of the Blum-Shub-Smale machine for the real numbers and ordered rings [BSS]. One can think of this machine as a program in a simple programming language which includes assumptions $x := t(x_1, \dots, x_n)$, where t is a term in the signature σ, conditional operators IF $\varphi(x_1, \dots, x_n)$ THEN GOTO n, where φ is an atomic formula, n is the number of a command, nonconditional jumps, and the stop operator. Variables are interpreted by elements of the structure $\mathcal{A}$.

In the second step, one has to consider a list superstructure over $\mathcal{A}$. It was introduced in [GS]. For a set M, let $F(M)$ be a set of all lists (finite sequences) over M. Now define

$$L_0 = A, \quad L_{n+1} = F(L_n) \cup L_n, \quad HL(A) = \bigcup_{n=0}^{\infty} L_n.$$

The elements of $HL(A)$ are called *inherently finite lists over* A. The elements of A are called *urelements*. Extend the signature σ to $\sigma_{HL} = \sigma \cup \{\text{nil}, \text{head}^{(1)}, \text{tail}^{(1)}, \text{cons}^{(2)}\}$. Here the constant nil means the empty list, head and tail are the operations of getting the first element of a list and the remaining elements, respectively. cons(α, β) adds a list β to α as the last element. All operations and predicates from σ are defined on $HL(A) \setminus A$ by nil. We obtain a structure $HL(\mathcal{A}) = \langle HL(A), \sigma_{HL}\rangle$ which is called an *inherent list superstructure over* $\mathcal{A}$.

Note that the list superstructure of any structure contains the set $HL(\emptyset)$ containing all lists generated by nil. We call it the *constructive part* of the superstructure. Denote $\sigma_{HL(\emptyset)} = \{\text{nil}, \text{head}^{(1)}, \text{tail}^{(1)}, \text{cons}^{(2)}\}$.

In the third step, we apply the machine to the list superstructure of $\mathcal{A}$ to obtain a notion of computable function $f : \Omega \longrightarrow HL(A)$, $\Omega \subseteq HL(A)^n$.

For more details see [ABM]. Note that performing only the first step is not sufficient: the class of computable functions will be too small (see examples in [ABM, KSU]). Our computability is equivalent to effective definitional schemes from [FM], machines with stacks and counters (see review [KSU]), BSS-computability [BSS] (over ordered rings and fields) and Turing computability (when $\mathcal{A} = \langle \mathbb{N}, +, \cdot, 0, 1 \rangle$).

Having a machine, one can define the usual notions of algorithm theory, such as computable functions, recursive and recursively enumerable (r. e.) sets, etc. The main feature here is the difference between two generalizations of the notion of an r. e. set. We shall call a set $X \subseteq HL(A)^n$ *recursively enumerable* iff it is a domain of a computable function. A set $X \subseteq HL(A)^n$ is an *output* iff it is the range of a computable function. These classes are different for arbitrary structures, but coincide, for example, for fields of the real and complex numbers and other structures with quantifier elimination (see [ABM]).

The natural numbers can be interpreted in $HL(\emptyset)$ as follows:

$$0 = \text{nil}, \quad n + 1 = \langle 0, 1, \dots, n \rangle = \text{cons}(n, n).$$

Denote the set of these lists by $\mathbb{N}_{HL}$. The set $\mathbb{N}_{HL}$ is recursive in any structure $HL(\mathcal{A})$ (see [ABM]).

There is a machine which, having an index of a term t or quantifier-free formula φ in the signature σ_{HL} and elements $\overline{\alpha} \in HL(A)^n$, computes the value $t(\overline{\alpha})$ or $\varphi(\overline{\alpha})$, so every quantifier-free definable set is recursive.

There is a description of r. e. sets by formulas of $L_{\omega_1 \omega}$ language:

Lemma 1.1. *Every r. e. set $X \subseteq HL(A)^n$ is definable in $HL(\mathcal{A})$ by a disjunction*

$$\bigvee_{i \in I} \varphi_i(\overline{x}), \tag{1}$$

where $\{\varphi_i \mid i \in I\}$ is a Turing r. e. set of quantifier-free pairwise inconsistent formulas in the signature σ_{HL}. Conversely, every Turing r. e. disjunction of quantifier-free formulas in the signature σ_{HL} defines an r. e. set.

Every r. e. set $X \subseteq A^n$ is definable in $HL(\mathcal{A})$ by disjunction (1) *where $\{\varphi_i \mid i \in I\}$ is a Turing r. e. set of quantifier-free pairwise inconsistent formulas in the signature σ.*

This is a modification of the Engeler Lemma for the list superstructure [ABM]. Output sets can be defined by r. e. disjunctions of $\exists$-formulas (see [ABM]).

For the definition of Turing reducibility (T-reducibility) one can introduce a machine with oracles as in classical algorithm theory. But in the general case there is a more convenient method from [BM, BMt]: *a set $X \subseteq HL(A)^m$ is T-reducible to $Y \subseteq HL(A)^n$ ($X \leq_T Y$) if X is recursive in the structure $\mathcal{A}_P = \langle \mathcal{A}, P^{(n)} \rangle$ where*

P is interpreted by the set Y. The *T-degree of a set X* is a class of sets Y, $X \leq_T Y$ and $Y \leq_T X$. The T-degree $\mathbf{a}$ is an *urelement* if it contains a set $X \subseteq A$, where $\mathbf{a}$ is *recursive (r. e.)* if it contains a recursive (r. e.) set.

From these definitions and Lemma 1.1 one can easily obtain the following result:

Lemma 1.2 (V. Belyaev). *For any $X, Y \subseteq HL(A)^n$, $X \leq_T Y$ iff X and $\overline{X}$ are definable by r. e. disjunctions $\bigvee_{i \in I} \varphi_i(\overline{x})$, where $\{\varphi_i \mid i \in I\}$ is a Turing r. e. set of quantifier-free pairwise inconsistent formulas of the following type:*

$$\varphi_i(\overline{x}) = \psi_i(\overline{x}) \ \& \ t_{i,1}(\overline{x}) \in Y \ \& \ \ldots \ \& \ t_{i,n_i}(\overline{x}) \in Y \ \&$$

$$\& \ t'_{i,1}(\overline{x}) \notin Y \ \& \ \ldots \ \& \ t'_{i,m_i}(\overline{x}) \notin Y.$$

One should understand expressions $t(\overline{x}) \in Y$ ($t'(\overline{x}) \notin Y$) as formulas $P(t(\overline{x}))$ ($\neg P(t'(\overline{x}))$) in the signature $\sigma_{HL} \cup \{P^{(1)}\}$ where P is interpreted by Y.

Assume that all quantifier-free formulas in the signatures used are enumerated effectively by natural numbers. The set $I \subseteq \mathbb{N}$ in the disjunction $\bigvee_{i \in I} \varphi_i(\overline{x})$ means a set of indices of φ_i.

Denote a list consisting of elements $\alpha_1, \ldots, \alpha_n$ as $\langle \alpha_1, \ldots, \alpha_n \rangle$, $\langle\ \rangle$ is the empty list.

The structure $\mathcal{N} = \langle \mathbb{N}, +, \cdot, 0, 1 \rangle$ is called arithmetic. We shall use the fields of the real, complex and p-adic numbers: $\mathcal{R} = \langle \mathbb{R}, +, \cdot, -, ^{-1}, 0, 1, \leq \rangle$, $\mathcal{C} = \langle \mathbb{C}, +, \cdot, -, ^{-1}, 0, 1 \rangle$, $\mathcal{Q}_p = \langle \mathbb{Q}_p, +, \cdot, -, ^{-1}, 0, 1, P \rangle$, where P is a valuation ring.

$\leq_T$ means Turing reducibility, $\equiv_T$ is Turing equivalence. The T-degree of a set A is denoted $d_T\, A$.

We use the notation $\overline{x}$ for a tuple $x_1, \ldots, x_n$ of variables or elements of a structure. For a tuple of terms $t_1, \ldots, t_n$ and a set X, we write $\overline{t}(x) \subseteq X$ ($\overline{t}(x) \not\subseteq X$) instead of $t_1(x) \in X \ \& \ \ldots \ \& \, t_n(x) \in X$ ($t_1(x) \notin X \ \& \ \ldots \ \& \, t_n(x) \notin X$).

2. Main result

First we need some useful functions over the list superstructure. Every list has two characteristics: *support* and *structure*. The support of α is a list of all the urelements that the list α is constructed with. A structure is a term in the signature $\sigma_{HL(\emptyset)}$ obtained from α by replacing all urelements by variables. More precisely, let $\mathrm{conc}(\alpha, \beta)$ the concatenation of α, β, $\mathrm{diff}(\alpha)$ be the result of deleting from α all repeated elements. Then

$\mathrm{supp}(\mathrm{nil}) = \mathrm{nil}$;
$\mathrm{supp}(a) = a, \ \ a \in A$;
$\mathrm{supp}(\langle \alpha_1, \ldots, \alpha_n \rangle) = \mathrm{diff}(\mathrm{conc}(\mathrm{supp}(\alpha_1), \ldots, \mathrm{conc}(\mathrm{supp}(\alpha_{n-1}), \mathrm{supp}(\alpha_n)) \ldots)$.

Let $T(\sigma_{HL(\emptyset)})$ be the set of all terms in the signature $\sigma_{HL(\emptyset)}$. Define a function $F : HL(A)^2 \longrightarrow T(\sigma_{HL(\emptyset)})$:

$$F(\langle a_1, \dots, a_n\rangle, \mathrm{nil}) = \mathrm{nil}, \qquad a_1, \dots, a_n \in A;$$
$$F(\langle a_1, \dots, a_n\rangle, a_i) = x_i, \qquad i = 1, \dots, n,$$
$$\begin{aligned} &F(\langle a_1, \dots, a_n\rangle, \langle \alpha_1, \dots, \alpha_k\rangle) \\ &\quad = \langle F(\langle a_1, \dots, a_n\rangle, \alpha_1), \dots, F(\langle a_1, \dots, a_n\rangle, \alpha_n)\rangle \\ &\quad = \mathrm{cons}(\dots \mathrm{cons}(\mathrm{nil}, F(\langle a_1, \dots, a_n\rangle, \alpha_1), \dots, F(\langle a_1, \dots, a_n\rangle, \alpha_k)). \end{aligned}$$

Then set $\mathrm{struct}(\alpha) = F(\mathrm{supp}(\alpha), \alpha)$.

Note that $\alpha = t(a_1, \dots, a_n)$ where $\mathrm{supp}(\alpha) = \langle a_1, \dots, a_n\rangle$, $\mathrm{struct}(\alpha) = t(x_1, \dots, x_n)$. An effective enumeration of the terms in the signature $\sigma_{HL(\emptyset)}$ induces an enumeration of all list structures. Denote it by ν. The proof of the following facts can found in [A1, A2].

Lemma 2.1. 1. *For any $k \in \mathbb{N}$ there are terms $s_j^{(k)}$ in the signature $\sigma_{HL(\emptyset)}$ such that $\forall \alpha\ k = \nu(\mathrm{struct}(\alpha)) \to \mathrm{supp}(\alpha) = \langle s_1^{(k)}(\alpha), \dots, s_n^{(k)}(\alpha)\rangle$. The indices of $s_j^{(k)}$ can be found effectively by k.*

2. *For any $k \in \mathbb{N}_{HL}$ the relation $R(x) = (\nu(\mathrm{struct}(x)) = k)$ is definable by a quantifier-free formula in the signature $\sigma_{HL(\emptyset)}$ whose index can be found effectively by k.*

3. *For any $k \in \mathbb{N}_{HL}$ and any quantifier-free formula $\varphi(x)$ in the signature σ_{HL} one can effectively construct a quantifier-free formula $\varphi^{(k)}(x_1, \dots, x_n)$ in the signature σ such that*

$$\forall \alpha\ \ \nu(\mathrm{struct}(\alpha)) = k,\ \mathrm{supp}(\alpha) = \langle a_1, \dots, a_n\rangle \longrightarrow HL(\mathcal{A}) \models \varphi(\alpha)$$
$$\Longleftrightarrow$$
$$\mathcal{A} \models \varphi^{(k)}(a_1, \dots, a_n).$$

The main result is the proof of two analogues of the Sacks Splitting Theorem for certain classes of algebraic structures, including the fields $\mathcal{R}$, $\mathcal{C}$, $\mathcal{Q}_p$ as well as arithmetic, hence the classical Sacks Theorem is a partial case. The Splitting Theorem for a structure $\mathcal{A}$ is the following statement.

Splitting Theorem. *Let $B, C \subseteq HL(A)$ be r. e. sets such that C is not recursive. Then there exist r. e. sets A_0, A_1 such that*

1) $A_0 \cup A_1 = B$, $A_0 \cap A_1 = \emptyset$;

2) $C \not\leq_T A_\xi$, $\xi = 0, 1$.

The proof is stated using the finite injury priority method in spite of the nonconstructive nature of the structures considered. It is possible because any r. e. set consists of countably many "pieces" (see Lemma 1.1). In fact, we construct r. e. disjunctions which define A_ξ instead of sets A_ξ.

A formula $\alpha(\overline{x})$ is called an *atom in a structure* $\mathcal{A}$ if $\mathcal{A} \models \exists\overline{x}\ \alpha(\overline{x})$ and there are pairwise inconsistent formulas β_1, β_2 such that $\mathcal{A} \models \exists\overline{x}\ (\alpha(\overline{x})\ \&\ \beta_i(\overline{x})), i = 1, 2$.

We are interested in algebraic structures satisfying the following condition.

Condition ($\star$). *There is an r. e. sequence of quantifier-free atom formulas* $\alpha_1, \alpha_2, \alpha_3, \ldots$ *in the signature* σ *such that for any quantifier-free formula* $\varphi(\overline{x})$ *such that* $\mathcal{A} \models \exists\overline{x}\ \varphi(\overline{x})$*, there exists an atom* $\alpha_i(\overline{x})$ *with the property* $\forall\overline{x}\ (\alpha_i(\overline{x}) \to \varphi(\overline{x}))$ *(or equivalently* $\exists\overline{x}\ (\alpha_i(\overline{x})\ \&\ \varphi(\overline{x}))$*). The truth of the relation* $\exists\overline{x}\ (\alpha_i(\overline{x})\ \&\ \varphi(\overline{x}))$ *can be determined effectively from indices of* α_i *and* φ.

If for an atom α and a formula φ, $\mathcal{A} \models \exists\overline{x}\ (\alpha(\overline{x})\ \&\ \varphi(\overline{x}))$, we say that α *is under* φ.

Proposition 2.1. *The structures* $\mathcal{R}$, $\mathcal{C}$, $\mathcal{Q}_p$ *and* $\mathcal{N}$ *satisfy the condition* ($\star$).

Proof. The atomic formulas in arithmetic are $\alpha(x_1, \ldots, x_n) = (x_1 = k_1)\ \&\ \ldots\ \&\ (x_n = k_n)$, where $k_1, \ldots, k_n \in \mathbb{N}$. The atomic formulas in the fields are $(f_1(x_1) = 0)\ \&\ \ldots\ \&\ (f_n(x_n) = 0)$, where $f_1, \ldots, f_n \in \mathbb{Q}[x]$ are irreducible over $\mathbb{Q}$. □

Lemma 2.2. *Let* $\alpha(\overline{x})$ *be an atom from* ($\star$)*, where* $t(\overline{x})$ *is a term. Then a formula* $\varphi(y) = \exists\overline{x}\ ((t(\overline{x}) = y)\ \&\ \alpha(\overline{x}))$ *is equivalent to an atom in* $\mathcal{A}$ *whose index can be found effectively from* α *and* t*. For any quantifier-free formula* ψ *one can effectively from* α *and* ψ *determine the truth value of the relation* $\exists\overline{x}\ \psi(\overline{x})\ \&\ \varphi(\overline{x})$.

Proof. Assume that there are pairwise inconsistent formulas θ_1 and θ_2 such that pairs φ, θ_1 and φ, θ_2 are consistent in $\mathcal{A}$. Then both formulas $\beta_i(\overline{x}) = \theta_i(t(\overline{x})))$ are inconsistent and split a set defined by α. This is a contradiction.

The relation $\exists\overline{x}\ \psi(\overline{x})\ \&\ \varphi(\overline{x})$ is equivalent in $\mathcal{A}$ to $\exists\overline{y}\ \alpha(\overline{y})\ \&\ \psi(t(\overline{y}))$. The last can be determined effectively from indices α and ψ since ($\star$). □

Lemma 2.3. *Let* $B \subseteq HL(A)$ *be an r. e. set,* $B^0 \subseteq B^1 \subseteq \cdots \subseteq B^n \subseteq \cdots$ *a sequence of r. e. sets and* $B = \bigcup_{n=0}^{\infty} B^n$*. Let also*

$$\theta(\overline{x}) = \chi(\overline{x})\ \&\ (\overline{t}^{(0)}(\overline{x}) \subseteq B)\ \&\ (\overline{t}^{(1)}(\overline{x}) \not\subseteq B),$$

$$\theta_s(\overline{x}) = \chi(\overline{x})\ \&\ (\overline{t}^{(0)}(\overline{x}) \subseteq B^s)\ \&\ (\overline{t}^{(1)}(\overline{x}) \not\subseteq B^s),$$

where $\chi(\overline{x})$ *is a quantifier-free formula,* $\overline{t}^{(0)} = (t_1^{(0)}, \ldots, t_n^{(0)})$ *and* $\overline{t}^{(1)} = (t_1^{(1)}, \ldots, t_m^{(1)})$ *are tuples of terms in the signature* σ_{HL}*. Then for any* $\alpha \in HL(\mathcal{A})^m$

$$\theta(\alpha) = 1 \iff \exists s_0\ \forall s \geq s_0\ \theta_s(\alpha) = 1,$$

$$\theta(\alpha) = 0 \iff \exists s_0\ \forall s \geq s_0\ \theta_s(\alpha) = 0.$$

Proof. The truth of $\theta(\overline{x})$ means that $\chi(\overline{x}) = 1$, $\{t_1^{(0)}(\overline{x}), \dots, t_n^{(0)}(\overline{x})\} \subseteq B$, and $\{t_1^{(1)}(\overline{x}), \dots, t_m^{(1)}(\overline{x})\} \subseteq \overline{B}$.

We have

$$\exists s_0\ \forall s \geq s_0\ \{t_1^{(0)}(\overline{x}), \dots, t_n^{(0)}(\overline{x})\} \subseteq B^s, \quad \forall s\ \{t_1^{(1)}(\overline{x}), \dots, t_m^{(1)}(\overline{x})\} \subseteq \overline{B}^s.$$

Thus

$$\exists s_0\ \forall s \geq s_0\ \theta_s(\overline{x}) = 1.$$

If $\theta(\overline{x}) = 0$ then either $\chi(\overline{x}) = 0$ or $\{t_1^{(0)}(\overline{x}), \dots, t_n^{(0)}(\overline{x})\} \not\subseteq B$ or $\{t_1^{(1)}(\overline{x}), \dots, t_m^{(1)}(\overline{x})\} \not\subseteq \overline{B}$. If $\chi(\overline{x}) \neq 0$ then we obtain

either $\forall s\ \{t_1^{(0)}(\overline{x}), \dots, t_n^{(0)}(\overline{x})\} \not\subseteq B^s$, or $\exists s_0\ \forall s \geq s_0\ \{t_1^{(1)}(\overline{x}), \dots, t_m^{(1)}(\overline{x})\} \not\subseteq \overline{B}^s$.

Therefore

$$\exists s_0\ \forall s \geq s_0\ \theta_s(\overline{x}) = 0.$$

Conversely, if $\exists s_0\ \forall s \geq s_0\ \theta_s(\overline{x}) = 1$, but $\theta(\overline{x}) = 0$ then $\exists s_0\ \forall s \geq s_0\ \theta_s(\overline{x}) = 0$ as proved, a contradiction. Similarly in the second case. □

Theorem 2.1. *If a structure $\mathcal{A}$ satisfies the condition $(\star)$ then its list superstructure $HL(\mathcal{A})$ also does.*

Proof. Consider a formula $\varphi(x)$ in the signature σ_{HL} which is not false in $HL(\mathcal{A})$. Then there is a list β, $\mathrm{supp}(\beta) = \langle b_1, \dots, b_n \rangle$, $\nu(\mathrm{struct}(\beta)) = k$ $\varphi(\beta) = 1$. By Lemma 2.1 there is a formula $\varphi^{(k)}(x_1, \dots, x_n)$ in the signature σ that $\mathcal{A} \models \varphi^{(k)}(b_1, \dots, b_n)$. Choose an atom $\alpha(x_1, \dots, x_n)$ under $\varphi^{(k)}$. Then $HL(\mathcal{A}) \models \exists x\ \alpha^\star(x) \,\&\, \varphi(x)$, where

$$\alpha^\star(x) = \alpha(s_1^{(k)}(x), \dots, s_n^{(k)}(x)) \,\&\, \nu(\mathrm{struct}(x)) = k).$$

The formula $\exists x\ (\alpha^\star(x) \,\&\, \varphi(x))$ is equivalent to

$$\exists x_1 \dots x_n\ \alpha(x_1, \dots, x_n) \,\&\, \varphi^{(k)}(x_1, \dots, x_n).$$

Hence $\alpha^\star(x)$ is as atom as is α.

Similarly, the question of the truth of $\exists x\ (\alpha^\star(x) \,\&\, \varphi(x))$ can be reduced to the case of the expression $\exists x_1 \dots x_n\ \alpha(x_1, \dots, x_n) \,\&\, \varphi^{(k)}(x_1, \dots, x_n)$, which can be decided effectively. □

Now we prove the main theorems.

Theorem 2.2. *Let $\mathcal{A}$ be an algebraic structure satisfying the condition $(\star)$ and having decidable $\exists$-theory. Then the Splitting Theorem is true in $\mathcal{A}$.*

Proof. Assume that B is not recursive. If not then $A_0 = \emptyset$, $A_1 = B$ is the required splitting.

By Lemma 1.1, the sets B and C are definable by r. e. disjunctions $\bigvee_{i\in I} \varphi_i$ and $\bigvee_{k\in K} \psi_k$, where $I, K \subseteq \mathbb{N}$ are Turing r. e. sets and φ_i (ψ_k) are pairwise inconsistent formulas in the signature σ_{HL}.

Construct the sets A_ξ in steps. Denote by A_ξ^s the part constructed in the s-th step. Actually, we construct r. e. disjunctions defining A_ξ instead of the sets A_ξ. This gives the recursive enumerability of A_ξ. These disjunctions will be constructed from the formulas φ_i, $i \in I$. The disjunctions for A_ξ^s will be finite, so the sets A_ξ^s will be recursive.

In each step, perform one of the following two actions:

1) choose a formula φ_i, $i \in I$ and add it to the disjunction for one of the set A_ξ, or

2) create a requirement for one of A_ξ.

A requirement for A_ξ is a finite set of atoms. It means that elements of $HL(\mathcal{A})$, satisfying these atoms, should not be placed in A_ξ. Every requirement has an index and an argument. No more than one requirement with index n (n-requirement) can exist in each step.

The condition $C \not\leq_T A_\xi$, $\xi = 0, 1$ means that $\overline{C}$ is not r. e. relative to the sets A_ξ, i.e., $\overline{C}$ is not definable by an r. e. disjunction $\bigvee_{j\in J} \theta_j$, where $\theta_j(x)$ are

$$\chi_j(x) \mathbin{\&} \bar{t}_j(x) \subseteq A_\xi \mathbin{\&} \bar{t}'_j(x) \not\subseteq A_\xi .$$

All such expressions are $L_{\omega_1\omega}$ formulas in the signature $\sigma_{HL} \cup \{P^{(1)}\}$. Enumerate effectively all these formulas. Let Φ_n be a disjunction with index n. Thus $\overline{C}$ is not r. e. relative to $A_\xi \iff$

$$\forall n \, \exists x \;\; (x \in C \Leftrightarrow \Phi_n(x) = 1).$$

Finally, we have two groups of conditions:

$$(1_n) \qquad \exists x \, \Big(\bigvee_{j\in J} \theta_j(x) \leftrightarrow \bigvee_{k\in K} \psi_k(x)\Big), \;\; \text{where } \Phi_n = \bigvee_{j\in J} \theta_j$$

$$(2_i) \qquad \{x \mid \varphi_i(x) = 1\} \subseteq A_0 \cup A_1$$

Let $\{R_s\}_{s\in\mathbb{N}}$ be an effective enumeration of these conditions containing every condition countably many times.

Now we describe the construction of A_ξ.

Step 0. Let $A_\xi^0 = \emptyset$, $\xi = 0, 1$.

Step $s + 1$. Consider a condition R_s.

1) R_s is (1_n), $\Phi_n = \bigvee_{j\in J} \theta_j$. Do nothing if there exists an n-requirement. Otherwise, using the decidability of the $\exists$-theory of $\mathcal{A}$ and Lemma 2.1, find a structure of a list

with the least index λ, $0 \leq \lambda \leq s$, such that the following condition is true:

$$\exists x_1 \ldots x_m \Big(\bigvee_{k \in K, k \leq s} \psi_k \ \& \ \bigvee_{j \in J, j \leq s} \theta_j^s \Big)^{(\lambda)} (\overline{x}). \qquad (2)$$

The number m is determined effectively from an index of λ. An expression (2) is a formula in the signature σ, therefore its truth value can be determined effectively. The condition (2) implies the following:

$$\exists x \ \nu(\mathrm{struct}(x)) \leq s \ \& \Big(\bigvee_{k \in K, k \leq s} \psi_k(x) \ \& \ \bigvee_{j \in J, j \leq s} \theta_j^s(x) \Big).$$

If there is no such λ then go to the next step. Otherwise, enumerating formulas from ($\star$), find the first atom $\alpha(x)$ under $\Big(\bigvee_{k \in K, k \leq s} \psi_k \ \& \ \bigvee_{j \in J, j \leq s} \theta_j^s \Big)(x)$.

Choose the least $j_0 \in J$, $j_0 \leq s$ that

$$\exists x \ \ \nu(\mathrm{struct}(x)) = \lambda \ \& \ \alpha(x) \ \& \ \theta_{j_0}^s(x) \ \& \ \bigvee_{k \in K, k \leq s} \psi_k(x).$$

Let $\theta_{j_0}(x) = \chi_{j_0}(x) \ \& \ \overline{t}_{j_0}(x) \subseteq A_\xi \ \& \ \overline{t}'_{j_0}(x) \nsubseteq A_\xi$, where $\overline{t}'_{j_0} = (t'_1, \ldots, t'_m)$.

According to Lemma 2.2, formulas $\exists y \ (\alpha(y) \ \& \ t'_l(y) = x)$, $l = 1, \ldots, m$, are equivalent to atoms. A set of these atoms is an n-requirement with argument α for A_ξ.

The sets A_ξ are not changed: $A_\xi^{s+1} = A_\xi^s$, $\xi = 0, 1$.

For a set M and a requirement τ, write $\tau \cap M = \emptyset$ if $\forall x \in M \ \forall \alpha \in \tau \ \neg\alpha(x)$ and $\tau \cap M \neq \emptyset$ otherwise. By Lemma 2.2, it can be determined effectively if M is definable by a quantifier-free formula. Note that if a requirement τ for A_ξ was created in step $s+1$ then $\tau \cap A_\xi^{s+1} = \emptyset$.

2) R_s is (2_i). Let $A_\xi^{s+1} = A_\xi^s$ and go to the next step if $\{x \mid \varphi_i(x) = 1\} \subseteq A_0 \cup A_1$. It can be determined effectively from the index of φ_i and s using finite disjunctions which define A_ξ.

Else consider all existing requirements (there are finitely many of them in each step) and find the n-requirement τ with the least n such that $\tau \cap \{x \mid \varphi_i(x)\} \neq \emptyset$. When such τ found and it is a requirement for the set A_ζ, let $A_{1-\zeta}^{s+1} = A_{1-\zeta}^s \cup \{x \mid \varphi_i(x) = 1\}$, $A_\zeta^{s+1} = A_\zeta^s$.

After extending the sets A_ξ, some requirement can be injured. More precisely, *a requirement τ for A_ξ is injured in a step* $s+1$ if

$$\tau \cap A_\xi^{s+1} \neq \emptyset, \ \text{but} \ \tau \cap A_\xi^s = \emptyset.$$

The $(s+1)$-th step is completed after deleting all injured requirements.

Having constructed all A_ξ^s, set $A_\xi = \bigcup_{s=0}^{\infty} A_\xi^s$, $\xi = 0, 1$.

Define the *injury set of the n-requirement:*

$$I_n = \{s \in \mathbb{N} \mid \text{the } n\text{-requirement is injured in step } s\}.$$

Note that if a requirement is injured by R_s then R_s is a condition of the type (2_i).

Lemma 2.4. *All injury sets I_n are finite.*

Proof. Induction on n. It is clear that $I_0 = \emptyset$. Assume that we prove the finiteness of the sets $I_0, \dots, I_{n-1}$. Consider I_n. Now $s \in I_n$ iff we consider a condition (2_i) in step s and there are n-requirement τ_n for A_ξ and m- requirement τ_m, $m < n$, for $A_{1-\xi}$ and

$$\tau_n \cap \{x \mid \varphi_i(x) = 1\} \neq \emptyset, \quad \tau_m \cap \{x \mid \varphi_i(x) = 1\} \neq \emptyset.$$

By our induction hypothesis, the requirements $\tau_0, \dots, \tau_{n-1}$ with indices $0, \dots, n-1$ are not changed in any step $s > s_0 = \max(I_0 \cup \dots \cup I_{n-1})$. But $\tau_0 \cup \dots \cup \tau_{n-1}$ is a finite set of atoms and the formulas φ_i, $i \in I$ are pairwise inconsistent; therefore $(\tau_0 \cup \dots \cup \tau_{n-1}) \cap \{x \mid \varphi_i(x) = 1\} = \emptyset$, starting with some step $s_1 \geq s_0$, so the n-requirement τ_n is not injured. □

Lemma 2.5. *Let* $s > \max I_n$, $\Phi_n = \bigvee_{j \in J} \theta_j$, *$R_s$ be a condition (1_n) and*

$$\exists x \ \nu(\mathrm{struct}(x)) \leq s \ \& \Big(\bigvee_{k \in K, k \leq s} \psi_k(x) \ \& \bigvee_{j \in J, j \leq s} \theta_j^s(x) \Big). \tag{3}$$

Then

$$\exists x \Big(\bigvee_{k \in K} \psi_k(x) \ \& \bigvee_{j \in J} \theta_j(x) \Big).$$

Proof. Consider step $s + 1$. If there are no n-requirements in this step, then such a requirement will be created since (3). In any case we have an active n-requirement τ with argument $\alpha(x)$ for A_ξ after proceeding the step $s + 1$.

By induction on t, prove that

$$\forall t \geq s \ \exists x \ \Big(\alpha(x) \ \& \bigvee_{k \in K, k \leq s} \psi_k(x) \ \& \bigvee_{j \in J, j \leq s} \theta_j^t(x) \Big). \tag{4}$$

The case $t = s$ follows from the lemma's condition and the algorithm. Assume that the statement (4) is proved for all $t' \leq t$. If the set A_ξ was not changed in step $t + 1$ then (4) is true. Otherwise since τ is not injured, $\tau \cap A_\xi^{t+1} = \emptyset$ and (4) is also true.

Now according to (4) and Lemma 2.3, $\exists x \ (\bigvee_{k \in K} \psi_k(x) \ \& \bigvee_{j \in J} \theta_j(x))$. □

Lemma 2.6. *The set $\overline{C}$ is not defined by any formula Φ_n.*

Proof. Assume that disjunction $\Phi_n = \bigvee_{j\in J} \theta_j$ defines $\overline{C}$. Consider the relation

$$\Theta(x) = \bigvee_{s\in\mathbb{N}_{HL}} \Big(\Gamma(s,x) \,\&\, \bigvee_{j\in J, j\le s} \theta_j^s(x)\Big),$$

where

$$\Gamma(s,x) = 1$$

$$\Longleftrightarrow$$

$$(s \in \mathbb{N}_{HL}) \,\&\, (R_s \text{has type } (1_n)) \,\&\, (s > \max I_n) \,\&\, (\nu(\text{struct}(x)) \le s).$$

The relation $\Gamma(s,x)$ is recursive, so $\Theta(x)$ defines an r. e. set.

Prove that $\Theta(x)$ defines $\overline{C}$, then C must be recursive (an analogue of Post's Theorem is true [ABM]), contradicting the lemma's hypotheses. We have $\forall x\ (\Phi_n(x) = 1 \to \Theta(x) = 1)$. Indeed, if for some x $\Phi_n(x) = 1$, then there is $j_0 \in J\ \theta_{j_0}(x) = 1$. By Lemma 2.3 $\exists s_0\ \forall s \ge s_0\ \theta^s_{j_0}(x) = 1$. Since for any x $\{s \mid \Gamma(s,x)\}$ is infinite, $\Theta(x) = 1$. Conversely, assume $\exists x\ (\Phi_n(x) = 0 \,\&\, \Theta(x) = 1)$, i.e., $\exists x\ (x \in C \,\&\, \Theta(x))$ or equivalently that

$$\exists x\Big(\bigvee_{k\in K} \psi_k(x) \ \&\ \Theta(x)\Big).$$

This means that for some x and sufficiently large s, satisfying $\Gamma(s,x)$, the following condition is true:

$$\bigvee_{k\in K, k\le s} \psi_k(x) \ \&\ \bigvee_{j\in J, j\le s} \theta_j^s(x).$$

Since $\Gamma(s,x)$ is true then $\nu(\text{struct}(x)) \le s$. By the previous lemma,

$$\exists x\Big(\bigvee_{k\in K} \psi_k(x) \,\&\, \bigvee_{j\in J} \theta_j(x)\Big),$$

that is $C \cap \overline{C} \ne \emptyset$. This contradiction proves the lemma. □

Thus $C \not\le_T A_\xi$, $\xi = 0, 1$. The conditions $A_0 \cup A_1 = B$, $A_0 \cap A_1 = \emptyset$ are true by our construction. The proof is complete. □

Since the function mapping every set $A \subseteq \mathbb{N}$ to itself induces an embedding of the semilattice of the r. e. T-degrees over arithmetic to the semilattice of the r. e. T-degrees over $\mathcal{R}$ which preserves $\le_T$ and $\mathbf{0}$ [A2], Theorem 2.2 implies the classical Sacks Theorem. Indeed, the Sacks Splitting Theorem can be deduced more easily from Theorem 2.3 below. But this Splitting Theorem for $\mathcal{R}$ cannot be proved more easily in spite of the decidability of its theory.

We say that a set of formulas $\{\alpha_0, \alpha_1, \dots,\}$ is *dense in a structure* $\mathcal{A} = \langle A, \sigma\rangle$, if for any $a_1, \dots, a_n \in A$ there is a formula $\alpha_j(x_1, \dots, x_n)$ that $\mathcal{A} \models \alpha_j(\overline{a})$.

Theorem 2.3. *Let $\mathcal{A}$ be an algebraic structure satisfying the condition ($\star$) and the set of atoms from ($\star$) is dense in $\mathcal{A}$. Then the Splitting Theorem is true in $\mathcal{A}$.*

Proof. According to Theorem 2.1, a structure $HL(\mathcal{A})$ also satisfies the condition ($\star$). Let $\alpha_0, \alpha_1, \ldots$ be the dense sequence of atoms from ($\star$) for $\mathcal{A}$ and $\alpha_0^\star, \alpha_1^\star, \ldots$ be the sequence of atoms from ($\star$) for $HL(\mathcal{A})$. We prove that $\{\alpha_j^\star\}$ is a dense set in $HL(\mathcal{A})$. Consider a list $\beta \in HL(A)$ with $\nu(\text{struct}(\beta)) = k$, $\text{supp}(\beta) = \langle b_1, \ldots, b_n\rangle$. Find an atom $\alpha(\overline{x})$ such that $\mathcal{A} \models \alpha(\overline{b})$. The elements of support are values of terms in the signature $\sigma_{HL(\emptyset)}$ on β: $b_j = s_j^{(k)}(\beta)$, $j = 1, \ldots, n$. Then $HL(\mathcal{A}) \models \alpha^\star(\beta)$, where $\alpha^\star(x) = \alpha(s_1^{(k)}(x), \ldots, s_n^{(k)}(x))$ & $(\nu(\text{struct}(x) = k)$, according to Theorem 2.1 this is an atom from ($\star$) in $HL(\mathcal{A})$.

The rest of the proof is similar to that of Theorem 2.2 except for the following minor changes. The algorithm of constructing A_ξ, $\xi = 0, 1$ will be as follows:

Step 0. Let $A_\xi^0 = \emptyset$, $\xi = 0, 1$.

Step $s + 1$. Choose R_s.

1) R_s is (1_n), $\Phi_n = \bigvee_{j\in J} \theta_j$. Go to the next step if there exists an n-requirement. Otherwise, look at the first s atoms from ($\star$) and find an atom α with least index such that

$$\exists x \ \alpha(x) \ \& \left(\bigvee_{k\in K, k\le s} \psi_k \ \& \bigvee_{j\in J, j\le s} \theta_j^s \right)(x).$$

According to ($\star$), this can be determined effectively. If such α exists then create an n-requirement as in Theorem 2.2, otherwise do nothing.

The sets A_ξ are not changed: $A_\xi^{s+1} = A_\xi^s$, $\xi = 0, 1$.

2) R_s is (2_i). As in Theorem 2.2, extend the sets A_ξ^s, taking into account existing requirements.

Similarly to Lemmas 2.4 and 2.5, we can prove the following statements:

All injury sets I_n are finite.

Let $s > \max I_n$, $\Phi_n = \bigvee_{j\in J} \theta_j$, R_s be a condition (1_n) and

$$\exists x \left(\bigvee_{k\in K, k\le s} \psi_k(x) \ \& \bigvee_{j\in J, j\le s} \theta_j^s(x) \right) \& \ \alpha_j(x),$$

for $1 \le j \le s$. Then

$$\exists x \left(\bigvee_{k\in K} \psi_k(x) \ \& \bigvee_{j\in J} \theta_j(x) \right).$$

The final result follows from the following lemma:

Lemma 2.7. *The set $\overline{C}$ is not defined by any formula Φ_n.*

Proof. Assume that $\overline{C}$ is defined by $\Phi_n = \bigvee_{j \in J} \theta_j$. Consider the relation

$$\Theta(x) = \bigvee_{s \in \mathbb{N}_{HL}} \Big(\Gamma(s,x) \;\&\; \bigvee_{j \in J, j \leq s} \theta_j^s(x) \Big),$$

where

$$\Gamma(s,x) = 1 \Leftrightarrow$$

$$(s \in \mathbb{N}_{HL}) \;\&\; (R_s \text{has type } (1_n)) \;\&\; (s > \max I_n) \;\&\; \exists j \leq s \; \alpha_j(x) = 1.$$

Again Γ is recursive, $\Theta(x)$ defines $\overline{C}$, so C is recursive, contradicting the lemma's hypotheses. □

The proof of Theorem 2.3 is complete. □

The following statements can be proved from the Splitting Theorem as in the case of arithmetic. The notation $\mathbf{a}|\mathbf{b}$ means that degrees $\mathbf{a}$ and $\mathbf{b}$ are incomparable.

Corollary 2.1. *Under the conditions of Theorems* 2.2 *or* 2.3,

1) *for any r. e. T-degree* $\mathbf{c} \neq \mathbf{0}$ *there are r. e.* $\mathbf{a}$ *and* $\mathbf{b}$, *such that* $\mathbf{0} < \mathbf{a}, \mathbf{b} < \mathbf{c}$ *and* $\mathbf{a}|\mathbf{b}$;

2) *for any r. e.* $\mathbf{0} < \mathbf{a} < \mathbf{0}'$ *there is r. e. degree* $\mathbf{b}$ *such that* $\mathbf{a}|\mathbf{b}$;

3) *there is no minimal r. e. Turing degree.*

References

[ABM] I. V. Ashaev, V. Y. Belyaev, A. G. Myasnikov, Approaches to the theory of generalized computability, Algebra i Logika 32 (4) (1993), 349–386; Algebra and Logic 32 (4) (1993), 183–205.

[A1] I. V. Ashaev, Splitting theorem in generalized computability, Prepr. 96-01, Omsk: Omsk State Univ., 1996., 24 pp. (in Russian).

[A2] I. V. Ashaev, Algorithmic computability over arbitrary algebraic structures, PhD dissertation, Omsk, 1996, 72 pp. (in Russian).

[BM] V. Y. Belyaev, A. G. Myasnikov, On hierarchy theorems in generalized computability, to appear.

[BMt] V. Y. Belyaev, A. G. Myasnikov, Hierarchy theorems in generalized recursion theory, Proc. III Internat. Conf. on Algebra, Krasnoyarsk, 1993, 42–43 (in Russian).

[BSS] L. Blum, M. Shub, S. Smale, On a theory of computation and complexity over the real numbers: NP-completeness, recursive functions and universal machines, Bull. Amer. Math. Soc. 212 (1989), 1–46.

[GS] S. S. Goncharov, D. I. Sviridenko, Σ-programming, Vychisl. Sistemy 107 (1985), 3–29; Amer. Math. Soc. Transl. Ser. II 142 (1989), 101–121.

[FM] H. Friedman, R. Mansfield, Algorithmic Procedures, Trans. Amer. Math. Soc. 332 (1992), 297–312.

[KSU] A. J. Kfoury, A. P. Stolboushkin, P. Uzhichin, Some open questions in theory of program schemes and dynamic logics, Uspekhi Mat. Nauk. 44 (1) (1989), 35–55; Russ. Math. Surveys 44 (1) (1989), 43–68.

Polynomial-time versus computable Boolean algebras

Douglas Cenzer and Jeffrey B. Remmel

The main goal of this paper is to study when a computable Boolean algebra $\mathcal{B}$ is computably isomorphic to a polynomial-time Boolean algebra over either the binary representation of the natural numbers or the tally representation of the natural numbers. Here a polynomial-time structure consists of a polynomial-time subset A of $\{0, 1\}^*$ together with polynomial-time relations and polynomial-time functions. More formally, let $\omega = \{0, 1, \dots\}$ denote the set of natural numbers. Let $[,]$ denote the usual quadratic-time pairing function $[m, n] = m + \frac{1}{2}(m + n)(m + n + 1)$, which maps $\omega \times \omega$ onto ω. Let $\phi_{e,n}$ denote the n-ary partial function on $(\{0, 1\}^*)^n$ computed by the e-th Turing machine. Then we say that a structure

$$\mathcal{A} = \left(A, \{R_i^{\mathcal{A}}\}_{i\in S}, \{f_i^{\mathcal{A}}\}_{i\in T}, \{c_i^{\mathcal{A}}\}_{i\in U}\right),$$

(where the universe A of $\mathcal{A}$ is a subset of $\{0, 1\}^*$) is *computable* if A is a computable subset of $\{0, 1\}^*$, S, T, and U are initial segments of ω, the set of relations $\{R_i^{\mathcal{A}}\}_{i\in S}$ is uniformly computable in the sense that there is a computable function G such that for all $i \in S$, $G(i) = [n_i, e_i]$ where $R_i^{\mathcal{A}}$ is an n_i-ary relation and ϕ_{e_i,n_i} computes the characteristic function of $R_i^{\mathcal{A}}$, the set of functions $\{f_i^{\mathcal{A}}\}_{i\in T}$ is uniformly computable in the sense that there is a computable function F such that for all $i \in T$, $F(i) = [n_i, e_i]$ where $f_i^{\mathcal{A}}$ is an n_i-ary function and ϕ_{e_i,n_i} restricted to A^{n_i} computes $f_i^{\mathcal{A}}$, and there is a computable function interpreting the constant symbols in the sense that there is a computable function H such that for all $i \in U$, $H(i) = c_i^{\mathcal{A}}$. Note that if $\mathcal{A}$ is a computable structure, then the atomic diagram of $\mathcal{A}$ is computable. We say that a computable structure $\mathcal{A} = (A, \{R_i^{\mathcal{A}}\}_{i\in S}, \{f_i^{\mathcal{A}}\}_{i\in T}, \{c_i^{\mathcal{A}}\}_{i\in U})$, is *polynomial-time* if A is a polynomial-time subset of $\{0, 1\}^*$ and the set of relations $\{R_i^{\mathcal{A}}\}_{i\in S}$ and the set of functions $\{f_i^{\mathcal{A}}\}_{i\in T}$ are uniformly polynomial-time in the sense that, in addition to the functions G and F defined above, there are computable functions G' and F' such that for $i \in S$, $G'(i) = m_i$ where for all $(x_1, \dots, x_{n_i})$ in $(\{0, 1\}^*)^{n_i}$, it takes at most $(\max\{2, |x_1|, \dots, |x_{n_i}|\})^{m_i}$ steps to compute $\phi_{e_i,n_i}(x_1, \dots, x_{n_i})$ and for all $i \in T$, $F'(i) = q_i$ where for all $(x_1, \dots, x_{n_i})$ in $(\{0, 1\}^*)^{n_i}$, it takes at most $(\max\{2, |x_1|, \dots, |x_{n_i}|\})^{q_i}$ steps to compute $\phi_{e_i,n_i}(x_1, \dots, x_{n_i})$. Note that if $\mathcal{A}$ is a polynomial-time structure with infinitely many relation symbols or with infinitely many function symbols, then our definition of a polynomial-time structure does not ensure that the atomic diagram of $\mathcal{A}$ is polynomial-time. Thus we say $\mathcal{A}$ is *uniformly polynomial-time* if the atomic diagram of $\mathcal{A}$ is polynomial-time. Note that the fact that $\mathcal{A}$ is uniformly polynomial-time implies, among other things, that the sequence of run times $\{x^{m_i} : i \in S\}$ and $\{x^{q_i} : i \in T\}$ are bounded by some fixed polynomial. Of course, if $\mathcal{A}$ is a structure over a finite language, then $\mathcal{A}$ is a polynomial-time structure if and only if $\mathcal{A}$ is a uniformly polynomial-time structure. Similar definitions may be given for other resource-bounded classes.

In the study of polynomial-time model theory and algebra, the fundamental problems for a class C of computable structures include the classification of those structures which are either

isomorphic to a polynomial-time structure;

computably isomorphic to a polynomial-time structure;

isomorphic to a polynomial-time structure with a specified universe such as the binary or tally representation of the natural numbers;

computably isomorphic to a polynomial-time structure with specified universe.

For example, in [2] it was shown that every torsion Abelian group G is computably isomorphic to a polynomial-time group and that if there is a finite bound on the order of the elements of G, then the universe may be specified as either the standard binary representation $\mathrm{Bin}(\omega)$ or the tally representation $\mathrm{Tal}(\omega) = \{1^n : n \in \omega\}$, of the natural numbers. In contrast, computable torsion groups are constructed which are not even isomorphic to any polynomial-time groups with universe $\mathrm{Bin}(\omega)$ or $\mathrm{Tal}(\omega)$. The recent paper [8] gives general conditions under which various structures, including graphs, trees and equivalence structures, will be computably isomorphic to polynomial-time structures with standard universe. The forthcoming [10] gives a similar analysis of torsion-free Abelian groups.

It is a theorem of Cenzer and Remmel [1] that every computable Boolean algebra is computably isomorphic to a polynomial-time Boolean algebra. This result easily follows from three basic facts. First it is well known that every computable Boolean algebra is computably isomorphic to the interval algebra of some computable linear ordering with a least element, see [20]. Next Grigorieff [11] proved that every computable linear ordering is computably isomorphic to a quasi-real-time linear ordering. Finally it can be checked that the interval algebra of a polynomial-time linear ordering is computably isomorphic to a polynomial-time Boolean algebra. However the Cenzer and Remmel result does not specify the universe of the resulting polynomial-time Boolean algebra. In this paper, we consider a more restricted kind of existence question, i.e. whether a given computable Boolean algebra is computably isomorphic to a polynomial time model which has a standard universe such $\mathrm{Bin}(\omega)$ or $\mathrm{Tal}(\omega)$.

A similar question has been studied for computable linear orderings and the results in that case are related to the main results of this paper.

First Grigorieff [11] proved that every computable linear ordering is isomorphic to a quasi-real-time linear ordering which has universe $\mathrm{Bin}(\omega)$. However Grigorieff's result can not be improved to the result that every computable linear ordering is *computably isomorphic* to a linear-time linear ordering over Bin(ω). That is, Cenzer and Remmel [1] showed that for any infinite polynomial-time set $A \subseteq \{0,1\}^*$, there exists a computable copy of the linear ordering $\omega + \omega^*$ which is not computably isomorphic to any polynomial-time linear ordering which has universe A. Here $\omega+\omega^*$ is the ordering obtained by taking a copy of $\omega = \{0, 1, 2, \dots\}$ under the usual ordering followed by a copy of the negative integers under the usual ordering. In fact, Grigorieff proved a bit more, namely he showed that every computable linear ordering $\mathcal{L} = (L, <_L)$ for which either

1. $\mathcal{L}$ has a computable $<_L$ increasing sequence which is either cofinal in $\mathcal{L}$ or has a limit in $\mathcal{L}$ or

2. $\mathcal{L}$ has a computable $<_L$ decreasing sequence which is either co-initial in $\mathcal{L}$ or has a limit in $\mathcal{L}$,

is computably isomorphic to a linear-time linear ordering with universe Bin(ω). One can then show that any computable linear ordering which does not have either property (1) or property (2) must have order type $\omega + \lambda \cdot Z + \omega^*$ where Z is the order type of the integers and λ is a countable linear ordering. Remmel [21] showed that for any computable linear ordering $\mathcal{L}$ which is of order type $\omega+\lambda \cdot Z+\omega^*$ for some countable linear ordering λ, there exists a computable linear ordering $\mathcal{M}$ which is isomorphic to $\mathcal{L}$ such that $\mathcal{M}$ is not computably isomorphic to any polynomial-time linear ordering with universe Bin(ω) or Tal(ω).

There are three main results in this paper. First we prove that if $\mathcal{B}$ is a countably infinite computable Boolean algebra such that both the set of atoms of $\mathcal{B}$ and the ideal generated by the atoms of $\mathcal{B}$ are computable, then $\mathcal{B}$ is computably isomorphic to a polynomial-time Boolean algebra with universe Bin(ω) and to a polynomial-time Boolean algebra with universe Tal(ω). Second, we prove that if $\mathcal{B}$ is a countably infinite computable Boolean algebra such that the set of atoms of $\mathcal{B}$ is computable, then $\mathcal{B}$ is computably isomorphic to a polynomial-time Boolean algebra with universe Bin(ω) and to a polynomial-time Boolean algebra with universe Tal(ω) unless $\mathcal{B}$ is isomorphic to the Boolean algebra generated by the closed intervals of the rationals. We show that this second result is best possible by constructing a computable Boolean algebra $\mathcal{B}$ which is isomorphic to the Boolean algebra generated by the closed intervals of the rationals such that set of atoms of $\mathcal{B}$ is computable, the ideal generated by the atoms of $\mathcal{B}$ is not computable, and $\mathcal{B}$ is not computably isomorphic to any polynomial-time Boolean algebra with universe Bin(ω) or with universe Tal(ω). Finally we prove that if $\mathcal{B}$ is a countably infinite computable Boolean algebra, then $\mathcal{B}$ is computably isomorphic to a polynomial-time Boolean algebra with universe Bin(ω) and to a polynomial-time Boolean algebra with universe Tal(ω) unless $\mathcal{B}$ is atomic. Again we prove that our third result is best possible by proving that for any countably infinite computable atomic Boolean algebra $\mathcal{B}$, there is a computable Boolean algebra $\mathcal{C}$ which is isomorphic to $\mathcal{B}$ but which is not computably isomorphic to any polynomial-time Boolean algebra with universe Bin(ω) or with universe Tal(ω).

The outline of this paper is as follows. In section one, we shall provide the basic definitions of complexity theory. In section 2, we shall state some basic lemmas due to the authors on when polynomial-time structures are polynomial-time isomorphic to a polynomial-time structure with universe Bin(ω) or Tal(ω). These lemmas will be needed to prove our positive results. In section three, we shall give some basic results on computable and polynomial-time Boolean algebras. In particular, we shall introduce polynomial-time representations of three specific Boolean algebras which will play a key role in our main results, namely, the Boolean algebra of finite and cofinite subsets of the integers, $\tilde{\mathcal{N}}$, the countable atomless Boolean algebra, $\tilde{\mathfrak{Q}}$, and the Boolean algebra generated by the closed intervals of the rationals, $\tilde{\mathfrak{C}}$. Finally in section four, we shall prove our main results.

1. Preliminaries

In this section, we will give the basic definitions from complexity theory which will be needed for the rest of the paper.

Let Σ be a finite alphabet. Then Σ^* denotes the set of finite strings of letters from Σ and Σ^ω denotes the set of infinite strings of letters from Σ where $\omega = \{0, 1, 2, \dots\}$ is the set of natural numbers. For any natural number $n \neq 0$, $\text{tal}(n) = 1^n$ is the tally representation of n and $\text{bin}(n) = i_0 i_1 \dots i_e \in \{0,1\}^*$ is the (reverse) binary representation of n if $n = i_0 + 2 \cdot i_1 + \cdots + 2^e \cdot i_e$ and $i_e \neq 0$. In general, the k-ary representation $b_k(n) = i_0 i_1 \dots i_e$ if $n = i_0 + i_1 \cdot k + \cdots + i_e \cdot k^e$ and $i_e \neq 0$. We let $\text{tal}(0) = \text{bin}(0) = b_k(0) = 0$. Then we let $\text{Tal}(\omega) = \{\text{tal}(n) : n \in \omega\}$, $\text{Bin}(\omega) = \{\text{bin}(n) : n \in \omega\}$ and, for each $k \geq 3$, $B_k(\omega) = \{b_k(n) : n \in \omega\}$. Occasionally, we will want to say that $B_2(\omega) = \text{Bin}(\omega)$ and that $B_1(\omega) = \text{Tal}(\omega)$.

For a string $\sigma = (\sigma(0), \sigma(1), \dots, \sigma(n-1))$, $|\sigma|$ denotes the length n of σ. The empty string has length 0 and will be denoted by $\emptyset$. A constant string σ of length n will be denoted by k^n. For $m < |\sigma|$, $\sigma\lceil m$ is the string $(\sigma(0), \dots, \sigma(m-1))$ and we say σ is an *initial segment* of τ (written $\sigma \prec \tau$) if $\sigma = \tau\lceil m$ for some m. The *concatenation* $\sigma^\frown\tau$ (or sometimes just $\sigma\tau$) is defined by

$$\sigma^\frown\tau = (\sigma(0), \sigma(1), \dots, \sigma(m-1), \tau(0), \tau(1), \dots, \tau(n-1)),$$

where $|\sigma| = m$ and $|\tau| = n$; in particular we write $\sigma^\frown a$ for $\sigma \frown (a)$ and $a^\frown\sigma$ for $(a)^\frown\sigma$.

Our basic computation model is the standard multi-tape Turing machine of Hopcroft and Ullman [14]. Note that there are different heads on each tape and that the heads are allowed to move independently. This implies that a string σ can be copied in linear time.

Let $t(n)$ be a function on natural numbers. A Turing machine M is said to be $t(n)$ *time bounded* if each computation of M on inputs of length n where $n \geq 2$ requires at most $t(n)$ steps. A function $f(x)$ on strings is said to be in DTIME(t) if there is a $t(n)$-time bounded deterministic Turing machine M which computes $f(x)$. For a function f of several variables, we let the length of $(x_1, \ldots, x_n)$ be $|x_1| + \cdots + |x_n|$. A set of strings or a relation on strings is in DTIME(t) if its characteristic function is in DTIME(t). Similar definitions can be given for *space-bounded* computation and DSPACE(t).

We define the standard notions of feasibility as follows. We say that a function $f(x)$ is quasi-real-time computable if it is in DTIME$(n + c)$ for some c. We say that $f(x)$ is linear-time (space) computable if it is in DTIME(cn) (DSPACE(cn)) for some $c \geq 0$; the class of linear-time computable functions is denoted by L. We say that $f(x)$ is polynomial-time (space) computable if it is in DTIME$(n^c + c)$ (DSPACE$(n^c + c)$) for some $c \geq 0$; the class of functions computable in polynomial time (space) is denoted by P (PSPACE). We say that $f(x)$ is exponential-time (space) computable if it is in DTIME(2^{cn}) (DSPACE$(2^{cn})$for some $c \geq 0$; finally, $f(x)$ is EXPTIME (EXPSPACE) computable if it is in DTIME(2^{n^c}) (DSPACE(2^{n^c})) for some $c \geq 0$;

Observe that for a function $f(x_1, \ldots, x_k)$ of several variables, the above definitions are equivalent, if we declare the size of the input $(x_1, \ldots, x_k)$ to be the maximum of the sizes $|x_1|, \ldots, |x_k|$ since we allow multiple tapes. This occasionally simplifies the computation of the complexity of various functions.

We refer the reader to Odifreddi [17] for the basic definitions of computability theory. Let $\phi_{i,n}$ be the partial computable function of n variables computed by the ith Turing machine M_i. If $n = 1$, we will write ϕ_i instead of $\phi_{i,1}$. Given a string $\sigma \in \{0, 1\}^*$, we write $\phi_{i,s}(\sigma)\downarrow$ if M_i gives an output in s or fewer steps when started on input string σ. Thus the function $\phi_{i,s}$ is uniformly polynomial-time. We write $\phi_e(\sigma)\downarrow$ if $(\exists s)(\phi_{e,s}(\sigma)\downarrow)$ and $\phi_e(\sigma)\uparrow$ if not $\phi_e(\sigma)\downarrow$.

Definition 1.1. For any complexity class Γ and any structures

$$\mathcal{A} = \left(A, \{R_i^A\}_{i \in S}, \{f_i^A\}_{i \in T}, \{c_i\}_{i \in U}\right),$$

and

$$\mathcal{B} = \left(B, \{R_i^B\}_{i \in S}, \{f_i^B\}_{i \in T}, \{c_i\}_{i \in U}\right),$$

we say that $\mathcal{A}$ and $\mathcal{B}$ are Γ-*isomorphic* if there is an isomorphism f from $\mathcal{A}$ onto $\mathcal{B}$ and Γ-computable functions F and G such that $f = F \lceil A$ (the restriction of F to A) and $f^{-1} = G \lceil B$.

2. Polynomial-time sets and isomorphisms

In this section we shall give a number of useful lemmas about the relations between the various standard universes that we will consider in our study of feasible structures. The basic type of standard universe is a set Σ^* where Σ is a finite alphabet and in particular where $\Sigma = \{0, 1\}$, $\{1\}$, or $\{0\}$. Other standard universes include the set Tal(ω) of tally representations of natural numbers, the set $\mathrm{Bin}(\omega)$ of binary representations of natural numbers, and, for any k, the set $B_k(\omega)$ of k-ary representations of natural numbers. In computability theory, all of these sets are computably isomorphic and therefore interchangeable. For our purposes, we must consider carefully which of these isomorphisms are polynomial-time or even polynomial-time in one direction.

First we need to explicitly define a polynomial-time pairing function. For any finite alphabet Σ, there is a natural embedding ρ of Σ^* into $\mathrm{Bin}(\omega)$ given as follows. We may suppose that $\Sigma \subset \{0, 1, 2, \ldots, n\}$ for some n. Let $\rho(\emptyset) = 0$ and, for $\sigma = (i_1, \ldots, i_k)$, let

$$\rho(\sigma) = 0^{i_1}\frown 1\frown 0^{i_2}\frown 1\frown \cdots \frown 0^{i_k}\frown 1.$$

The function ρ is actually an isomorphism from ω^* onto $\mathrm{Bin}(\omega)$ and has an inverse ρ^{-1}. It is also clear that, for each n, the set $\rho(\{0, 1, \ldots, n\}^*)$ is linear-time (uniformly in n). Thus we can normally assume that an arbitrary structure has universe a subset of $\mathrm{Bin}(\omega)$.

The coding function $\langle\sigma_1, \sigma_2, \ldots, \sigma_k\rangle_k$ for $\sigma_1, \ldots, \sigma_k \in \{0, 1\}^*$ is now defined by

$$\langle\sigma_1, \sigma_2, \ldots, \sigma_k\rangle_k = \rho(\sigma_1\frown 2\frown \sigma_2\frown 2\frown \cdots\frown \sigma_{k-1}\frown 2\frown \sigma_k).$$

Let $Q_k = \{\langle\sigma_1, \sigma_2, \ldots, \sigma_k\rangle_k : \sigma_i \in \{0, 1\}^* \text{ for each i}\}$. For $i = 1, \ldots k$, the projection functions π_i^k from Q_k onto $\{0, 1\}^*$ are implicitly defined by the equation

$$\sigma = \langle\pi_1^k(\sigma), \pi_2^k(\sigma), \ldots, \pi_k^k(\sigma)\rangle_k.$$

The subscript k will normally be omitted. It is easy to see that the sets Q_k and $B_k(\omega)$ are all linear-time and that the functions π_i^k, and $\langle\ ,\ldots,\ \rangle_k$ are all computable in linear time.

Given two subsets A and B of $\{0, 1\}^*$, define

$$A \otimes B = \{\langle a, b\rangle : a \in A, b \in B\}$$

and

$$A \oplus B = \{\langle 0, a\rangle : a \in A\} \cup \{\langle 1, b\rangle : b \in B\}.$$

It is clear that if A and B are p-time, then both $A \oplus B$ and $A \otimes B$ will also be p-time.

Now, for each $k \geq 2$, a natural number in (reverse) k-ary form is simply a string $\sigma \in \{0, 1, \ldots, k-1\}^*$ which is either 0 or else ends with an element of $\{1, \ldots, k-1\}$. Thus the set $B_k(\omega)$ of k-ary representations of natural numbers is a linear-time subset of $\{0, 1, \ldots, k-1\}^*$. $\mathrm{Tal}(\omega) = \{0\} \cup \{1\}^*1$ is of course a linear-time subset of $\{0, 1\}^*$, but is not polynomial-time isomorphic to the whole set. For a string $\sigma = b_k(n) \in B_k(\omega)$ with $n > 0$, let $\mu_k(\sigma)$ be the unary representation, 1^n, of n and let $\mu_k(0) = 0$. We now state a sequence of lemmas which will be useful for our main results. Most of these lemmas are proved in [2, 5, 9].

The first lemma is stated for polynomial-time and exponential-time structures and isomorphisms but applies to other notions of complexity as well.

Lemma 2.1. *Suppose that $\mathcal{A}$ is a polynomial-time (exponential-time) structure with universe A and that ϕ is a polynomial-time (exponential-time) set isomorphism from A onto a set B. Then $\mathcal{B}$ is a polynomial-time (exponential-time) structure, where the functions and relations on B are defined to make ϕ an isomorphism of the structures.*

Next we state two lemmas which relate tally representation of a structure to its binary representation and, more generally, to its k-ary representation.

Lemma 2.2. *For each $k > 1$,*

(a) $\mathrm{Bin}(\omega)$ *is linear-time isomorphic to* $\{0, 1\}^*$.

(b) *there is a linear-time function p such that, for all n, both the computation of $\mu_k(b_k(n)) = 1^n$ and the inverse computation of $\mu_k^{-1}(1^n) = b_k(n)$ can be computed in time $p(n)$;*

For any subset M of ω, let $\mathrm{tal}(M) = \{\mathrm{tal}(n) : n \in M\}$, let $\mathrm{bin}(M) = \{\mathrm{bin}(n) : n \in M\}$, and for any finite $k > 1$, let $b_k(M) = \{b_k(n) : n \in M\}$.

For any structure $\mathcal{M}$ with universe $M \subseteq \omega$, let $\mathrm{tal}(\mathcal{M})$ be the tally representation of $\mathcal{M}$ with universe $\mathrm{tal}(\omega)$ and relations and functions defined so that the mapping taking n to $\mathrm{tal}(n)$ is an isomorphism from $\mathcal{M}$ onto $\mathrm{tal}(\mathcal{M})$; $\mathrm{bin}(\mathcal{M})$ and $b_k(\mathcal{M})$ are similarly defined. Of course, $\mathrm{Tal}(\omega)$ and $\mathrm{Bin}(\omega)$ are not p-time isomorphic, but the isomorphism μ_k may be used to compare the complexity of isomorphic structures with universes $\mathrm{Tal}(\omega)$ and $\mathrm{Bin}(\omega)$

Lemma 2.3. *Let $k \geq 2$, let $\mathcal{M}$ be a structure with universe $M \subseteq \omega$ and let $\mathcal{A} = \mathrm{tal}(\mathcal{M})$ and $\mathcal{B} = b_k(\mathcal{M})$. Then*

(a) *If $\mathcal{A}$ is p-time, then $\mathcal{B}$ is exponential-time.*

(b) *If $\mathcal{B}$ is exponential-time, then $\mathcal{A}$ is* EXPTIME.

(c) *If $\mathcal{B}$ is* EXPTIME *and there is a fixed constant c such that, for all functions $f^{\mathcal{M}}$, $f^{\mathcal{M}}(n_1, \ldots, n_k) \leq 2^{c(n_1+\cdots+n_k)}$ for all but finitely many k-tuples, then* $\mathrm{tal}(\mathcal{M})$ *is exponential-time.*

(d) *If $\mathcal{B}$ is exponential-time and, for all functions $f^{\mathcal{M}}$, $f^{\mathcal{M}}(n_1, \ldots, n_k) \leq (n_1 + \cdots + n_k)^c$ for some fixed constant c and all but finitely many k-tuples, then $\mathcal{A}$ is p-time.*

(e) *If $\mathcal{B}$ is polynomial-time and, for all functions $f^{\mathcal{M}}$, $f^{\mathcal{M}}(n_1, \ldots, n_k) \leq c(n_1 + \cdots + n_k)$ for some fixed constant c and all but finitely many k-tuples, then $\mathcal{A}$ is linear-time.*

We will frequently want to combine structures using disjoint unions and direct sums. The following lemma, which is due to Cenzer and Remmel [2], will be used repeated for our positive results.

Lemma 2.4. (a) *Let A be a p-time subset of* $\mathrm{Tal}(\omega)$. *Then* $A \oplus \mathrm{Tal}(\omega)$ *is p-time isomorphic to* $\mathrm{Tal}(\omega)$ *and* $A \oplus \mathrm{Bin}(\omega)$ *is p-time isomorphic to* $\mathrm{Bin}(\omega)$.

(b) *Let A be a nonempty p-time subset of* $\mathrm{Tal}(\omega)$. *Then* $A \otimes \mathrm{Tal}(\omega)$ *is p-time isomorphic to* $\mathrm{Tal}(\omega)$ *and* $A \otimes \mathrm{Bin}(\omega)$ *is p-time isomorphic to* $\mathrm{Bin}(\omega)$.

3. Computable Boolean algebras

Let $\tilde{\mathcal{B}}$ be a Boolean algebra. Let $0_{\tilde{\mathcal{B}}}$ and $1_{\tilde{\mathcal{B}}}$ denote the zero and unit of $\tilde{\mathcal{B}}$ respectively. We say $x \in \tilde{\mathcal{B}}$ is nontrivial if $x \in \tilde{\mathcal{B}} - \{0_{\tilde{\mathcal{B}}}, 1_{\tilde{\mathcal{B}}}\}$. We let $\preceq_{\tilde{\mathcal{B}}}$ denote the order relation on $\tilde{\mathcal{B}}$, that is , if $x, y \in \tilde{\mathcal{B}}$, then $x \preceq_{\tilde{\mathcal{B}}} y$ if and only if $x \wedge_{\tilde{\mathcal{B}}} y = x$. (We shall often drop the subscripts on $0_{\tilde{\mathcal{B}}}$, $1_{\tilde{\mathcal{B}}}$, $\wedge_{\tilde{\mathcal{B}}}$, $\vee_{\tilde{\mathcal{B}}}$, $\neg_{\tilde{\mathcal{B}}}$, and $\preceq_{\tilde{\mathcal{B}}}$ where there is no confusion.) We say $x \in \tilde{\mathcal{B}}$ is an atom if $x \neq 0$ and for all $y \in \tilde{\mathcal{B}}$ such that $y \preceq_{\tilde{\mathcal{B}}} x$ either $y = 0$ or $y = x$ and we say $x \in \tilde{\mathcal{B}}$ is atomless if $x \neq 0$ and there is no atom y of $\tilde{\mathcal{B}}$ such that $y \preceq_{\tilde{\mathcal{B}}} x$. $\tilde{\mathcal{B}}$ is called atomic if $\tilde{\mathcal{B}}$ contains no atomless elements. $\tilde{\mathcal{B}}$ is called atomless if $\tilde{\mathcal{B}}$ contains no atoms. We let $\mathrm{At}(\tilde{\mathcal{B}})$ denote the set of atoms in $\tilde{\mathcal{B}}$. If B is a subalgebra of $\tilde{\mathcal{B}}$, the atoms relative to B, $\mathrm{At}(B)$, are not necessarily atoms relative to $\tilde{\mathcal{B}}$. If $S \subseteq \tilde{\mathcal{B}}$, we let $\mathrm{sub}_{\tilde{\mathcal{B}}}(S)$ denote the subalgebra of $\tilde{\mathcal{B}}$ generated by S and $I(S)$ denote the ideal generated by S. If $S = \{z\}$ is a singleton, we write $[z]$ for $I(\{z\})$.

The basic definitions and terminologies on computable Boolean algebras are from [20, 18]. For a comprehensive discussion of computable Boolean algebras, see the recent book [13] of S. Goncharov. Let

$$\tilde{\mathcal{B}}_1 = (B_1, \vee_{\tilde{\mathcal{B}}_1}, \wedge_{\tilde{\mathcal{B}}_1}, \neg_{\tilde{\mathcal{B}}_1}, 0_{\tilde{\mathcal{B}}_1}, 1_{\tilde{\mathcal{B}}_1})$$

and

$$\tilde{\mathcal{B}}_2 = (B_2, \vee_{\tilde{\mathcal{B}}_2}, \wedge_{\tilde{\mathcal{B}}_2}, \neg_{\tilde{\mathcal{B}}_2}, 0_{\tilde{\mathcal{B}}_2}, 1_{\tilde{\mathcal{B}}_2}).$$

We write $\tilde{\mathcal{B}}_1 \cong \tilde{\mathcal{B}}_2$ if $\tilde{\mathcal{B}}_1$ and $\tilde{\mathcal{B}}_2$ are isomorphic and we write $\tilde{\mathcal{B}}_1 \cong_c \tilde{\mathcal{B}}_2$ if $\tilde{\mathcal{B}}_1$ and $\tilde{\mathcal{B}}_2$ are computable Boolean algebras which are computable isomorphic. The product of $\tilde{\mathcal{B}}_1$ and $\tilde{\mathcal{B}}_2$, $\tilde{\mathcal{B}}_1 \times \tilde{\mathcal{B}}_2$, is the set of all pairs (x_1, x_2) with $x_1 \in \tilde{\mathcal{B}}_1$ and $x_2 \in \tilde{\mathcal{B}}_2$ where the operations of $\wedge$, $\vee$, and $\neg$ are taken componentwise. If $\tilde{\mathcal{B}}_1$ and $\tilde{\mathcal{B}}_2$ are computable Boolean algebras where $B_1, B_2 \subseteq \{0, 1\}^*$, then by $\tilde{\mathcal{B}}_1 \times \tilde{\mathcal{B}}_2$ we mean the computable presentation of the product of $\tilde{\mathcal{B}}_1$ and $\tilde{\mathcal{B}}_2$ induced by the pairing function $\langle , \rangle$ defined in the previous section. That is, we let

$$|\tilde{\mathcal{B}}_1 \times \tilde{\mathcal{B}}_2| = \{\langle x_1, x_2 \rangle) : x_1 \in |\tilde{\mathcal{B}}_1| \ \& \ x_2 \in |\tilde{\mathcal{B}}_2|\}$$

and $\wedge_{\tilde{\mathcal{B}}_1 \times \tilde{\mathcal{B}}_2}$, $\vee_{\tilde{\mathcal{B}}_1 \times \tilde{\mathcal{B}}_2}$, and $\neg_{\tilde{\mathcal{B}}_1 \times \tilde{\mathcal{B}}_2}$ are the computable operations obtained by using the computable operations of $\tilde{\mathcal{B}}_1$ and $\tilde{\mathcal{B}}_2$ componentwise.

If $\tilde{\mathcal{B}}$ is a computable Boolean algebra, then the order relation on $\tilde{\mathcal{B}}$, $\preceq_{\tilde{\mathcal{B}}}$, is computable. Thus if $z \in \tilde{\mathcal{B}}$, then we may define a computable Boolean algebra $[z] = [z]_{\tilde{\mathcal{B}}}$

with universe

$$|[z]| = \{x \in |\tilde{\mathcal{B}}| : x \preceq_{\tilde{\mathcal{B}}} z\}$$

and operations $\wedge_{[z]}$ and $\vee_{[z]}$ which are just the restrictions of $\wedge_{\tilde{\mathcal{B}}}$ and $\vee_{\tilde{\mathcal{B}}}$ respectively and

$$\neg_{[z]}x = z - x = z \wedge_{\tilde{\mathcal{B}}} (\neg_{\tilde{\mathcal{B}}} x).$$

Moreover, if $\tilde{\mathcal{B}}$ is a computable Boolean algebra, then $\tilde{\mathcal{B}}$ is computably isomorphic to $[z] \times [\neg z]$ under the isomorphism which sends $x \in |\tilde{\mathcal{B}}|$ to $\langle x \wedge z, x \wedge \neg z \rangle$.

Next we introduce specific polynomial-time representations of three basic Boolean algebras which will be used in the proofs of our main results.

Let $\tilde{\mathcal{N}}$ be the Boolean algebra of finite and cofinite subsets of the set of natural numbers ω. We introduce two specific representations of $\tilde{\mathcal{N}}$ which we call the *binary* and *tally* representation of $\tilde{\mathcal{N}}$. That is, we shall define two maps $F_{\text{Bin}} : \tilde{\mathcal{N}} \to \text{Bin}(\omega)$ and $F_{\text{Tal}} : \tilde{\mathcal{N}} \to \text{Tal}(\omega)$ and let $\text{Bin}(\tilde{\mathcal{N}})$ denote the Boolean algebra with universe $\text{Bin}(\omega)$ whose operations are defined so that F_{Bin} is an isomorphism and $\text{Tal}(\tilde{\mathcal{N}})$ denote the Boolean algebra with universe $\text{Tal}(\omega)$ whose operations are defined so that F_{Tal} is an isomorphism.

First let $F_{\text{Bin}}(0_{\tilde{\mathcal{N}}}) = 0$ and $F_{\text{Bin}}(1_{\tilde{\mathcal{N}}}) = 1$.

Then for any finite set $S = s_1 < \cdots < s_n$, let σ_S denote the string of 0's and 1's of length $s_n + 1$ such that $\sigma_S(i) = 1 \iff i \in S$. Then we can let

$$F_{\text{Bin}}(S) = 1^\frown \sigma_S \tag{1}$$

$$F_{\text{Bin}}(\omega - S) = 0^\frown \sigma_S \tag{2}$$

It is then easy to see that $\text{Bin}(\tilde{\mathcal{N}})$ will be a linear-time Boolean algebra. It then easily follows from Lemma 2.3 that if $F_{\text{Tal}}(x) = \mu_2(F_{\text{Bin}}(x))$ for all $x \in \tilde{\mathcal{N}}$, then $\text{Tal}(\tilde{\mathcal{N}})$ will also be a linear-time Boolean algebra. Moreover it is also easy to see that $\text{At}(\text{Bin}(\tilde{\mathcal{N}}))$ and $\text{At}(\text{Tal}(\tilde{\mathcal{N}}))$ are also linear-time in our presentation.

The computable isomorphism type of $\text{Bin}(\tilde{\mathcal{N}})$ is characterized by the following result due to Remmel.

Theorem 3.1 (Remmel [18]). *Let $\tilde{\mathcal{B}}$ be a computable Boolean algebra such that $\tilde{\mathcal{B}} \cong \tilde{\mathcal{N}}$ and $\text{At}(\tilde{\mathcal{B}})$, the set of atoms of $\tilde{\mathcal{B}}$, is a computable set. Then $\tilde{\mathcal{B}} \cong_c \text{Bin}(\tilde{\mathcal{N}})$.*

Let $\tilde{\mathcal{Q}}$ be the Boolean algebra generated by the left closed-right open subsets of Q. Note that Cantor's basic back and forth argument which shows that the theory of a countable atomless Boolean algebra is $\aleph_0$ categorical is effective. This means that any two computable countable atomless Boolean algebras are computably isomorphic. Thus we have the following result.

Theorem 3.2. *Let $\tilde{\mathcal{B}}$ be a computable Boolean algebra such that $\tilde{\mathcal{B}} \cong \tilde{\mathcal{Q}}$, then $\tilde{\mathcal{B}} \cong_c \text{Bin}(\tilde{\mathcal{Q}})$.*

In particular, the Boolean algebra $\tilde{\mathcal{Q}}$ may be represented as generated by the left closed-right open dyadic intervals. We can now give a specific polynomial-time realization of $\tilde{\mathcal{Q}}$ as follows, using

Let $\mathcal{P}([0, 1))$ denote the Boolean algebra of all subsets of the rational left-closed right-open interval $[0, 1)$ in the dyadic rational numbers. The Boolean operations of meet, join, and complementation on $\mathcal{P}([0, 1))$ are respectively intersection, union, and relative complement in $[0, 1)$. Let $\mathcal{B}([0, 1))$ be the subalgebra of $\mathcal{P}([0, 1))$ generated by the left-closed right-open intervals of the form

$$\left[\tfrac{i}{2^n}, \tfrac{j}{2^n}\right)$$

with $n \geq 0$ and $0 \leq i < j \leq 2^n$.

Next we define a natural generating sequence $a_1, a_2, \ldots$ for $\mathcal{B}([0, 1))$ by induction:

$a_1 = [0, 1)$;

$a_{2^{n-1}+m+1} = \left[\frac{2m}{2^n}, \frac{2m+1}{2^n}\right)$ if $n \geq 1$ and $0 \leq m < 2^{n-1}$.

Thus

$$a_2 = \left[0, \tfrac{1}{2}\right), \quad a_3 = \left[0, \tfrac{1}{4}\right), \quad a_4 = \left[\tfrac{1}{2}, \tfrac{3}{4}\right), \quad a_5 = \left[0, \tfrac{1}{8}\right),$$

$$a_6 = \left[\tfrac{1}{4}, \tfrac{3}{8}\right), \quad a_7 = \left[\tfrac{1}{2}, \tfrac{5}{8}\right), \quad a_8 = \left[\tfrac{3}{4}, \tfrac{7}{8}\right)$$

Let $A_n = \mathrm{sub}_{\mathcal{B}([0,1))}(\{a_1, \ldots, a_n\})$. Then it is not difficult to see that $A_1, A_2, A_3, \ldots$ is a strictly increasing sequence of subalgebras such that for each $n \geq 1$, there is a unique atom $x_n \in \mathrm{At}(A_n)$ such that a_{n+1} splits x_n, i.e. $\emptyset \subset a_{n+1} \subset x_n$. In fact, one can easily show by induction that if k is of the form $2^{n-1} + m$ with $0 \leq m < 2^{n-1}$, then

$$\mathrm{At}(A_k) = \left\{\left[\tfrac{i}{2^n}, \tfrac{i+1}{2^n}\right) : 0 \leq i < 2m\right\} \cup \left\{\left[\tfrac{j}{2^{n-1}}, \tfrac{j+1}{2^{n-1}}\right) : m \leq j < 2^{n-1}\right\}.$$

Hence $a_{k+1} = \left[\frac{2m}{2^n}, \frac{2m+1}{2^n}\right)$ splits the atom $x_k = \left[\frac{m}{2^{n-1}}, \frac{m+1}{2^{n-1}}\right)$ of A_k. It follows that for each $n \geq 1$, A_n has exactly n atoms so that A_n has exactly 2^n elements.

We use this generating sequence and its corresponding sequence of subalgebras

$$A_1 \subset A_2 \subset A_3 \subset \cdots$$

to define the binary and tally representations of $\mathcal{B}([0, 1))$ as was done by Nerode and Remmel in [15].

The binary representation of $\mathcal{B}([0, 1))$

First we describe a coding of the elements of $\mathcal{B}([0, 1))$ for the binary representation of $\mathcal{B}([0, 1))$. Our idea is to use binary numbers of length n to code the elements of $A_n - A_{n-1}$ for $n > 1$. Formally, we define a 1:1 correspondence $\delta \to s_\delta$ between $\mathrm{Bin}(\omega)$ and $\mathcal{B}([0, 1))$ by induction.

For the base step, set

$$s_0 = \emptyset, s_1 = [0, 1).$$

For the induction step, assume that the correspondence $\text{bin}(k) \to s_k$ has been defined between $\{\text{bin}(k) : 0 < k < 2^n\}$ and A_n. We then extend our correspondence to A_{n+1} as follows. Given a binary number m of length $n+1$, let $m = k \cdot 2^i$ where k is odd, so that $\text{bin}(m) = 0^i \frown \text{bin}(k)$, and let

$$s_m = \begin{cases} s_k \cup a_{n+1} & \text{if } s_k \cap a_{n+1} = \emptyset \\ s_k - a_{n+1} & \text{if } s_k \supseteq a_{n+1}. \end{cases} \tag{3}$$

Now let $\text{At}(A_n) = \{x_1, \ldots, x_{n-1}, x_n\}$ where x_n is the atom of A_n which is split by a_{n+1}. Then it is easy to see that every element of $A_{n+1} - A_n$ is either of the form

$$a_{n+1} \cup \bigcup_{i \in S} x_i$$

or

$$(x_n - a_{n+1}) \cup \bigcup_{i \in S} x_i$$

for some set $S \subseteq \{1, \ldots, n-1\}$. Thus (3) defines a 1:1 correspondence between $\{k : 2^n \leq k < 2^{n+1}\}$ and $A_{n+1} - A_n$.

Indeed it is quite easy to use (3) to computably construct s_n. We write $s_{\text{bin}(n)}$ for s_n in the following.

Example 3.1. Suppose $\text{bin}(n) = 0101101$. Then s_n can be constructed as follows.

$s_0 = \emptyset$.

$s_{01} = a_2 = [0, \frac{1}{2})$ since $a_2 \cap s_0 = \emptyset$.

$s_{0101} = s_{01} \cup a_4 = [0, \frac{1}{2}) \cup [\frac{1}{2}, \frac{3}{4}) = [0, \frac{3}{4})$ since $a_4 \cap s_{01} = \emptyset$.

$s_{01011} = s_{0101} - a_5 = [0, \frac{3}{4}) - [0, \frac{1}{8}) = [\frac{1}{8}, \frac{3}{4})$ since $a_4 \subseteq s_{0101}$.

$s_{0101101} = s_{01011} - a_7 = [0, \frac{3}{4}) - [\frac{1}{2}, \frac{5}{8}) = [\frac{1}{8}, \frac{1}{2}) \cup [\frac{5}{8}, \frac{3}{4})$ since $a_7 \subseteq s_{01011}$.

It is not difficult to show that given two elements, σ and τ, in $\text{Bin}(\omega)$ with $|\sigma| \leq |\tau|$, we can find α, β and γ in $\text{Bin}(\omega)$ such that

$$s_\alpha = s_\sigma \cup s_\tau, \quad s_\beta = s_\sigma \cap s_\tau, \quad s_\gamma = [0, 1) - s_\tau$$

in polynomial-time in $|\tau|$. Furthermore, note that each of α, β and γ has length $\leq 2|\tau|$, since each of s_α, s_β and s_γ belongs to A_n if $s_\tau \in A_n$. See [15] for details. It follows that if we then define

$$\sigma \wedge_s \tau = \alpha \tag{4}$$

$$\sigma \vee_s \tau = \beta \tag{5}$$

$$\neg_s \tau = \gamma \tag{6}$$

Then $\mathrm{Bin}(\mathcal{B}) = (\mathrm{Bin}(\omega), \wedge_s, \vee_s, \neg_s)$ is a polynomial-time representation of the countable atomless Boolean algebra $\mathcal{B}([0, 1))$ which we call the binary representation of $\mathcal{B}([0, 1))$.

The tally representation of $\mathcal{B}([0, 1))$

The tally representation $\mathrm{Tal}(\mathcal{B})$ of $\mathcal{B}([0, 1))$ can easily be defined from the binary representation $\mathrm{Bin}(\mathcal{B}([0, 1))$ to be the isomorphic image under the map μ_2 and is therefore a p-time structure by Lemma 2.3 in light of the note above.

Our third Boolean algebra $\tilde{\mathfrak{C}}$ is the one generated by the closed intervals of Q, or equivalently, the one generated by $\tilde{\mathfrak{Q}}$ together with the set of singletons $\{q\}$ for $q \in Q$. Let

$$\theta = \{\{q\} : q \in Q\} \cup \{[q_1, q_2) : q_1, q_2 \in Q \;\&\; q_1 < q_2\} \cup \{\phi, Q\}.$$

Again we could give specific polynomial-time realization of $\tilde{\mathfrak{C}}$, $\mathrm{Bin}(\tilde{\mathfrak{C}})$ with universe $\mathrm{Bin}(\omega)$ and $\mathrm{Tal}(\tilde{\mathfrak{C}})$ with universe $\mathrm{Tal}(\omega)$. Instead we shall give a more general construction which will show that such polynomial-time realizations of $\tilde{\mathfrak{C}}$ exist.

As mentioned in the introduction, Cenzer and Remmel [1] showed that the interval algebra $\mathrm{Intalg}(\mathcal{L})$ of any p-time linear ordering $\mathcal{L}$ with a least element is a p-time Boolean algebra. Here we make the convention that for any such linear ordering ∞ is an element which is larger than any element of $\mathcal{L} = (L, <_L)$ and the elements of the interval algebra are either the empty set, unions of left-closed right-open intervals of the form

$$[l_1, l_2) \cup \cdots \cup [l_{2n-1}, l_{2n}),$$

or unions of the form

$$[l_1, l_2) \cup \cdots \cup [l_{2n-1}, l_{2n}) \cup [l_{2n+1}, \infty)$$

where $l_1 <_L \cdots <_L l_{2n+1}$ and

$$[l_{2i-1}, l_{2i}) = \{l \in L : l_{2i-1} <_L l \leq_L l_{2i}\}$$

$$[l_{2n+1}, \infty) = \{l \in L : l_{2n+1} \leq_L l\}$$

Thus the one of $\mathrm{Intalg}(\mathcal{L})$ is $[l_0, \infty)$ where l_0 is the least element of $\mathcal{L}$ and the zero of $\mathrm{Intalg}(\mathcal{L})$ is the empty set. We now strengthen that result in two ways. First, we show that if $\mathcal{L}$ has a standard universe, either $\mathrm{Tal}(\omega)$ or $\mathrm{Bin}(\omega)$, then $\mathrm{Intalg}(\mathcal{L})$ can be defined to have the same standard universe. Second, we obtain similar results for other notions of complexity, that is, linear-time and exponential-time as well as space versions of all three notions.

Lemma 3.3. *Suppose that the linear ordering $\mathcal{L} = (B, <^L)$ has a least element and has standard universe, either* $\mathrm{Tal}(\omega)$ *or* $\mathrm{Bin}(\omega)$. *Then* $\mathrm{Intalg}(\mathcal{L})$ *has a representation*

$\mathcal{B}$ as a Boolean algebra with universe Bin(ω) *and a representation $\mathcal{C}$ as a Boolean algebra with universe* Tal(ω) *such that*

(i) *If $\mathcal{L}$ is linear space, then $\mathcal{B}$ and $\mathcal{C}$ are both linear space.*

(ii) *If $\mathcal{L}$ is polynomial-time/space, then $\mathcal{B}$ is polynomial-time/space and $\mathcal{C}$ is linear-time/space.*

(iii) *If $\mathcal{L}$ is exponential-time/space, then $\mathcal{B}$ is exponential-time/space and $\mathcal{C}$ is polynomial-time/space.*

(iv) *If $\mathcal{L}$ is* EXPTIME/SPACE, *then $\mathcal{B}$ is* EXPTIME/SPACE *and $\mathcal{C}$ is exponential-time/space.*

Proof. Suppose that $\mathcal{L} = (B, <^L)$ is a polynomial-time linear ordering with a least element and $B = \text{Tal}(\omega)$. There is no loss in generality in assuming that the least element of $\mathcal{L}$ is $u_0 = 0$. For each n, let $L_n = \{\text{tal}(m) : m < n\}$, let $\mathcal{L}_n = (L_n, <^L)$. Now suppose that L_n is ordered as $u_0 = u_0^n <_L u_1^n <_L \cdots <_L u_{n-1}^n <_L u_n^n = \infty$. Then Intalg($\mathcal{L}_n$) has n atoms, $a_i^n = [u_i^n, u_{i+1}^n)$ for $0 \leq i \leq n-1$, and 2^n elements.

Now an arbitrary nontrivial element b of Intalg($\mathcal{L}$) is expressed as a finite union of disjoint intervals of the form

$$[\text{tal}(l_1), \text{tal}(l_2)) \cup \cdots \cup [\text{tal}(l_{2n-1}), \text{tal}(l_{2n}))$$

or of the form

$$[\text{tal}(l_1), \text{tal}(l_2)) \cup \cdots \cup [\text{tal}(l_{2n-1}), \text{tal}(l_{2n})) \cup [\text{tal}(l_{2n+1}), \infty)$$

where $\text{tal}(l_1) <_L \text{tal}(l_2) <_L \text{tal}(l_{2n-1}) <_L \text{tal}(l_{2n}) <_L \text{tal}(l_{2n+1})$. Thus such elements may be viewed as an element of Intalg($\mathcal{L}_n$), where n is the least such that all $\text{tal}(l_i) <_L \text{tal}(n)$.

We let 0 represent the zero of Intalg($\mathcal{L}$) and 1 represent the one of Intalg($\mathcal{L}$). Any element b of Intalg($\mathcal{L}_n$) may be represented by the binary number $\text{bin}(k) = (e(0)e(1)\ldots e(n)1)$, where $e(i) = 1$ if and only if $a_i^n \subseteq b$. We define a representation $\mathcal{B}_n$ of Intalg($\mathcal{L}_n$) to have universe the set of binary numbers of length $n+1$ and operations $\neg_n$, $\wedge_n$ and $\vee_n$ defined as follows:

(i) $\neg_n(e(0)^\frown e(1)^\frown \cdots {}^\frown e(n)^\frown 1) = (1 - e(0)^\frown 1 - e(1)^\frown \cdots {}^\frown 1 - e(n)^\frown 1)$,

(ii) $(e(0)^\frown e(1)^\frown \cdots {}^\frown e(n)^\frown 1) \vee_n (f(0)^\frown f(1)^\frown \cdots {}^\frown f(n)^\frown 1)$
$= (\max\{e(0), f(0)\}^\frown \max\{e(1), f(1)\}^\frown \cdots {}^\frown \max\{e(n), f(n)\}^\frown 1)$, and

(iii) $(e(0)^\frown e(1)^\frown \cdots {}^\frown e(n)^\frown 1) \wedge_n (f(0)^\frown f(1)^\frown \cdots {}^\frown f(n)^\frown 1)$
$= (\min\{e(0), f(0)\}^\frown \min\{e(1), f(1)\}^\frown \cdots {}^\frown \min\{e(n), f(n)\}^\frown 1)$.

These operations are clearly real-time computable in $\mathcal{B}_n$. Next we define a Boolean algebra $\mathcal{A}$ with universe $A \subseteq \text{Bin}(\omega)$ which is isomorphic to Intalg($\mathcal{L}$) and which for each $b \in$ Intalg($\mathcal{L}$) contains only the shortest representation in $\bigcup_n \mathcal{B}_n$, that is, the representation from $\mathcal{B}_n$ where n is the least such that the end points of every interval in b are $\leq n$. We call this n the *height* of b. In particular, the shortest representation of the zero of Intalg($\mathcal{L}$) is 0 and the short representation of the one of Intalg($\mathcal{L}$) is 1. We will show later that $\mathcal{A}$ is p-time isomorphic to a structure with universe Bin(ω).

The complement $\neg a$ of any nontrivial element of $\mathcal{A}$ has the same set of end points from B and hence $\neg a$ may be computed as in (i) above. Since the other operations are so easily computable with $\mathcal{B}_n$, we have the following plan for computing $a \vee b$ (or $a \wedge b$) where a and b are nontrivial. Let n be the maximum height of a and b and find representations $a' \in \mathcal{B}_n$ for a and $b' \in \mathcal{B}_n$ for b. Compute $c' = a' \vee b'$ (or $a' \wedge b'$) in $\mathcal{B}_n$. Then compute from c' the shortest representation c of c' in some $\mathcal{B}_p$ with $p \leq n$.

Given $a \in \mathcal{B}_m$ and $b \in \mathcal{B}_n$ with $m \leq n$, the procedure for computing $a' \in \mathcal{B}_n$ and equivalent to a is the following. First use the input b of length $n+1$ to compute the ordering $u_0 = u_0^n < u_1^n < \cdots < u_n^n$. Here we are just sorting $\{\text{tal}(i) : i < n\}$ under the p-time ordering $<^L$. It is well-known that such sorting takes $o(n \log n)$ comparisons, so that, since each of the elements u_i^n has length $\leq n$, this can be done in polynomial time. Suppose now that $a = (e(0)^\frown e(1)^\frown \cdots ^\frown e(m)^\frown 1)$. For each i, count the number k_i of the elements u_j^n not in $\mathcal{B}_m$ which fall into the interval $a_i = [u_i^m, u_{i+1}^m)$. Then let

$$a' = ((e(0)^{k_0})^\frown (e(1)^{k_1})^\frown \cdots ^\frown (e(m)^{k_m})^\frown 1).$$

Given $c' = (g(0)^\frown g(1)^\frown \cdots ^\frown g(m)^\frown 1) \in \mathcal{B}_n$, the procedure for computing the shortest representation $c \in \mathcal{B}_p$ is the following. First check to see whether or not the interval $[u_i^n, u_{i+2}^n)$, where $u_{i+1}^n = \text{tal}(n-1)$, is split in c', that is, whether $g(i) \neq g(i+1)$. If the interval is not split, that is if $g(i) = g(i+1)$, then we can reduce c' to $c_{n-1} \in \mathcal{B}_{n-1}$ by deleting $g(i)$ from c'. This computation may be performed in linear time given the sorted ordering $u_0 = u_0^n < u_1^n < \cdots < u_n^n$ previously computed. Then we proceed computably to see whether c_{n-1} may be reduced to c_{n-2} and so on. This whole procedure clearly may be done in polynomial time.

It follows that $\mathcal{A}$ is a polynomial-time Boolean algebra which is computably isomorphic to Intalg($\mathcal{L}$). Furthermore, it follows from the above discussion that $|\neg a| = |a|$ and that $|a \vee b|$ and $|a \wedge b|$ are both $\leq \max\{|a|, |b|\}$.

It remains to see that the set A is p-time isomorphic to Bin(ω). The isomorphism is very simply defined. Given $a = (e(0)^\frown e(1)^\frown \cdots ^\frown e(n)^\frown 1) \in A$ of height $n \geq 2$ and $u_{i+1} = \text{tal}(n-1)$, define $\phi(a)$ of length n by

$$\phi(a) = (e(0)^\frown e(1)^\frown \cdots ^\frown e(i)^\frown e(i+2)^\frown \cdots ^\frown e(n)^\frown 1),$$

We may omit $e(i+1)$ because we must have $e(i+1) = 1 - e(i)$ given that $a \in A$ and therefore has height n. Observe that the elements in A of height n are mapped onto the elements of Bin(ω) of length n. The inverse map simply inserts $e(i+1) = 1 - e(i)$ after $e(i)$ to get a back from $\phi(a)$. Note that $|\phi(a)| = |a| - 1$. The map ϕ and its inverse are both computable in polynomial time, since we may compute the $<_L$-increasing list of $\{\text{tal}(i) : i < n\}$ in polynomial-time in $|a|$.

It now follows from Lemma 2.1 that there is a p-time structure $\mathcal{B}$ with universe $\mathrm{Bin}(\omega)$ which is computably isomorphic to $\mathrm{Intalg}(\mathcal{L})$. Furthermore, the operations in $\mathcal{B}$ never increase the length. It therefore follows from Lemma 2.3 that there is a *linear-time* structure $\mathcal{C}$ with universe $\mathrm{Tal}(\omega)$ which is isomorphic to $\mathrm{Intalg}(\mathcal{L})$.

It is easily checked that if the linear ordering $\mathcal{L}$ was linear space, exponential time/space or EXPTIME/SPACE computable, then the resulting structure $\mathcal{B}$ with universe $\mathrm{Bin}(\omega)$ is also computable in linear space, exponential time, etc. It then follows from the methods of Lemma 2.3 that if $\mathcal{L}$ is computable in polynomial space, then the structure $\mathcal{C}$ with universe $\mathrm{Tal}(\omega)$ is computable in linear space, if $\mathcal{L}$ is computable in exponential time/space, then $\mathcal{C}$ is computable in polynomial time/space, and if $\mathcal{L}$ is computable in EXPTIME/SPACE, then $\mathcal{C}$ is computable in exponential time/space.

Finally we should observe that if $\mathcal{L}$ has universe $\mathrm{Bin}(\omega)$, then we can replace simply replace $\mathcal{L}$ by the ordering isomorphic to $\mathcal{L}$ which induced by the map μ_2 which sends $\mathrm{bin}(n)$ to $\mathrm{tal}(n)$ and derive the same conclusions. □

Remark. It is easy to see that an atom of $\mathrm{Intalg}(\mathcal{L})$ corresponds to an interval $[u_i^n, u_{i+1}^n)$ where $\mathrm{tal}(n-1) = u_i^n$ or $\mathrm{tal}(n-1) = u_{i+1}^n$ and u_{i+1}^n is an immediate successor of u_i^n in $\mathcal{L}$. Similarly elements in the ideal generated by the atoms of $\mathrm{Intalg}(\mathcal{L})$ correspond to unions of intervals of $\mathcal{L}$ which are finite. It is then easy to see that the same proof shows that if $\mathcal{L}$ is polynomial-time linear ordering with a least element which has universe $\mathrm{Tal}(\omega)$ or $\mathrm{Bin}(\omega)$ and whose immediate successor relation polynomial time, then $\mathrm{Intalg}(\mathcal{L})$ is computably isomorphic to polynomial-time Boolean algebra $\mathcal{B}$ with universe $\mathrm{Bin}(\omega)$ whose set of atoms is polynomial-time. Similarly if $\mathcal{L}$ is polynomial-time linear ordering with a least element which has universe $\mathrm{Tal}(\omega)$ or $\mathrm{Bin}(\omega)$ and whose finite interval relation is polynomial time, then $\mathrm{Intalg}(\mathcal{L})$ is computably isomorphic to polynomial-time Boolean algebra $\mathcal{B}$ with universe $\mathrm{Bin}(\omega)$ whose ideal generated by the atoms is polynomial-time.

It is easy to see that $\tilde{\mathcal{C}}$ is isomorphic to the interval algebra of $1+\eta\cdot\omega$ where η is the order type of the rationals. Moreover it follows from Grigorieff's results that we can construct a polynomial-time linear ordering $L = (\mathrm{Tal}(\omega), <_L)$ isomorphic to $1+\eta\cdot\omega$ with the property that the immediate successor relation and the relation Finite, where $\mathrm{Finite}(x, y)$ holds if and only if $x <_L y$ and the interval $[x, y]_L = \{z \in \mathrm{Tal}(\omega) : x \leq_L z \leq_l y\}$ is finite, are both polynomial-time. We can then apply Lemma 3.3 to conclude that there exists a polynomial-time $\mathrm{Bin}(\tilde{\mathcal{C}})$ such that the universe of $\mathrm{Bin}(\tilde{\mathcal{C}})$ is $\mathrm{Bin}(\omega)$, the set of atoms and the ideal generated by the atoms of $\mathrm{Bin}(\tilde{\mathcal{C}})$ are polynomial-time, and $\mathrm{Bin}(\tilde{\mathcal{C}})$ is isomorphic to $\tilde{\mathcal{C}}$. Similarly there is a polynomial-time $\mathrm{Tal}(\tilde{\mathcal{C}})$ such that the universe of $\mathrm{Tal}(\tilde{\mathcal{C}})$ is $\mathrm{Tal}(\omega)$, the set of atoms and the ideal generated by the atoms of $\mathrm{Tal}(\tilde{\mathcal{C}})$ are polynomial-time, and $\mathrm{Tal}(\tilde{\mathcal{C}})$ is isomorphic to $\tilde{\mathcal{C}}$.

The computable isomorphism type of $\mathrm{Bin}(\tilde{\mathcal{C}})$ is characterized by the following result of Remmel.

Theorem 3.4 (Remmel [18]). *Let $\tilde{\mathcal{B}}$ a computable Boolean algebra such that $\tilde{\mathcal{B}} \cong \tilde{\mathcal{C}}$ and $\mathrm{At}(\tilde{\mathcal{B}})$ and $I(\mathrm{At}(\tilde{\mathcal{B}}))$ are computable subsets of $|\tilde{\mathcal{B}}|$, then $\tilde{\mathcal{B}} \cong_c \mathrm{Bin}(\tilde{\mathcal{C}})$.*

The importance of our three Boolean algebras $\tilde{\mathcal{N}}$, $\tilde{\mathcal{Q}}$, and $\tilde{\mathcal{C}}$ for our purposes is based on the following result of Remmel.

Theorem 3.5 (Remmel [18]). *Let $\tilde{\mathcal{B}}$ be an infinite computable Boolean algebra. Then either*

1. $\tilde{\mathcal{B}} \cong \tilde{\mathcal{C}}$,
2. *there is a computable Boolean algebra, B, such that* $\tilde{\mathcal{B}} \cong \tilde{\mathcal{N}} \times B$, *or*
3. *there is a computable Boolean algebra, B, such that* $\tilde{\mathcal{B}} \cong \tilde{\mathcal{Q}} \times B$.

This theorem say that we can divide the set of countably infinite computable Boolean algebras into three classes, i.e. Boolean algebras of the form $\tilde{\mathcal{C}}$, or $\tilde{\mathcal{N}} \times B$ or $\tilde{\mathcal{Q}} \times B$ where B is some computable Boolean algebra. Note that this theorem doesn't say that any computable Boolean algebra, $\tilde{\mathcal{B}}$, is computably isomorphic to one of $\tilde{\mathcal{C}}$, $\tilde{\mathcal{N}} \times B$, or $\tilde{\mathcal{Q}} \times B$ where B is a computable Boolean algebra. For example, Remmel [20] constructed a computable Boolean algebra $\mathcal{B}$ which is isomorphic to $\tilde{\mathcal{N}}$ but for which $I(\mathrm{At}(\mathcal{B}))$ is immune and hence is certainly not computably isomorphic to $\tilde{\mathcal{N}}$. Indeed it is a result of Goncharov and LaRoche that any countable Boolean algebra which contains a single computable isomorphism type within its classical isomorphism type has only finitely many atoms; see [20]

4. Main results

In this section, we shall prove our three main results. That is, we shall consider three types of computable Boolean algebras, namely,

(i) computable Boolean algebras $\tilde{\mathcal{B}}$ where both $\mathrm{At}(\tilde{\mathcal{B}})$ and $I(\mathrm{At}(\tilde{\mathcal{B}}))$ are computable,

(ii) computable Boolean algebras $\tilde{\mathcal{B}}$ where $\mathrm{At}(\tilde{\mathcal{B}})$ is computable but $I(\mathrm{At}(\tilde{\mathcal{B}}))$ is not computable, and

(iii) computable Boolean algebras $\tilde{\mathcal{B}}$ where neither $\mathrm{At}(\tilde{\mathcal{B}})$ nor $I(\mathrm{At}(\tilde{\mathcal{B}}))$ are computable.

In case (i), we shall show that $\tilde{\mathcal{B}}$ is computably isomorphic to a polynomial-time Boolean algebra $\mathrm{Bin}(\tilde{\mathcal{B}})$ such that the universe of $\mathrm{Bin}(\tilde{\mathcal{B}})$ is $\mathrm{Bin}(\omega)$ and the set of atoms and the ideal generated by the atoms of $\mathrm{Bin}(\tilde{\mathcal{B}})$ are polynomial-time and $\tilde{\mathcal{B}}$ is computably isomorphic to a polynomial-time Boolean algebra $\mathrm{Tal}(\tilde{\mathcal{B}})$ such that the universe of $\mathrm{Tal}(\tilde{\mathcal{B}})$ is $\mathrm{Tal}(\omega)$ and the set of atoms and the ideal generated by the atoms of $\mathrm{Tal}(\tilde{\mathcal{B}})$ are polynomial-time.

In case (ii), we shall show that if $\tilde{\mathcal{B}}$ is not isomorphic to $\tilde{\mathcal{C}}$, then $\tilde{\mathcal{B}}$ is computably isomorphic to a polynomial-time Boolean algebra $\mathrm{Bin}(\tilde{\mathcal{B}})$ such that the universe of $\mathrm{Bin}(\tilde{\mathcal{B}})$ is $\mathrm{Bin}(\omega)$ and the set of atoms of $\mathrm{Bin}(\tilde{\mathcal{B}})$ is polynomial-time and $\tilde{\mathcal{B}}$ is computably isomorphic to a polynomial-time $\mathrm{Tal}(\tilde{\mathcal{B}})$ such that the universe of $\mathrm{Tal}(\tilde{\mathcal{B}})$ is

$\mathrm{Tal}(\omega)$ and the set of atoms of $\mathrm{Tal}(\tilde{\mathcal{B}})$ is polynomial-time. However we shall show that for any infinite polynomial-time set $A \subseteq \{0,1\}^*$, there exists a computable Boolean algebra $\mathcal{C}$ which is isomorphic to $\tilde{\mathcal{C}}$ and has a computable set of atoms but which is not computably isomorphic to any polynomial-time Boolean algebra with universe A.

Finally in case (iii), we shall show that if $\tilde{\mathcal{B}}$ not atomic, then $\tilde{\mathcal{B}}$ is computably isomorphic to a polynomial-time Boolean algebra $\mathrm{Bin}(\tilde{\mathcal{B}})$ such that the universe of $\mathrm{Bin}(\tilde{\mathcal{B}})$ is $\mathrm{Bin}(\omega)$ and $\tilde{\mathcal{B}}$ is computably isomorphic to a polynomial-time $\mathrm{Tal}(\tilde{\mathcal{B}})$ such that the universe of $\mathrm{Tal}(\tilde{\mathcal{B}})$ is $\mathrm{Tal}(\omega)$. However we shall show that if $\tilde{\mathcal{B}}$ is an atomic Boolean algebra, then for any infinite polynomial-time set $A \subseteq \{0,1\}^*$, there exists a computable Boolean algebra $\mathcal{D}$ which is isomorphic to $\tilde{\mathcal{B}}$ but which is not computably isomorphic to any polynomial-time Boolean algebra with universe A.

Before proving these results, we should point out two facts. First it is a result of Goncharov [12] that there exist computable atomic Boolean algebras $\mathcal{B}$ with the property that $\mathcal{B}$ is not isomorphic to any computable Boolean algebra $\mathcal{C}$ such that $\mathrm{At}(\mathcal{C})$ is computable. Thus there are computable Boolean algebras $\mathcal{B}$ which are covered by case (iii) but such that the classical isomorphism type of $\mathcal{D}$ does not intersect the computable Boolean algebras in case (ii). Similarly Remmel [19] showed that there exist computable Boolean algebras $\mathcal{D}$ such that $\mathrm{At}(\mathcal{D})$ is computable, but $\mathcal{D}$ is not isomorphic to any computable Boolean algebra $\mathcal{E}$ such that both $\mathrm{At}(\mathcal{E})$ and $I(\mathrm{At}(\mathcal{E}))$ are computable. Thus there are computable Boolean algebras $\mathcal{D}$ which are covered by case (ii) but such that the classical isomorphism type of $\mathcal{D}$ does not intersect the computable Boolean algebras in case (i). Second we note that it is not the case that every infinite polynomial-time subset A of $\{0,1\}^*$ can be the universe of a polynomial-time Boolean algebra. That is, we have the following result.

Theorem 4.1. *There exists an infinite polynomial-time subset A of* $\mathrm{Tal}(\omega)$ *such that there are no polynomial-time Boolean algebras with universe A.*

Proof. The polynomial-time set A is just the sequence of iterated exponentials of powers of 2 in the tally notation. That is, inductively define $m_0 = 1$ and $m_{n+1} = 2^{(m_n)}$. It is easy to see that $A = \{\mathrm{tal}(m_n) : n \in \omega\}$ is a polynomial-time subset of $\mathrm{Tal}(\omega)$. Now suppose there were a polynomial-time Boolean algebra $\mathcal{B} = (A, \vee_{\mathcal{B}}, \wedge_{\mathcal{B}}, \neg_{\mathcal{B}}, 0_{\mathcal{B}}, 1_{\mathcal{B}})$. Then consider the subalgebra generated by $\{\mathrm{tal}(m_0), \dots, \mathrm{tal}(m_{n-1})\}$ where $n \neq 2^k$ for any k. Since all finite Boolean algebras have cardinality 2^k for some k, this means that for each such n, there must be some Boolean combination of $\mathrm{tal}(m_0), \dots, \mathrm{tal}(m_{n-1})$ which lies in $A - \{\mathrm{tal}(m_0), \dots, \mathrm{tal}(m_{n-1})\}$. But then for some $i_0, i_1 < n$, we must have one of $\mathrm{tal}(m_{i_0}) \vee_{\mathcal{B}} \mathrm{tal}(m_{i_1})$, $\mathrm{tal}(m_{i_0}) \wedge_{\mathcal{B}} \mathrm{tal}(m_{i_1})$, and $\neg_{\mathcal{B}} \mathrm{tal}(m_{i_1})$ in $A - \{\mathrm{tal}(m_0), \dots, \mathrm{tal}(m_{n-1})\}$. Now if $\vee_{\mathcal{B}}$, $\wedge_{\mathcal{B}}$, and $\neg_{\mathcal{B}}$ are polynomial-time operations, there will be some q large enough so that

1. $|\,\mathrm{tal}(m_{i_0}) \vee_{\mathcal{B}} \mathrm{tal}(m_{i_1})| < 2^{\max(|\mathrm{tal}(m_{i_0})|, |\mathrm{tal}(m_{i_1})|)}$,

2. $|\,\mathrm{tal}(m_{i_0}) \wedge_{\mathcal{B}} \mathrm{tal}(m_{i_1})| < 2^{\max(|\mathrm{tal}(m_{i_0})|, |tal(m_{i_1})|)}$, and

3. $\neg_{\mathcal{B}} \mathrm{tal}(m_{i_1})| < 2^{|\mathrm{tal}(m_{i_1})|}$

unless both i_0 and i_1 are less than q. Now pick p is large enough so that p is not a power of 2 and $\{\text{tal}(m_0), \ldots, \text{tal}(m_{p-1})\}$ contains the subalgebra of $\mathcal{B}$ generated by $\{\text{tal}(m_0), \ldots, \text{tal}(m_{q-1})\}$. Then the $i_0, i_1 < p$ such that one of $\text{tal}(m_{i_0}) \vee_{\mathcal{B}} \text{tal}(m_{i_1})$, $\text{tal}(m_{i_0}) \wedge_{\mathcal{B}} \text{tal}(m_{i_1})$, and $\neg_{\mathcal{B}} tal(m_{i_1})$ is in $A - \{\text{tal}(m_0), \ldots, \text{tal}(m_{p-1})\}$ cannot have the property that both i_0 and i_1 are $\leq m$. This would then yield a contradiction since every element of $A - \{\text{tal}(m_0), \ldots, \text{tal}(m_{p-1})\}$ has length greater than or equal to $2^{\max(|\text{tal}(m_{i_0})|, |\text{tal}(m_{i_1})|)}$. □

4.1. Computable Boolean algebras $\tilde{\mathcal{B}}$ with $\text{At}(\tilde{\mathcal{B}})$ and $I(\text{At}(\tilde{\mathcal{B}}))$ computable

Suppose that $\tilde{\mathcal{B}}$ is a countably infinite computable Boolean algebra such that $\text{At}(\tilde{\mathcal{B}})$ and $I(\text{At}(\tilde{\mathcal{B}}))$ are computable. Then by Theorem 3.5, there are three cases that we have to consider.

Case 1. $\tilde{\mathcal{B}}$ is isomorphic to $\tilde{\mathfrak{C}}$. In this case, we can use Theorem 3.4 to conclude that $\tilde{\mathcal{B}}$ is computably isomorphic to $\text{Bin}(\tilde{\mathfrak{C}})$ which is turn is computably isomorphic to $\text{Tal}(\tilde{\mathfrak{C}})$.

Case 2. $\tilde{\mathcal{B}}$ is isomorphic to $\tilde{\mathcal{N}} \times \mathcal{B}$ where $\mathcal{B}$ is some computable Boolean algebra. This means that there is an $x \in \tilde{\mathcal{B}}$ such that $[x]_{\tilde{\mathcal{B}}}$ is isomorphic to $\tilde{\mathcal{N}}$. However since $\text{At}(\tilde{\mathcal{B}})$ is computable it follows that $\text{At}([x]_{\tilde{\mathcal{B}}})$ is also computable and hence, by Theorem 3.1, $[x]_{\tilde{\mathcal{B}}}$ is computably isomorphic to $\text{Bin}(\tilde{\mathcal{N}})$ and $[x]_{\tilde{\mathcal{B}}}$ is computably isomorphic to $\text{Tal}(\tilde{\mathcal{N}})$. Next consider $[\neg_{\tilde{\mathcal{B}}} x]_{\tilde{\mathcal{B}}}$. Clearly $[\neg_{\tilde{\mathcal{B}}} x]_{\tilde{\mathcal{B}}}$ is also a computable Boolean algebra. It well known that every countable Boolean algebra is isomorphic to the interval algebra of some countable linear ordering with a least element. The proof of this fact is effective so that every computable Boolean algebra is computably isomorphic to the interval algebra of some computable linear ordering with a least element, see [20]. Thus there is a computable linear ordering $\mathcal{L}$ with a least element such that $\text{Intalg}(\mathcal{L})$ is computably isomorphic to $[\neg_{\tilde{\mathcal{B}}} x]_{\tilde{\mathcal{B}}}$. Since the set of atoms and the ideal generated by the set of atoms of $[\neg_{\tilde{\mathcal{B}}} x]_{\tilde{\mathcal{B}}}$ are computable, the immediate successor relation and the finite interval relations are computable for $\mathcal{L}$.

Cenzer and Remmel [9] showed that any computable relational structure is computably isomorphic to a polynomial-time structure whose universe is a subset of $\text{Tal}(\omega)$. Thus we can assume that $\mathcal{L}$ is a polynomial-time linear ordering where the immediate successor relation and the finite interval relation are polynomial-time and where the universe is a polynomial-time subset of $\text{Tal}(\omega)$. It then follows from Lemma 4.2 below that the interval algebra of $\mathcal{L}$ is computably isomorphic to a polynomial-time Boolean algebra $\mathcal{D}$ whose universe is a subset of $\text{Tal}(\omega)$ and where the set of atoms and the ideal generated by the set of atoms are polynomial-time. Thus $\tilde{\mathcal{B}}$ is computably isomorphic to $[x]_{\tilde{\mathcal{B}}} \times [\neg_{\tilde{\mathcal{B}}} x]_{\tilde{\mathcal{B}}}$ which in turn is computably isomorphic to $\text{Bin}(\tilde{\mathcal{N}}) \times \mathcal{D}$ and to

$\mathrm{Tal}(\tilde{\mathcal{N}}) \times \mathcal{D}$. However the universe of $\mathrm{Bin}(\tilde{\mathcal{N}}) \times \mathcal{D}$ is $\mathrm{Bin}(\omega) \otimes |\mathcal{D}|$. Since $|\mathcal{D}|$ is a polynomial-time subset of $\mathrm{Tal}(\omega)$, it follows from Lemma 2.4 that $\mathrm{Bin}(\omega) \otimes |\mathcal{D}|$ is polynomial-time isomorphic to $\mathrm{Bin}(\omega)$. It then follows from Lemma 2.1, that $\mathrm{Bin}(\tilde{\mathcal{N}}) \times \mathcal{D}$ is polynomial time isomorphic to polynomial-time Boolean algebra $\mathcal{C}_1$ whose universe is $\mathrm{Bin}(\omega)$, whose set of atoms is polynomial-time, and whose ideal generated by the set of atoms is polynomial-time.

Similarly the universe of $\mathrm{Tal}(\tilde{\mathcal{N}}) \times \mathcal{D}$ is $\mathrm{Tal}(\omega) \otimes |\mathcal{D}|$. Again it follows from Lemma 2.4 that $\mathrm{Tal}(\omega) \otimes |\mathcal{D}|$ is polynomial-time isomorphic to $\mathrm{Tal}(\omega)$. It then follows from Lemma 2.1, that $\mathrm{Tal}(\tilde{\mathcal{N}}) \times \mathcal{D}$ is polynomial-time isomorphic to a polynomial-time Boolean algebra $\mathcal{C}_2$ whose universe is $\mathrm{Tal}(\omega)$, whose set of atoms is polynomial-time, and whose ideal generated by the set of atoms is polynomial-time. Thus once we have proved the lemma below, we will have shown that $\tilde{\mathcal{B}}$ is computably isomorphic to a polynomial-time Boolean algebra with universe $\mathrm{Bin}(\omega)$ and is computably isomorphic to a polynomial-time Boolean algebra with universe $\mathrm{Tal}(\omega)$.

Lemma 4.2. *Suppose that $\mathcal{L} = (L, <_L)$ is a polynomial-time linear ordering with a least element. Then* $\mathrm{Intalg}(\mathcal{L})$ *is computably isomorphic to a polynomial-time Boolean algebra $\mathcal{B} = (B, \vee_B, \wedge_B, \neq_B, 0_B, 1_B)$ where $B \subseteq \mathrm{Tal}(\omega)$. Moreover if the immediate successor relation on $\mathcal{L}$ is polynomial-time, then the set of atoms of $\mathcal{B}$ is polynomial time and if the finite interval relation on $\mathcal{L}$ is polynomial-time, the ideal generated by the atoms of $\mathcal{B}$ is polynomial-time.*

Proof. There is no loss in generality in assuming that $0, 1 \notin L$. Suppose $L = \{\mathrm{tal}(l_0) < \mathrm{tal}(l_1) < \cdots\}$ where $<$ is the usual order on $\mathrm{Tal}(\omega)$ and that $\mathrm{tal}(l_0)$ is the $<_L$-least element of L. Then we can a define map $\psi : \mathrm{Intalg}(\mathcal{L}) \to \mathrm{Bin}(\omega)$ as follows:

(i) If x is the zero of $\mathrm{Intalg}(\mathcal{L})$, $\emptyset$, then we set let $\psi(x) = 0$,

(ii) if $x = [\mathrm{tal}(l_{i_1}), \mathrm{tal}(l_{i_2})) \cup \cdots \cup [\mathrm{tal}(l_{i_{2n+1}}), \mathrm{tal}(l_{i_{2n+2}}))$ where $\mathrm{tal}(l_{i_1}) <_L \mathrm{tal}(l_{i_2}) <_L \cdots <_L \mathrm{tal}(l_{i_{2n+1}}) <_L \mathrm{tal}(l_{i_{2n+2}})$, then we set

$$\psi(x) = \mathrm{tal}(l_{i_1})^\frown 0^\frown \mathrm{tal}(l_{i_2})^\frown 0^\frown \cdots {}^\frown 0^\frown \mathrm{tal}(l_{i_{2n+1}})^\frown 0^\frown \mathrm{tal}(l_{i_{2n+2}}),$$

(iii) if $x = [\mathrm{tal}(l_{i_1}), \mathrm{tal}(l_{i_2})) \cup \cdots \cup [\mathrm{tal}(l_{i_{2n+1}}), \infty)$ where $\mathrm{tal}(l_{i_1}) <_L \mathrm{tal}(l_{i_2}) <_L \cdots <_L \mathrm{tal}(l_{i_{2n+1}})$, then we set

$$\psi(x) = \mathrm{tal}(l_{i_1})^\frown 0^\frown \mathrm{tal}(l_{i_2})^\frown 0^\frown \cdots {}^\frown 0^\frown \mathrm{tal}(l_{i_{2n+1}}).$$

It is easy to see that image of $\mathrm{Intalg}(\mathcal{L})$ under ψ is a polynomial-time set of $\mathrm{Bin}(\omega)$. Moreover if we define the operations on $\psi(\mathrm{Intalg}(\mathcal{L}))$ so that ψ is a Boolean algebra isomorphism, then the operations will also be polynomial-time. This makes $\psi(\mathrm{Intalg}(\mathcal{L}))$ a polynomial-time Boolean algebra which is computably isomorphic to $\mathrm{Intalg}(\mathcal{L})$. It is also easy to see that if the immediate successor relation of $\mathcal{L}$ is polynomial-time, the set of atoms of $\psi(\mathrm{Intalg}(\mathcal{L}))$ is polynomial-time. Similarly if the finite interval relation of $\mathcal{L}$ is polynomial-time, the ideal generated by the atoms is polynomial-time. Finally it is not difficult to check that the operation $\vee$, $\wedge$, and $\neg$ on $\psi(\text{Intalg}(\mathcal{L}))$ have the property that $|x \vee y| \leq 3(|x|+|y|)$, $|x \wedge y| \leq 3(|x|+|y|)$, and $|\neg y| \leq 3(|y|)$ for all $x, y \in \psi(\text{Intalg}(\mathcal{L}))$. It then follows from Lemma 2.3, that if μ_2 is the map which sends $\text{bin}(n)$ to $\text{tal}(n)$ for all $n \in \omega$, then $\mu_2(\psi(\text{Intalg}(\mathcal{L}))) \subseteq \text{Tal}(\omega)$ and if we define the operations on $\mu_2(\psi(\text{Intalg}(\mathcal{L})))$ to ensure that μ_2 is a Boolean algebra isomorphism from $\psi(\text{Intalg}(\mathcal{L}))$ onto $\mu_2(\psi(\text{Intalg}(\mathcal{L})))$, then $\mu_2(\psi(\text{Intalg}(\mathcal{L}))$ will be a polynomial-time Boolean algebra with universe a subset of $\text{Tal}(\omega)$ which is computably isomorphic to $\text{Intalg}(\mathcal{L})$. Moreover if the set of atoms of $\psi(\text{Intalg}(\mathcal{L}))$ is polynomial-time, then the set of atoms of $\mu_2(\psi(\text{Intalg}(\mathcal{L}))$ will also be polynomial-time and if the ideal generated set of atoms of $\psi(\text{Intalg}(\mathcal{L}))$ is polynomial-time, then the ideal generated by set of atoms of $\mu_2(\psi(\text{Intalg}(\mathcal{L}))$ will also be polynomial-time. □

Case 3. $\tilde{\mathcal{B}}$ is isomorphic to $\tilde{\mathcal{Q}} \times \mathcal{B}$ where $\mathcal{B}$ is some computable Boolean algebra. This case is essentially identical to Case 2. That is, there is an $x \in \tilde{\mathcal{B}}$ such that $[x]_{\tilde{\mathcal{B}}}$ is isomorphic to $\tilde{\mathcal{Q}}$. By the same argument that we used in case 2, there exists a polynomial-time Boolean algebra $\mathcal{D}$ whose universe is a polynomial-time subset of $\text{Tal}(\omega)$ and which is computably isomorphic to $[\neg_{\tilde{\mathcal{B}}} x]_{\tilde{\mathcal{B}}}$. Thus $\tilde{\mathcal{B}}$ is computably isomorphic to $[x]_{\tilde{\mathcal{B}}} \times [\neg_{\tilde{\mathcal{B}}} x]_{\tilde{\mathcal{B}}}$ which in turn is computably isomorphic to $\text{Bin}(\tilde{\mathcal{Q}}) \times \mathcal{D}$ and to $\text{Tal}(\tilde{\mathcal{Q}}) \times \mathcal{D}$. However the universe of $\text{Bin}(\tilde{\mathcal{Q}}) \times \mathcal{D}$ is $\text{Bin}(\omega) \otimes |\mathcal{D}|$. Since $|\mathcal{D}|$ is a polynomial-time subset of $\text{Tal}(\omega)$, it follows from Lemma 2.4 that $\text{Bin}(\omega) \otimes |\mathcal{D}|$ is polynomial-time isomorphic to $\text{Bin}(\omega)$. It then follows from Lemma 2.1, that $\text{Bin}(\tilde{\mathcal{Q}}) \times \mathcal{D}$ is polynomial time isomorphic to a polynomial-time Boolean algebra $\mathcal{C}_1$ whose universe is $\text{Bin}(\omega)$. Similarly the universe of $\text{Tal}(\tilde{\mathcal{Q}}) \times \mathcal{D}$ is $\text{Tal}(\omega) \otimes |\mathcal{D}|$. Again it follows from Lemma 2.4 that $\text{Tal}(\omega) \otimes |\mathcal{D}|$ is polynomial-time isomorphic to $\text{Tal}(\omega)$ and it follows from Lemma 2.1, that $\text{Tal}(\tilde{\mathcal{Q}}) \times \mathcal{D}$ is polynomial time isomorphic to a polynomial-time Boolean algebra $\mathcal{C}_2$ whose universe is $\text{Tal}(\omega)$. Thus $\tilde{\mathcal{B}}$ is computably isomorphic to a polynomial-time Boolean algebra with universe $\text{Bin}(\omega)$ and $\tilde{\mathcal{B}}$ is computably isomorphic to a polynomial-time Boolean algebra with universe $\text{Tal}(\omega)$.

Our analysis of the three cases has proved the following.

Theorem 4.3. *Suppose $\tilde{\mathcal{B}}$ is a computable Boolean algebra.*

(a) *If $\tilde{\mathcal{B}}$ is isomorphic to $\tilde{\mathcal{C}}$ and* $\text{At}(\tilde{\mathcal{B}}))$ *and* $I(\text{At}(\tilde{\mathcal{B}})$ *are computable, then $\tilde{\mathcal{B}}$ is computably isomorphic to a polynomial-time Boolean algebra* $\text{Bin}(\tilde{\mathcal{B}})$ *with universe* $\text{Bin}(\omega)$ *such that* $\text{At}(\text{Bin}(\tilde{\mathcal{B}}))$ *and* $I(\text{At}(\text{Bin}(\tilde{\mathcal{B}})))$ *are polynomial-time and $\tilde{\mathcal{B}}$ is computably isomorphic to a polynomial-time Boolean algebra* $\text{Tal}(\tilde{\mathcal{B}})$ *with universe* $\text{Tal}(\omega)$ *such that* $\text{At}(\text{Tal}(\tilde{\mathcal{B}}))$ *and* $I(\text{At}(\text{Tal}(\tilde{\mathcal{B}})))$ *are polynomial-time.*

(b) *If $\tilde{\mathcal{B}}$ is isomorphic to $\tilde{\mathcal{N}} \times \mathcal{B}$ for some computable Boolean algebra and* $\text{At}(\tilde{\mathcal{B}})$ *is computable, then $\tilde{\mathcal{B}}$ is computably isomorphic to a polynomial-time Boolean algebra* $\text{Bin}(\tilde{\mathcal{B}})$ *with universe* $\text{Bin}(\omega)$ *such that the set of atoms of* $\text{Bin}(\tilde{\mathcal{B}})$ *are polynomial-time. Similarly $\tilde{\mathcal{B}}$ is computably isomorphic to a polynomial-time*

Boolean algebra Tal($\tilde{\mathcal{B}}$) *with universe* Tal(ω) *such that the set of atoms of* Tal($\tilde{\mathcal{B}}$) *are polynomial-time.*

(c) *If $\tilde{\mathcal{B}}$ has an atomless element, i.e. if $\tilde{\mathcal{B}}$ is isomorphic to $\tilde{\mathcal{Q}} \times \mathcal{B}$ for some computable Boolean algebra $\mathcal{B}$, then $\tilde{\mathcal{B}}$ is computably isomorphic to a polynomial-time Boolean algebra with universe* Bin(ω) *and is computably isomorphic to a polynomial-time Boolean algebra with universe* Tal(ω).

If we combine Theorem 4.3 with Theorem 3.5, then we immediately get the following.

Theorem 4.4. *If $\tilde{\mathcal{B}}$ is a countably infinite computable Boolean algebra such that* At($\tilde{\mathcal{B}}$) *and* I(At($\tilde{\mathcal{B}}$)) *are computable, then $\tilde{\mathcal{B}}$ is computably isomorphic to a polynomial-time Boolean algebra* Bin($\tilde{\mathcal{B}}$) *with universe* Bin(ω) *where both the set of atoms and the ideal generated by the set of atoms of* Bin($\tilde{\mathcal{B}}$) *are polynomial-time. Also $\tilde{\mathcal{B}}$ is computably isomorphic to a polynomial-time Boolean algebra* Tal($\tilde{\mathcal{B}}$) *with universe* Tal(ω) *where both the set of atoms and the ideal generated by the set of atoms of* Tal($\tilde{\mathcal{B}}$) *are polynomial-time.*

Next we consider the other extreme case of our three cases, namely, when we assume that neither At($\tilde{\mathcal{B}}$) nor I(At($\tilde{\mathcal{B}}$)) are necessarily computable.

4.2. General computable Boolean algebras $\tilde{\mathcal{B}}$

Suppose that $\tilde{\mathcal{B}}$ is a countably infinite computable Boolean algebra. Now if $\tilde{\mathcal{B}}$ has an atomless element so that $\tilde{\mathcal{B}}$ is isomorphic to $\tilde{\mathcal{Q}} \times \mathcal{B}$ where $\mathcal{B}$ is a computable Boolean algebra, then by part (c) of Theorem 4.3, $\tilde{\mathcal{B}}$ is computably isomorphic to a polynomial-time Boolean algebra with universe Bin(ω) and is computably isomorphic to a polynomial-time Boolean algebra with universe Tal(ω). Thus we can assume that $\tilde{\mathcal{B}}$ is an atomic Boolean algebra. In that case, we cannot guarantee that that $\tilde{\mathcal{B}}$ is computably isomorphic to a polynomial-time Boolean algebra with universe Bin(ω) and is computably isomorphic to a polynomial-time Boolean algebra with universe Tal(ω). Indeed, we have the following.

Theorem 4.5. *If $\mathcal{B}$ is an infinite atomic computable Boolean algebra and G is any infinite polynomial-time subset of* $\{0, 1\}^*$, *then there exists a computable Boolean algebra $\mathcal{C}$ such that $\mathcal{C}$ is isomorphic to $\mathcal{B}$, but $\mathcal{C}$ is not computably isomorphic to any polynomial-time Boolean algebra with universe G.*

Proof. We shall construct our desired computable Boolean algebra $\mathcal{C}$ as a subalgebra of some fixed computable copy of the Boolean algebra of left-closed right-open interval of the rationals Q which we denote by $\tilde{\mathcal{Q}}$.

Let $\mathcal{G}_0, \mathcal{G}_1, \ldots$ be a list of all structures of the form

$$\mathcal{G} = \langle G, 0_{\mathcal{G}}, 1_{\mathcal{G}}, \wedge_{\mathcal{G}}, \vee_{\mathcal{A}}, \neg_{\mathcal{A}} \rangle$$

where $0_{\mathcal{G}}, 1_{\mathcal{G}} \in G$, $\wedge_{\mathcal{G}}$ and $\vee_{\mathcal{G}}$ are binary polynomial-time functions which map $\{0,1\}^* \times \{0,1\}^*$ into $\{0,1\}^*$, and $\neg_{\mathcal{G}}$ is a unary polynomial-time function which maps $\{0,1\}^*$ into $\{0,1\}^*$. We think of each such structure $\mathcal{G}_i$ as potentially defining a polynomial-time Boolean algebra with universe G,

$$\mathcal{D}_i = \langle G, 0_{\mathcal{D}_i}, 1_{\mathcal{D}_i}, \wedge_{\mathcal{D}_i}, \vee_{\mathcal{D}_i}, \neg_{\mathcal{D}_i} \rangle,$$

by setting

1. $\wedge_{\mathcal{D}_i}$ equal to $\wedge_{\mathcal{G}_i}$ restricted to $G \times G$,

2. $\vee_{\mathcal{D}_i}$ equal to $\vee_{\mathcal{G}_i}$ restricted to $G \times G$, and

3. $\neg_{\mathcal{D}_i}$ equal to $\neg_{\mathcal{G}_i}$ restricted to G.

Thus $\mathcal{D}_0, \mathcal{D}_1, \ldots$ contains all polynomial Boolean algebras with universe G. Of course, not every $\mathcal{D}_i$ will be a Boolean algebra. Indeed, as we saw in Proposition 4.1, there exists polynomial-time subsets A of $\{0,1\}^*$ for which there are no polynomial-time Boolean algebras with universe A. Nevertheless, to meet the conditions of the theorem, we need only construct $\mathcal{C}$ isomorphic to $\mathcal{B}$ so that we meet the following requirements.

R_e: If $e = [n,k]$, then ϕ_n restricted to $|\mathcal{C}|$ is not a computable Boolean algebra isomorphism from $\mathcal{C}$ onto $\mathcal{D}_i$.

The first thing that we shall do at stage s of the construction is to check the first s elements of G, $g_1, \ldots, g_s$, under the lexicographic order of $\{0,1\}^*$ and attempt to compute the subalgebra of $\mathcal{D}_i$ generated by $g_1, \ldots, g_s$ for $i = 0, \ldots, s$. That is, for each $i \leq s$, we shall compute all Boolean combinations using the functions $\wedge_{\mathcal{G}_i}$, $\vee_{\mathcal{G}_i}$, and $\neg_{\mathcal{G}_i}$ to see if the operations on $\mathcal{D}_i$ are consistent with being a polynomial-time Boolean algebra with universe G. If we find any inconsistency, e.g. a given Boolean combination which is not in G, two equivalent Boolean combinations which do yield the same answer, etc., then we will say that any requirement R_j where $j = [i,k]$ is *canceled* at stage s.

Let $b_0 = 1_{\mathcal{B}}, b_1, \ldots$ be an computable generating sequence for $\mathcal{B}$. That is, if $\mathcal{B}_s = \mathrm{sub}_{\mathcal{B}}(\{b_0, b_1, \ldots b_s\})$, then

1. $\mathcal{B}_0 \subset \mathcal{B}_1 \subset \cdots \subset \mathcal{B}_s \subset \cdots$,

2. for each s, $\mathcal{B}_s$ is a Boolean algebra with exactly $s+1$ atoms, say $\mathrm{At}(\mathcal{B}_s) = \{d_{1,s}, \ldots, d_{s+1,s}\}$, and

3. for each s, b_{s+1} splits exactly one atom of $\mathcal{B}_s$, that is, there is a $k \leq s+1$ such that $0_{\mathcal{B}} <_{\mathcal{B}} b_{s+1} <_{\mathcal{B}} d_{k,s}$.

At each stage s of our construction, we shall specify a finite subalgebra C_s of $\tilde{\mathcal{Q}}$ and a sequence of isomorphisms $F_k^s : \mathcal{B}_k \to E_k^s$ for $k = 1, \ldots, s$ where E_k^s is a subalgebra of C_s. Our sequence of isomorphisms will be consistent in the sense that for all $0 < k < l \leq s$,

$$F_l^s \lceil B_k = F_k^s.$$

Let $F_s = F_s^s : B_s \to E_s^s$. Now each E_k^s will be a Boolean algebra with exactly $k+1$ atoms. We let $e_{j,k}^s = F_k^s(d_{j,k})$ so that $\mathrm{At}(E_k^s) = \{e_{1,k}^s, \ldots, e_{k+1,k}^s\}$.

Our construction will ensure that we meet the following set of requirements for all $e \geq 1$.

Q_e : $(\forall x \in B_e) \lim_s F_s(x) = F(x)$ exists.

We rank our requirements in decreasing order of priority as

$$Q_0, R_0, Q_1, R_1, \ldots.$$

In addition, our construction will ensure that $C_0 \subset C_1 \subset \cdots$ is a computable sequence of subalgebras of $\tilde{\mathcal{Q}}$ and our desired Boolean algebra $\mathcal{C}$ equals $\bigcup_s C_s$. We let $E_e = F_e(B_e)$. It will then follow that $F = \lim_s F_s$ will be an isomorphism from $\mathcal{B}$ onto $E = \cup_e E_e$ and E will be a subalgebra of $\mathcal{C}$. As we shall see, it will not necessarily be the case that $E = \mathcal{C}$. However our construction will ensure that $\mathcal{C}$ is an atomic Boolean algebra with the property $\mathcal{C}/I(\mathrm{At}(\mathcal{C}))$ is isomorphic to $E/I(\mathrm{At}(E))$. We can then appeal to the following theorem of Vaught that $\mathcal{C}$ is isomorphic to E and hence to $\mathcal{B}$.

Theorem 4.6 ([23]). *Let A and B be countably infinite atomic Boolean algebras. Then $A/I(\mathrm{At}(A))$ is isomorphic to $B/I(\mathrm{At}(B))$ if and only if A is isomorphic to B.*

In addition to the structure given above at stage s, our construction will specify some additional splittings of the atoms of each E_k^s. That is, for each atom $e_{i,k}^s$ of E_k^s, we shall specify a pair of nonzero elements of C^s, $e_{i,k,0}^s$ and $e_{i,k,1}^s$, with the following properties.

1. $e_{i,k}^s = e_{i,k,0}^s \vee_{\tilde{\mathcal{Q}}} e_{i,k,1}^s$ and $0_{\tilde{\mathcal{Q}}} = e_{i,k,0}^s \wedge_{\tilde{\mathcal{Q}}} e_{i,k,1}^s$,
2. if $e_{i,k}^s = e_{j,l}^s$ for some $k \neq l$, then $e_{i,k,0}^s = e_{j,l,0}^s$ and $e_{i,k,1}^s = e_{j,l,1}^s$,
3. if $e_{i,k}^s$ is not an atom of E_s^s and hence $k < s$, then $e_{i,k,0}^s$ and $e_{i,k,1}^s$ are atoms of E_j^s where j is the least $q > k$ such that $e_{i,k}^s$ is not an atom in E_q^k.

Moreover each $e_{i,k,j}^s$ will be further split into two elements of C^s, $f_{i,k,j}^s$ and $g_{i,k,j}^s$ with the following properties.

1. $e_{i,k,j}^s = f_{i,k,j}^s \vee_{\tilde{\mathcal{Q}}} g_{i,k,j}^s$ and $0_{\tilde{\mathcal{Q}}} = f_{i,k,j}^s \wedge_{\tilde{\mathcal{Q}}} g_{i,k,j}^s$,
2. if $e_{i,k,j}^s = e_{t,l,q}^s$ for some $k \neq l$, then $f_{i,k,j}^s = f_{t,l,q}^s$ and $g_{i,k,j}^s = g_{t,l,q}^s$,
3. $g_{i,k,j}^s$ is a nonzero element of C_s,

4. $f^s_{i,k,j}$ is a join of a finite number of atoms of C_s, $f^s_{i,k,j,t}$ for $1 \leq t \leq p^s_{i,k,j}$, and

5. if $e^s_{i,j,k} = e^s_{r,t}$ for some $r > j$, then either $f^s_{i,k,j} \leq_{C_s} f^s_{r,t,0}$ or $f^s_{i,k,j} \leq_{C_s} f^s_{r,t,1}$. (This condition says later splittings of $e^s_{i,j,k}$ will be accomplished by splitting $g^s_{i,k,j}$ rather than $f^s_{i,k,j}$.)

At any given stage s, C_s will simply be the subalgebra of $\tilde{\mathcal{Q}}$ which is generated by E^s_s plus all the elements of the form $g^s_{i,k,j}$, $f^s_{i,k,j}$, and $f^s_{i,k,j,t}$ where $k \leq s$. The role of these extra specified splittings will become clear when we explain our strategy to meet the requirements R_e.

Our basic strategy for meeting a requirement R_e where $e = [n, k]$ which is never canceled is to wait until s is large enough so that $\phi^s_n(x) \downarrow$ for all $x = e^s_{i,e,j}$ where $1 \leq i \leq e+1$ and $j \in \{0, 1\}$. In that case, we let $h^s_{i,e,j} = \phi_n(e^s_{i,e,j})$. We then consider the elements $h^s_{i,e,j}$ in the structure $\mathcal{D}_k$. If ϕ_n were an isomorphism from $\mathcal{C}$ onto $\mathcal{D}_k$, then the elements $h^s_{i,e,j}$ where $1 \leq i \leq e+1$ and $j \in \{0, 1\}$ would be a set of pairwise disjoint elements of $\mathcal{D}_k$ whose join is $1_{\mathcal{D}_k}$. If this is not the case, then we will cancel requirement R_e. If it is the case, then we start to compute the sequence of subalgebras of $\mathcal{D}_k$, $\mathcal{D}_{k,t} = \text{sub}_{\mathcal{D}_k}(\{g_0, \ldots, g_t\})$ in order for $t = 0, 1, 2, \ldots$. We then wait for one of two things to happen. Either we find some inconsistency with $\mathcal{D}_k$ being a Boolean algebra in which case we cancel requirement R_e or there will be a t large enough so that the cardinality of

$$[h^s_{i,e,j}]_{\mathcal{D}_{k,t}} = \{x \in \mathcal{D}_{k,t} : x \leq_{\mathcal{D}_k} h^s_{i,e,j}\}$$

is strictly larger than $|C_s|^2$.

In the latter case, we know that if we can restrict any further splittings of $e^s_{k,e,j}$ in $\mathcal{C}$, then it can not be the case that ϕ_n is an isomorphism from $\mathcal{C}$ onto $\mathcal{D}_k$. That is we would like to *freeze* $e^s_{i,e,j}$ from every being split at any latter stages of the construction. We note that if $\mathcal{D}_k$ is actually a p-time Boolean algebra, then eventually we must find such a t and $h^s_{i,e,j}$ because $\mathcal{D}_k$ is an infinite Boolean algebra. Once we find such a t and $h^s_{i,e,j}$, we will do one of two things. First suppose that $e^s_{i,e,j}$ is not an atom of E^s_s. In that case, $e^s_{i,e,j}$ will be an atom of some E^s_k where $e < k \leq s$. In this case, we let $e^s_{i,e,1-j} = f^s_{i,e,1-j} \vee g^s_{i,e,1-j}$. In that case, it may be that $f^s_{i,e,1-j}$ has been frozen by some other requirement so that we do not want to split $f^s_{i,e,1-j}$. However, we assume that $g^s_{i,e,1-j}$ or at least some atom under $g^s_{i,e,1-j}$ is free to split. Thus we will look for the least atom C_s of the form $g^s_{u,v,w}$ which is contained in $g^s_{i,e,1-j}$. We will then split it into two nonzero elements of $\tilde{\mathcal{Q}}$, a and b and let

1. $e^{s+1}_{i,e,j} = e^s_{i,e,j} \vee a$,

2. $f^{s+1}_{i,e,j} = e^s_{i,e,j}$ and $g^{s+1}_{i,e,j} = a$,

3. $e^{s+1}_{i,e,1-j} = e^s_{i,e,1-j} - a$,

4. $f^{s+1}_{i,e,1-j} = e^{s}_{i,e,1-j} - (a \vee b)$ and $g^{s+1}_{i,e,1-j} = b$.

We will also put a Δ_e marker on $f^{s}_{i,e,j}$ to record the fact that we have taken an action to satisfy requirement R_e. Note that we will still have $e^{s+1}_{i,e} = e^{s+1}_{i,e,0} \vee e^{s+1}_{i,e,1} = e^{s}_{i,e}$ and hence we can let $E^s_u = E^{s+1}_u$ and $F^s_u = F^{s+1}_u$ for $u \leq e$. We may have to split $e^{s+1}_{i,e,0}$ and $e^{s+1}_{i,e,1}$ further to reconstruct the required isomorphisms F^{s+1}_k for $k > e$ but we will ensure that all such splittings are accomplished by splitting $g^{s+1}_{i,e,0}$ and $g^{s+1}_{i,e,1}$ but not splitting $f^{s+1}_{i,e,0}$ and $f^{s+1}_{i,e,1}$. If $e^{s}_{i,e}$ is an atom of E^s, then we will simply put a Δ_e marker on $e^{s}_{i,e,j}$. The idea is that it is possible that $e^{s}_{i,e}$ is really the image of an atom of $\mathcal{B}$ and hence can not allow it to be split infinitely often. Now if at some latter stage t, we discover that $e^{s}_{i,e}$ is not an atom of E^t_t, then we would like to take the action as described above to freeze $e^{t-1}_{i,e,j}$ and create new versions of $e^{t}_{i,e,0}$ and $e^{t}_{i,e,1}$. However at stage t, it may be that both $e^{t-1}_{i,e,0}$ and $e^{t-1}_{i,e,1}$ have some Δ_i markers on them. In such a situation, we will have taken an action to freeze $e^{t-1}_{i,e,j}$ if it has a Δ_i marker on it of highest priority. Thus we may cause an injury to those requirements which want to freeze $e^{t-1}_{i,e,1-j}$, but the usual priority argument will show that we eventually get a chance to take an action to satisfy every requirement which will not be injured.

We will ensure that C is computable by ensuring that for all s, $s \in C \iff s \in C_s$. This is easily accomplished for subalgebras $\tilde{\mathcal{Q}}$ by the following mechanism. Suppose that we are given a finite subalgebra, C_s, of $\tilde{\mathcal{Q}}$ and the set of atoms of C_s is $\{c^s_1, \ldots, c^s_n\}$. Then when we extend the subalgebra C_s to C_{s+1}, we may need to split certain c^s_i into finitely many elements $c^s_{i,1}, \ldots, c^s_{i,t_i}$. There are infinitely many splittings of c^s_i in $\tilde{\mathcal{Q}}$. Moreover it is easy to show that if $x_1, x_2, y_1, y_2 \in \tilde{\mathcal{Q}}$ are such that

1. $x_1 \vee_{\tilde{\mathcal{Q}}} x_2 = c^s_i$,

2. $x_1 \wedge_{\tilde{\mathcal{Q}}} x_2 = 0_{\tilde{\mathcal{Q}}}$,

3. $y_1 \vee_{\tilde{\mathcal{Q}}} y_2 = c^s_i$,

4. $y_1 \wedge_{\tilde{\mathcal{Q}}} y_2 = 0_{\tilde{\mathcal{Q}}}$, and

5. $x_1 \notin \{y_1, y_2\}$

then $\mathrm{sub}_{\tilde{\mathcal{Q}}}(C_s \cup x_1, x_2) \cap \mathrm{sub}_{\tilde{\mathcal{Q}}}(C_s \cup \{y_1, y_2\}) = C_s$. Since there are infinitely many splittings of c^s_i in $\tilde{\mathcal{Q}}$, it follows that for any finite set $S \subseteq \tilde{\mathcal{Q}} - C_s$, there exists $x_1, x_2 \in \tilde{\mathcal{Q}}$ such that $x_1 \vee_{\tilde{\mathcal{Q}}} x_2 = c^s_i, x_1 \wedge_{\tilde{\mathcal{Q}}} x_2 = 0_{\tilde{\mathcal{Q}}}$, and $\mathrm{sub}_{\tilde{\mathcal{Q}}}(C_s \cup \{x_1, x_2\}) \cap S = \emptyset$. Thus in particular, if $S = \{0, \ldots s\} - C_s$, then for any $j \leq s$, $j \in C_s \iff j \in \mathrm{sub}_{\tilde{\mathcal{Q}}}(C_s \cup \{x_1, x_2\})$. Thus we can extend C_s by splitting the atom c^s_i in such a way that we do not add any new elements which are less than s. We can then repeat the process and extend $\mathrm{sub}_{\tilde{\mathcal{Q}}}(C_s \cup \{x_1, x_2\})$ by splitting x_1 or x_2 into two elements and again ensure that we do add new elements which are less than or equal to s, etc. It follows that we can split each of the atoms C_s into any finite number of pieces in such a way that we do not add any new elements which are less than or equal to s to the extension. In this way we can guarantee that $s \in \mathcal{C}$ if and only if $s \in C_s$ and hence $\mathcal{C}$ will be a computable Boolean algebra.

Construction

Stage 0: Let $E_0^0 = \{0_{\tilde{\mathcal{Q}}}, 1_{\tilde{\mathcal{Q}}}\}$, $F_0^0(1_{\mathcal{B}}) = 1_{\tilde{\mathcal{Q}}}$ and $F_0^0(0_{\mathcal{B}}) = 0_{\tilde{\mathcal{Q}}}$. Let x be the least nonzero element of $\tilde{\mathcal{Q}}$ and let $e_{1,0} = 1_{\tilde{\mathcal{Q}}}$, $e_{1,0,0}^0 = x$ and $e_{1,0,1}^0 = \neg_{\tilde{\mathcal{Q}}}(x)$. Finally let x_1, x_2, y_1, y_2 be the least quadruple of elements in the lexicographic order of $\tilde{\mathcal{Q}}$ such that

$$x_1 \vee_{\tilde{\mathcal{Q}}} x_2 = x \text{ and } x_1 \wedge_{\tilde{\mathcal{Q}}} x_2 = 0_{\tilde{\mathcal{Q}}} \text{ and}$$

$$y_1 \vee_{\tilde{\mathcal{Q}}} y_2 = \neg_{\tilde{\mathcal{Q}}}(x) \text{ and } y_1 \wedge_{\tilde{\mathcal{Q}}} y_2 = 0_{\tilde{\mathcal{Q}}}.$$

Then let $f_{1,0,0}^0 = x_1$, $f_{1,0,1}^0 = y_1$, let $g_{1,0,0}^0 = x_2$, and $g_{1,0,1}^0 = y_2$.

Stage $s+1$: Assume that we have defined the subalgebra C_s, the sequence of subalgebras $E_0^s \subset \cdots \subset E_s^s \subseteq C_s$, the sequence of isomorphisms $F_k^s : \mathcal{B}_k \to E_k^s$, the set of specified splittings $e_{i,k}^s = e_{i,k,0}^s \vee e_{i,k,1}^s$, $e_{i,k,0}^s = f_{i,k,0}^s \vee g_{i,k,0}^s$, and $e_{i,k,1}^s = f_{i,k,1}^s \vee g_{i,k,1}^s$, for $k \leq s$ and $1 \leq i \leq k+1$, as described in our remarks preceding the construction. Moreover assume that if any of the elements $e_{i,k,j}^s$ or $f_{i,k,j}^s$ has a Δ_e marker on it, then $e < j$ where j is the least i such that $e_{i,k}^s$ is not an atom of E_j^s.

First for each $j \leq s+1$, check the consistency of $\mathcal{D}_j$ being a Boolean algebra by attempting to compute the subalgebra generated by $g_0, \ldots, g_{s+1}$. Cancel all requirements R_e where $e = [n, j]$ where $j \leq s+1$ and $\mathcal{D}_j$ is found to be not consistent with being a Boolean algebra by this consistency check.

Next look for a requirement R_e where $e = [n, k] \leq s$ such that R_e is not canceled, there does not exists a Δ_e marker which is currently resting on some element of C_s, and $\phi_n(e_{i,e,j}^s)\downarrow$ for all $i \leq e+1$ and $j \in \{0, 1\}$.

If there is no such e, let $d_{i,s}$ be that atom of $\mathcal{B}_s$ which is split by b_{s+1}. This means that we must split $e_{i,s}^s$ in E_s^s. Then for $p < s$, $1 \leq k \leq p$ and $j \in \{0, 1\}$, we let $E_p^s = E_p^{s+1}$, $F_p^s = F_p^{s+1}$, $e_{k,p}^s = e_{k,p}^{s+1}$, $e_{k,p,j}^s = e_{k,p,j}^{s+1}$, $f_{k,p,j}^s = f_{k,p,j}^{s+1}$, and $g_{k,p,j}^s = g_{k,p,j}^{s+1}$. Similarly we let $E_s^s = E_s^{s+1}$ and $F_s^s = F_s^{s+1}$ and for $k \neq i$ and $j \in \{0, 1\}$, we set $e_{k,s}^s = e_{k,s}^{s+1}$, $e_{k,s,j}^s = e_{k,s,j}^{s+1}$, $f_{k,s,j}^s = f_{k,s,j}^{s+1}$, and $g_{k,s,j}^s = g_{k,s,j}^{s+1}$. We also let $e_{i,s}^s = e_{i,s}^{s+1}$. Next consider $e_{i,s,0}^s = f_{i,s,0}^s \vee_{\tilde{\mathcal{Q}}} g_{i,s,0}^s$ and $e_{i,s,1}^s = f_{i,s,1}^s \vee_{\tilde{\mathcal{Q}}} g_{i,s,1}^s$. Look for any Δ_r markers which lie on either $e_{i,s,0}^s$ or $e_{i,s,1}^s$. Pick $j \in \{0, 1\}$ so that either there is no Δ_j marker on $e_{i,s,j}^s$ or there is a Δ_t marker on $e_{i,s,1-j}^s$ which has higher priority than any marker on $e_{i,s,j}^s$. If neither $e_{i,s,0}^s$ nor $e_{i,s,1}^s$ have Δ_r markers on them for any r, then let $j = 0$. Then find the least nonzero elements in $\tilde{\mathcal{Q}}$, $a_1, a_2, a_3, b_1, b_2, b_3, b_4$, such that

1. $a_1, a_2, a_3, b_1, b_2, b_3, b_4$ are pairwise disjoint in $\tilde{\mathcal{Q}}$,

2. $a_1 \vee_{\tilde{\mathcal{Q}}} a_2 \vee_{\tilde{\mathcal{Q}}} a_3 \vee_{\tilde{\mathcal{Q}}} b_1 \vee_{\tilde{\mathcal{Q}}} b_2 \vee_{\tilde{\mathcal{Q}}} b_3 \vee_{\tilde{\mathcal{Q}}} b_4 = g^s_{i,s,j}$, and

3. $\text{sub}_{\tilde{\mathcal{Q}}}(C_s \cup \{a_1, a_2, a_3, b_1, b_2, b_3, b_4\}) \cap \{0, \ldots, s+1\} = C_s \cap \{0, \ldots, s+1\}$.

Let $a = a_1 \vee_{\tilde{\mathcal{Q}}} a_2 \vee_{\tilde{\mathcal{Q}}} a_3$ and $b = b_1 \vee_{\tilde{\mathcal{Q}}} b_2 \vee_{\tilde{\mathcal{Q}}} b_3 \vee_{\tilde{\mathcal{Q}}} b_4$. Then we let $e^{s+1}_{i,s,1-j} = e^{s+1}_{i,s} -_{\tilde{\mathcal{Q}}} b$ and let $f^{s+1}_{i,s,1-j} = e^{s+1}_{i,s,1-j} -_{\tilde{\mathcal{Q}}} a$ and $g^{s+1}_{i,s,1-j} = a$. We then let $e^{s+1}_{i,s,j} = b$ and let $f^{s+1}_{i,s,1-j} = b_1$ and $g^{s+1}_{i,s,1-j} = b_2 \vee_{\tilde{\mathcal{Q}}} b_3 \vee_{\tilde{\mathcal{Q}}} b_4$. (Note that this has the effect of incorporating $e^s_{i,s,1-j}$ into $f^{s+1}_{i,s,1-j}$ and hence this will ensure that we don't split $e^s_{i,s,1-j}$ later in the construction.) We remove all the Δ_r markers which where on $e^s_{i,s,j}$. The markers which were on the element $e^s_{i,s,1-j}$ do not move but now they will now rest on elements which are $\leq_C f^{s+1}_{i,s,1-j}$. We next define F^{s+1}_{s+1} by defining it on the atoms $\mathcal{B}_{s+1}$ and extending to be a Boolean algebra isomorphism by setting

1. $F^{s+1}_{s+1}(d_{k,s}) = e^{s+1}_{k,s}$ if $k \neq i$,

2. $F^{s+1}_{s+1}(d_{i,s} - b_{s+1}) = e^{s+1}_{i,s,1-j}$, and

3. $F^{s+1}_{s+1}(b_{s+1}) = e^{s+1}_{i,s,j}$.

It follows that if $d_{k,s+1} = d_{i,s} - b_{s+1}$ and $d_{l,s+1} = b_{s+1}$, then we need only define $e^{s+1}_{q,s+1,r}$, $f^{s+1}_{q,s+1,r}$, and $g^{s+1}_{q,s+1,r}$ for $q \in \{k, l\}$ and $r \in \{0, 1\}$. Now since $F^{s+1}_{s+1}(d_{i,s} - b_{s+1}) = F^{s+1}_{s+1}(d_{k,s+1}) = e^{s+1}_{k,s+1} = e^{s+1}_{i,s,1-j}$, we let $e^{s+1}_{k,s+1,0} = f^{s+1}_{i,s,1-j} \vee_{\tilde{\mathcal{Q}}} a_1$, $f^{s+1}_{k,s+1,0} = f^{s+1}_{i,s,1-j}$ and $g^{s+1}_{k,s+1,0} = a_1$. Similarly we let $e^{s+1}_{k,s+1,1} = a_2 \vee_{\tilde{\mathcal{Q}}} a_3$, $f_{k,s+1,1^{s+1}} = a_2$ and $g^{s+1}_{k,s+1,1} = a_3$. Since $F^{s+1}_{s+1}(b_{s+1}) = F^{s+1}_{s+1}(d_{l,s+1}) = e^{s+1}_{k,s+1} = e^{s+1}_{i,s,j}$, we let $e^{s+1}_{l,s+1,0} = f^{s+1}_{i,s,j} \vee_{\tilde{\mathcal{Q}}} b_2$, $f^{s+1}_{k,s+1,0} = f^{s+1}_{i,s,j} = b_1$ and $g^{s+1}_{k,s+1,0} = b_2$ and we let $e^{s+1}_{k,s+1,1} = b_3 \vee_{\tilde{\mathcal{Q}}} b_4$, $f^{s+1}_{k,s+1,1} = b_3$ and $g^{s+1}_{k,s+1,1} = b_4$. It is then easy to see that our construction ensures that all $e^{s+1}_{x,y,z}$, $f^{s+1}_{x,y,z}$ and $g^{s+1}_{x,y,z}$ are nonzero elements of $\tilde{\mathcal{Q}}$ which satisfy the conditions specified in our discussion preceding the constructions

Next suppose that there is a requirement R_e where $e = [n, k] \leq s$ such that R_e is not canceled, there does not exists a Δ_e marker which is currently resting on some element of C_s and $\phi_n(e^s_{i,e,j}) \downarrow$ for all $i \leq e + 1$ and $j \in \{0, 1\}$. Let $f = [u, w]$ be the least such requirement. Let $h^s_{i,e,j} = \phi_{u,s}(e^s_{i,f,j})$ for all $1 \leq i \leq f$ and $j \in \{0, 1\}$. Then begin to compute the subalgebras $\mathcal{D}_{w,t} = \text{sub}_{\mathcal{D}_w}(\{g_0, \ldots, g_t\})$ in order until either we find that $\mathcal{D}_w$ is not consistent with being a Boolean algebra with universe G, in which case we cancel all requirements of the form R_e where $e = [v, w]$, or we find t, k and j such that the cardinality of $[h^s_{k,f,j}]_{\mathcal{D}_{w,t}}$ is strictly greater than $|C_s|^2$.

If we find that R_f is canceled, then we extend C_s to C_{s+1} as described above. Otherwise assume that t, k are j are the least triple such that $[h^s_{k,f,j}]_{\mathcal{D}_{w,t}}$ is strictly greater than $|C_s|^2$. We now have two cases.

The first case is when $e^s_{k,f}$ is an atom of E^s_s. In this case, we simply put a Δ_f marker on $e^s_{k,f,j}$ and proceed to extend C_s to C_{s+1} as described above. Note that one of two things will happen in this case. It could be that $e^s_{k,f}$ is the atom that must be split in order to extend E^s_s to E^{s+1}_{s+1}. In that case, the fact that the Δ_f marker is on $e^s_{k,f,j}$ will be taken into account when we decide which of $e^s_{k,f,j}$ and $e^s_{k,f,1-j}$ will be incorporated into a frozen element in the next stage. If $e^s_{k,f,j}$ is incorporated into a frozen element at the next stage, then we will have satisfied requirement R_f since $e^s_{k,f,j}$ will never be split in C. If $e^s_{k,f,1-j}$ is not incorporated into a frozen element at the next stage, the Δ_f will be abandoned, but we will have taken an action to satisfy a higher priority requirement.

The second case is when $e^s_{k,f}$ is not an atom in E^s_s. Then we proceed as follows. Our first task is to reconstruct E^{s+1}_s and then we shall extend E^{s+1}_s to E^{s+1}_{s+1}. For $p < f$, $1 \leq r \leq p+1$, and $q \in \{0,1\}$, we let $E^s_p = E^{s+1}_p$, $F^s_p = F^{s+1}_p$, $e^s_{r,p} = e^{s+1}_{r,p}$, $e^s_{r,p,q} = e^{s+1}_{r,p,q}$, $f^s_{r,p,q} = f^{s+1}_{r,p,q}$, and $g^s_{r,p,q} = g^{s+1}_{r,p,q}$. Similarly we let $E^s_f = E^{s+1}_f$ and $F^s_f = F^{s+1}_f$ and for $i \neq k$ and $q \in \{0,1\}$, we set $e^s_{i,f} = e^{s+1}_{i,f}$, $e^s_{i,f,q} = e^{s+1}_{i,f,q}$, $f^s_{i,f,q} = f^{s+1}_{i,f,q}$, and $g^s_{i,f,q} = g^{s+1}_{k,f,q}$. We also let $e^s_{k,s} = e^{s+1}_{k,s}$. Next consider $e^s_{k,f,0} = f^s_{k,f,0} \vee_{\tilde{Q}} g^s_{k,f,0}$ and $e^s_{k,f,1} = f^s_{i,f,1} \vee_{\tilde{Q}} g^s_{i,f,1}$. By our conditions on the $e^s_{a,b}$, $f^s_{a,b,c}$ and $g^s_{a,b,c}$, we know that there is some atom $e^s_{x,s}$ of E^s_s such that $e^s_{x,s} \leq_{C_s} e^s_{k,f,j}$ and there is some q such that $g^s_{x,s,q} \wedge_{\tilde{Q}} f^s_{k,f,j} = 0_{\tilde{Q}}$. Then find the least elements in $\tilde{Q}$, $a_1, a_2, a_3, b_1, b_2, b_3, b_4$ such that

1. $a_1, a_2, a_3, , b_1, b_2, b_3, b_4$ are pairwise disjoint in $\tilde{Q}$,
2. $a_1 \vee_{\tilde{Q}} a_2 \vee_{\tilde{Q}} a_3 \vee_{\tilde{Q}} b_1 \vee_{\tilde{Q}} b_2 \vee_{\tilde{Q}} b_3 \vee_{\tilde{Q}} b_4 = g^s_{x,s,q}$, and
3. $\mathrm{sub}_{\tilde{Q}}(C_s \cup \{a_1, a_2, a_3, b_1, b_2, b_3, b_4\}) \cap \{0, \ldots, s+1\} = C_s \cap \{0, \ldots, s+1\}$.

Let $a = a_1 \vee_{\tilde{Q}} a_2 \vee_{\tilde{Q}} a_3$ and $b = b_1 \vee_{\tilde{Q}} b_2 \vee_{\tilde{Q}} b_3 \vee_{\tilde{Q}} b_4$. Then we let $e^{s+1}_{k,f,j} = e^{s+1}_{k,f} -_{\tilde{Q}} b$ and let $f^{s+1}_{k,f,j} = e^{s+1}_{k,f,j} -_{\tilde{Q}} a$ and $g^{s+1}_{k,f,j} = a$. We then let $e^{s+1}_{k,f,1-j} = b$ and let $f^{s+1}_{k,f,1-j} = b_1$ and $g^{s+1}_{k,f,1-j} = b_2 \vee_{\tilde{Q}} b_3 \vee_{\tilde{Q}} b_4$. (Note that this has the effect of incorporating $e^s_{k,f,j}$ into $f^{s+1}_{k,f,j}$ and hence this will ensure that we don't split $e^s_{k,f,j}$ later in the construction.) We then place a Δ_f marker on $f^{s+1}_{k,f,j}$. We now have to reconstruct the $e^{s+1}_{x,y}$, $e^{s+1}_{x,y,z}$, $f^{s+1}_{x,y,z}$, and $g^{s+1}_{x,y,z}$ for $f < y \leq s$. Now if it is the case that $e^s_{x,y} \wedge_{\tilde{Q}} e^s_{k,f} = 0_{\tilde{Q}}$, then we simply define $e^{s+1}_{x,y} = e^s_{x,y}$, $e^{s+1}_{x,y,z} = e^s_{x,y,z}$, $f^{s+1}_{x,y,z} = f^s_{x,y,z}$, and $g^{s+1}_{x,y,z} = g^s_{x,y,z}$ for all z.

This leaves us only to define $e^{s+1}_{x,y}$, $e^{s+1}_{x,y,z}$, $f^{s+1}_{x,y,z}$, and $g^{s+1}_{x,y,z}$ for those x, y such that $e^s_{x,y} \leq_{C_s} e^s_{k,f}$. We define these values in stages according to the first stage these elements show up the sequence $E^s_0 \subset \cdots E^s_s$. That is, suppose that $t > f$ is such that $e^s_{k,f}$ is an atom of E^s_{t-1} but is not an atom of E^s_t. It follows that we have defined $e^{s+1}_{i,r}$ for all $f < r \leq t-1$ and $1 \leq i \leq r+1$ and hence we can let $E^{s+1}_r = E^s_r$ and

$F_r^{s+1} = F_r^s$ for $f < r \leq t-1$. We next define F_t^{s+1} by defining it on the atoms $\mathcal{B}_t$ and extending to be a Boolean algebra isomorphism. Now suppose that $e_{k,f}^s = e_{i,t-1}^s$ and that b_t splits the atom $d_{i,t-1}$ in $\mathcal{B}_{t-1}$. Moreover assume that $b_t = d_{n,t}$ and $d_{i,t-1} - b_t = d_{m,t}$. Then we define F_t as follows.

1. $F_t^{s+1}(d_{z,t}) = e_{z,t}^{s+1}$ if $z \notin \{m, n\}$,
2. $F_{s+1}^{s+1}(d_{i,s} - b_{s+1}) = e_{k,f,j}^{s+1}$, and
3. $F_{s+1}^{s+1}(b_{s+1}) = e_{k,f,1-j}^{s+1}$.

Next we define $e_{q,t,r}^{s+1}$, $f_{q,t,r}^{s+1}$, $g_{q,t,r}^{s+1}$ for $q \in \{m, n\}$ and $r \in \{0, 1\}$. Now since $F_{s+1}^{s+1}(d_{i,t-1} - b_t) = F_{s+1}^{s+1}(d_{n,t}) = e_{n,t}^{s+1} = e_{k,f,j}^{s+1}$, we let $e_{n,t,0}^{s+1} = f_{k,f,j}^{s+1} \vee_{\tilde{\mathcal{Q}}} a_1$, $f_{n,t,0}^{s+1} = f_{k,f,j}^{s+1}$ and $g_{n,t,0}^{s+1} = a_1$ and we let $e_{n,t,1}^{s+1} = a_2 \vee_{\tilde{\mathcal{Q}}} a_3$, $f_{n,t,1}^{s+1} = a_2$, and $g_{n,t,1}^{s+1} = a_3$. Next since $F_t^{s+1}(b_{s+1}) = F_t^{s+1}(d_{m,t}) = e_{m,t}^{s+1} = e_{k,f,1-j}^{s+1} = f_{k,f,1-j}^{s+1} \vee_{\tilde{\mathcal{Q}}} (b_2 \vee_{\tilde{\mathcal{Q}}} b_3 \vee_{\tilde{\mathcal{Q}}} b_4)$, we let $e_{m,t,0}^{s+1} = f_{k,f,j-1}^{s+1} \vee_{\tilde{\mathcal{Q}}} b_2$, $f_{m,t,0}^{s+1} = f_{k,f,j-1}^{s+1}$ and $g_{m,t,0} = b_2$ and we let $e_{m,t,1}^{s+1} = b_3 \vee_{\tilde{\mathcal{Q}}} b_4$, $f_{m,t,1}^{s+1} = b_3$ and $g_{m,t,1} = b_4$.

Any further splittings of either $e_{n,t}^{s+1}$ or $e_{m,t}^{s+1}$ are to be constructed in the following manner. Suppose, for example, that $t < t_1 \leq s$ is a stage such that $e_{n,t}^s$ is an atom of $E_{t_1-1}^s$ but is not an atom of $E_{t_1}^s$. Moreover suppose that $e_{n,t}^s = e_{n_1,t_1}^s \vee_{\tilde{\mathcal{Q}}} e_{n_2,t_1}^s$. That is e_{n_1,t_1}^s and e_{n_2,t_1}^s are the atoms of $E_{t_1-1}^s$ whose join is $e_{n,t}^s$. Then we must reconstruct atoms e_{n_1,t_1}^{s+1} and e_{n_2,t_1}^{s+1} of $E_{t_1}^{s+1}$ whose join is $e_{n,t}^{s+1}$. This is easy. We simply let $e_{n_1,t_1}^{s+1} = e_{n,t,0}^{s+1}$ and $e_{n_2,t_1}^{s+1} = e_{n,t,1}^{s+1}$. Then find least elements in $\tilde{\mathcal{Q}}$, $x_1, x_2, x_3, y_1, y_2, y_3$ such that

1. $x_1, x_2, x_3, y_1, y_2, y_3$ are pairwise disjoint in $\tilde{\mathcal{Q}}$,
2. $x_1 \vee_{\tilde{\mathcal{Q}}} x_2 \vee_{\tilde{\mathcal{Q}}} x_3 = g_{n,t,0}^{s+1}$,
3. $y_1 \vee_{\tilde{\mathcal{Q}}} y_2 \vee_{\tilde{\mathcal{Q}}} y_3 = g_{n,t,1}^{s+1}$ and
4. adding $x_1, x_2, x_3, y_1, y_2, y_3$ to the elements that we have already added to C_{s+1} will not generate any element which is in $\{0, \dots, s+1\} - C_s$.

Then we let

1. $e_{n_1,t_1,0}^{s+1} = f_{n,t,0}^{s+1} \vee_{\tilde{\mathcal{Q}}} x_1$,
2. $f_{n_1,t_1,0}^{s+1} = f_{n,t,0}^{s+1}$,
3. $g_{n_1,t_1,0}^{s+1} = x_1$,
4. $e_{n_1,t_1,1}^{s+1} = x_2 \vee_{\tilde{\mathcal{Q}}} x_3$,

5. $f^{s+1}_{n_1,t_1,1} = x_2$, and

6. $g^{s+1}_{n_1,t_1,1} = x_3$

Similarly we let

1. $e^{s+1}_{n_2,t_1,0} = f^{s+1}_{n,t,1} \vee_{\tilde{\mathcal{Q}}} y_1$,

2. $f^{s+1}_{n_2,t_1,0} = f^{s+1}_{n,t,1}$,

3. $g^{s+1}_{n_2,t_1,0} = y_1$,

4. $e^{s+1}_{n_2,t_1,1} = y_2 \vee_{\tilde{\mathcal{Q}}} y_3$,

5. $f^{s+1}_{n_1,t_1,1} = y_2$, and

6. $g^{s+1}_{n_1,t_1,1} = y_3$

In this way, we can reconstruct all $e^{s+1}_{x,y}$, $e^{s+1}_{x,y,z}$, $f^{s+1}_{x,y,z}$, and $g^{s+1}_{x,y,z}$ for those x, y such that $e^s_{x,y} \leq_{C_s} e^s_{k,f}$. It will then be the case that we have defined all $e^{s+1}_{x,y}$, $e^{s+1}_{x,y,z}$, $f^{s+1}_{x,y,z}$, and $g^{s+1}_{x,y,z}$ for $y \leq s$ and $1 \leq x \leq y+1$ and $z \in \{0, 1\}$. This allows us to define E^{s+1}_y as the subalgebra generated by $\{e^{s+1}_{1,y}, \ldots, e^{s+1}_{y+1,y}\}$ and define F^{s+1}_y be defining $F^{s+1}_y(d_{i,y}) = e^{s+1}_{i,y}$ for all $y \leq s$. Thus we are only left with the process of extending E^{s+1}_s to E^{s+1}_{s+1} which we do exactly as in the case where there was no requirement to satisfy.

End of construction

It is easy to see that each stage is completely effective, thus $\mathcal{C} = \bigcup_s C_s$ is a c.e. subalgebra of $\tilde{\mathcal{Q}}$. Moreover our construction ensures that $s \in C_s \iff s \in \mathcal{C}$ so that $\mathcal{C}$ is a computable subalgebra. We now prove a sequence of lemmas to show that $\mathcal{C}$ has the properties claimed in the theorem.

Lemma 4.7. *For any given requirement R_e, we act for R_e at most finitely often, that is, $\{s : e(s) = e\}$ is finite for all e.*

Proof. We proceed by induction on e. Thus we can assume that there is a stage s_0 such that $e(s) \geq e$ for all $s \geq s_0$.

If a requirement R_e is canceled at some stage s, then if will stay canceled at all later stages. It follows that $e(t) \neq e$ for all $t > s$ so that $\{s : e(s) = e\}$ will automatically be finite. Thus we can assume that R_e is never canceled.

Next observe that our construction ensures that we never split any element of $\mathcal{C}$ of the form $f^s_{x,y,a}$. That is, at stage s, it may be that $f^s_{x,y,a}$ is a finite union of atoms of C_s but we never introduce any additional $\mathcal{C}$-splittings of $f^s_{x,y,a}$ after stage s. Now suppose that we take an action for $R(e)$ at stage $s+1 > s_0$. Let $e = [n,k]$ and $e^s_{i,e,j}$ be the element such that $[\phi_{n,s}(e^s_{i,e,j})]_{\mathcal{D}_{k,t}}$ has cardinality greater than or equal $|C_s|^2$ which we use to satisfy R_e at stage $s+1$. If $e^s_{i,e,j}$ is not an atom of E^s_s, then $e^s_{i,e,j}$ will be $\leq_{C_{s+1}} f^{s+1}_{i,e,j}$ at stage $s+1$ and hence we place a Δ_e marker on $f^{s+1}_{i,e,j}$. This marker is never removed which means that $e(t) \neq e$ for all $t > s+1$. Thus once again $\{s : e(s) = e\}$ will be finite.

The other possibility is that $e^s_{i,e}$ is an atom of E^s_s in which case we place a Δ_e marker on $e^s_{i,e,j}$. The only way that this marker can be removed is if at some stage $t \geq s$, we split $e^s_{i,e}$ in E^t_t and there is a Δ_f marker on $e^t_{i,e,1-j}$ with a higher priority than Δ_e. However in this case $e^t_{i,e,1-j}$ will then become $\leq_{C_{t+1}} f^{t+1}_{i,e,1-j}$ and hence the Δ_f marker on $e^t_{i,e,1-j}$ will never be removed. It follows that a Δ_e marker can be removed at most e times, once for each of $R_0, \ldots, R_{e-1}$. But then if $\{s : e(s) = e\}$ were infinite, there would be a stage $t > s_0$ so that we place a Δ_e marker on an element at stage t and it is never removed there after. But this is a contradiction since that would mean that $\{s : e(s) = e\} \subseteq \{0, \ldots, t\}$. Thus $\{s : e(s) = e\}$ must be finite as claimed. □

Lemma 4.8. *For all k, $\lim_s E^s_k = E_k$ and $\lim_s F^s_k = F_k$ exists and hence $F = \bigcup_k F_k$ is an isomorphism from $\mathcal{B}$ onto $E = \bigcup_k E_k$.*

Proof. Our construction ensures that $E^s_k \neq E^{s+1}_k$ only if there $e(s+1) < k$ and the action that we take to meet requirement $R_{e(s+1)}$ at stage $s+1$ causes use to redefine $e^s_{i,e(s+1),0}$ and $e^s_{i,e(s+1),1}$ at stage $s+1$. However by Lemma 4.7, this can happen at most finitely often. Thus $\lim_s E^s_k = E_k$ and $\lim_s F^s_k = F_k$ exists for each k. It then follows that $F = \bigcup_k F_k$ is an isomorphism from $\mathcal{B}$ onto $\bigcup_k E_k$ since for each k, F_k is an isomorphism from $\mathcal{B}_k$ onto E_k. □

Lemma 4.9. *For all e, $1 \leq i \leq e+1$, and $j \in \{0,1\}$, $\lim_s e^s_{i,e,j} = e_{i,e,j}$, $\lim_s f^s_{i,e,j} = f_{i,e,j}$, and $\lim_s g^s_{i,e,j} = g_{i,e,j}$ exist.*

Proof. First observe that we need only prove that $\lim_s e^s_{i,e,j} = e_{i,e,j}$ exists for $1 \leq i \leq e+1$ and $j \in \{0,1\}$ since our construction ensures that if $e^s_{i,e,j} = e^{s+1}_{i,e,j}$, then $f^s_{i,e,j} = e^{s+1}_{i,e,j}$ and $g^s_{i,e,j} = e^{s+1}_{i,e,j}$.

It follows from Lemma 4.8, that $\lim_s e^s_{i,e} = e_{i,e}$ exists for all $1 \leq i \leq e+1$. There are two cases. If $e_{i,e}$ is not an atom of E, then there is a $k > e$, i_0, i_1, and s_e such that $e^t_{i,e,0} = e^t_{i_0,k}$ and $e^t_{i,e,1} = e^t_{i_1,k}$ for all $t > s_e$. Thus since $\lim_s e^s_{i_0,k} = e_{i_0,k}$ and $\lim_s e^s_{i_1,k} = e_{i_1,k}$ exist, it follows that $\lim_s e^s_{i,e,0} = e_{i,e,0}$ and $\lim_s e^s_{i,e,1} = e_{i,e,1}$ exist. Now if $e_{i,e}$ is an atom of E, then simply find a stage $t > e$ such that $e^s_{i,e} = e^t_{i,e}$ for all $s > t$. It is then easy to see that our construction ensures that $e^s_{i,e,j} = e^t_{i,e,j}$ for all $j \in \{0,1\}$ and $s > t$. □

Lemma 4.10. *For all e, requirement R_e is satisfied.*

Proof. Let $e = [n, k]$. There are three possibilities.

First, it could be that requirement R_e is canceled at some stage s. In that case, $\mathcal{D}_k$ is not a Boolean algebra with universe G so that requirement R_e is automatically satisfied.

The second possibility is that there is a stage s such that a Δ_e marker is placed and that marker is never removed. In that case, it should be clear that there is an element of $e^s_{i,e,j}$ of C_s such that $\phi_{n,s}(e^s_{i,e,j}) = h^s_{i,e,j}$ is defined and the cardinality of $[h^s_{i,e,j}]_{\mathcal{D}_k}$ is greater than $|C_s|^2$. Then either $e^s_{i,e,j}$ is incorporated into some $f^t_{x,y,z}$ at stage $t > s$ so that $|[e^s_{i,e,j}]_{\mathcal{C}}| = |[e^s_{i,e,j}]_{C_s}|$, which implies that $|[e^s_{i,e,j}]_{\mathcal{C}}| \neq |[h^s_{i,e,j}]_{\mathcal{D}_k}|$ and hence ϕ_n cannot be an isomorphism from $\mathcal{C}$ onto $\mathcal{D}_k$ or $e^s_{i,e}$ is an atom of $\mathcal{C}$ and once again, $|[e^s_{i,e,j}]_{\mathcal{C}}| = |[e^s_{i,e,j}]_{C_s}|$ which implies that ϕ_n cannot be an isomorphism from $\mathcal{C}$ onto $\mathcal{D}_k$.

The final case is when there is a stage t such that for all $s \geq t$, there is no element of C_s with a Δ_e marker on it. By Lemmas 4.7 and 4.9, we can assume that we choose such a $t > e$ is large enough so that $e(s) > e$, $e^s_{i,e,j} = e^t_{i,e,j}$, $f^s_{i,e,j} = f^t_{i,e,j}$, and $g^s_{i,e,j} = g^t_{i,e,j}$ for all $s \geq t$, $1 \leq i \leq e+1$, and $j \in \{0, 1\}$. But then it can never be the case that there is a stage $s \geq t$ such that $\phi_{n,s}(e^s_{i,e,j})$ is defined for all $1 \leq i \leq e+1$ and $j \in \{0, 1\}$. But this means that ϕ_n is not defined on all of $\mathcal{C}$ so once again requirement R_e is satisfied. □

Lemma 4.11. *$\mathcal{C}$ is isomorphic to $\mathcal{B}$.*

Proof. Our construction ensures that if $f^s_{i,e,j}$ is defined then we will never again split $f^s_{i,e,j}$ in the construction of $\mathcal{C}$. This means that if $f^s_{i,e,j,t}$, for $t = 1, \dots, p_{i,e,j,s}$, are the atoms of C_s which are under $f^s_{i,e,j}$, then the $f^s_{i,e,j,t}$ will be in fact atoms of $\mathcal{C}$. It is easy to see that our construction ensures $\mathcal{C} = \mathrm{sub}_{\tilde{\mathcal{Q}}}(E \cup \bigcup_{i,e,j,s}\{e^s_{i,e,j}, f^s_{i,e,j,t}, g^s_{i,e,j}\})$.

We need to analyze the elements in $\bigcup_{i,e,j}\{e^s_{i,e,j}, f^s_{i,e,j}, g^s_{i,e,j}\}$. In particular, we want to analyze these elements modulo the ideal generated by the atoms of $\mathcal{C}$, $I(\mathrm{At}(\mathcal{C}))$. By our observation above, each element $f^s_{i,e,j}$ is in $I(\mathrm{At}(\mathcal{C}))$. Moreover since $e^s_{i,e,j} = f^s_{i,e,j} \vee_{\mathcal{C}} g^s_{i,e,j}$, it follows that $e^s_{i,e,j} \equiv_{I(\mathrm{At}(\mathcal{C}))} g^s_{i,e,j}$ so that we need only analyze the elements of the form $e^s_{i,e,j}$. If $e^s_{i,e,j} \neq e^{s+1}_{i,e,j}$, then either $e^s_{i,e}$ is the atom of E^s_S that must be split in order to extend our isomorphism from $\mathcal{B}_s$ to $\mathcal{B}_{s+1}$ or $e^s_{i,e}$ is contained in some atom $e^s_{k,f}$ which is used to meet a requirement R_f and so $f \leq e$.

In the first case, it is easy to see that our construction ensures that $e^s_{i,e} = e^{s+1}_{i,e}$ and that either $e^s_{i,e,j}$ is incorporated into $f^{s+1}_{i,e,j}$ or the new $g^{s+1}_{i,e,0}$ and $g^{s+1}_{i,e,1}$ are contained in $g^s_{i,e,j}$ in which case

$$e^s_{i,e,j} \equiv_{I(\mathrm{At}(\mathcal{C}))} g^{s+1}_{i,e,0} \vee_{\mathcal{C}} g^{s+1}_{i,e,1} \equiv_{I(\mathrm{At}(\mathcal{C}))} e^{s+1}_{i,e,0} \vee_{\mathcal{C}} e^{s+1}_{i,e,1} = e^{s+1}_{i,e}.$$

The second case is similar. That is, either $e^s_{i,e,j}$ is incorporated into one of $f^{s+1}_{k,f,0}$ or $f^{s+1}_{k,f,1}$ or the the new $g^{s+1}_{f,k,0}$ and $g^{s+1}_{f,k,1}$ are contained in $g^s_{i,e,j}$ in which case

$$e^s_{i,e,j} \equiv_{I(\mathrm{At}(\mathcal{C}))} g^{s+1}_{f,k,0} \vee_{\mathcal{C}} g^{s+1}_{f,k,1} \equiv_{I(\mathrm{At}(\mathcal{C}))} e^{s+1}_{f,k,0} \vee_{\mathcal{C}} e^{s+1}_{f,k,1} = e^{s+1}_{f,k}.$$

Thus we have shown that if $e^s_{i,e,j} \neq e^{s+1}_{i,e,j}$, then $e^s_{i,e,j} \equiv_{I(\mathrm{At}(\mathcal{C}))} e^{s+1}_{r,u}$ where $u \leq e$. Now a similar remark holds for elements of the the form $e^t_{r,u}$. That is, if $e^t_{r,u} \neq e^{t+1}_{r,u}$, then it must be the case that $e^s_{r,u}$ is contained in some atom $e^s_{k,f}$ which is used to meet a requirement R_f. In that case, it is easy to see that our construction ensures that $e^t_{f,k} = e^{t+1}_{f,k}$ and that either $e^t_{r,u}$ is incorporated into one of $f^{t+1}_{k,f,0}$ or $f^{t+1}_{k,f,1}$ or the the new $g^{t+1}_{f,k,0}$ and $g^{t+1}_{f,k,1}$ are contained in $e^t_{r,u,p}$ for some p, in which case

$$e^t_{r,u} \equiv_{I(\mathrm{At}(\mathcal{C}))} g^{s+1}_{f,k,0} \vee_{\mathcal{C}} g^{s+1}_{f,k,1} \equiv_{I(\mathrm{At}(\mathcal{C}))} e^{s+1}_{f,k,0} \vee_{\mathcal{C}} e^{s+1}_{f,k,1} = e^{s+1}_{f,k}.$$

Thus once again we can conclude that if if $e^t_{r,u} \neq e^{t+1}_{r,u}$, then $e^t_{r,u} \equiv_{I(\mathrm{At}(\mathcal{C}))} e^{s+1}_{k,f}$ where $f \leq u$. Thus we have shown that either $e^s_{i,e,j} = e^t_{i,e,j}$ for all $t > s$ or for all $t > s$, $e^s_{i,e,j}$ is equivalent modulo $I(\mathrm{At}(\mathcal{C}))$ to some element of E^t_e. In the first case, it must be that $e^s_{i,e} = e^t_{i,e}$ for all $t > s$ and hence $e_{i,e}$ is an atom of E. In that case, it is easy to see that our construction ensures that for all $q \in 0, 1$ and $t > s$, $e^s_{i,e,q} = e^t_{i,e,q}$, $f^s_{i,e,q} = f^t_{i,e,q}$, and $g^s_{i,e,q} = g^t_{i,e,q}$. This means that $[e^s_{i,e}]_{\mathcal{C}_s} = [e^s_{i,e}]_{\mathcal{C}}$ so that $e^s_{i,e}$ is in the ideal generated by the atoms of $\mathcal{C}$. In the second case, we conclude that $e^s_{i,e,j}$ is equivalent modulo $I(\mathrm{At}(\mathcal{C}))$ to some element of E_e.

We get two important conclusions from our argument above. The first conclusion is that every element in $\mathcal{C}$ is equivalent modulo $I(\mathrm{At}(\mathcal{C}))$ to some element of E. The second is that every atom of E is in $I(\mathrm{At}(\mathcal{C}))$. This means that in fact $\mathcal{C}$ is an atomic Boolean algebra. That is, since $\mathcal{B}$ is atomic and E is isomorphic to $\mathcal{B}$, E is atomic. Thus for each element $x \in E$, there is an $y \in \mathrm{At}(E)$ such that $y \leq_E x$. But then $y \in I(\mathrm{At}(\mathcal{C}))$ so that there is some $a \in \mathrm{At}(\mathcal{C})$ such that $a \leq_{\mathcal{C}} x$. It follows that $\mathcal{C}$ is an atomic Boolean algebra and that $\mathcal{C}/I(\mathrm{At}(\mathcal{C}))$ is isomorphic to $E/I(\mathrm{At}(E))$. But then since E is isomorphic to $\mathcal{B}$, it follows that $\mathcal{C}/I(\mathrm{At}(\mathcal{C}))$ is isomorphic to $\mathcal{B}/I(\mathrm{At}(\mathcal{B}))$. Thus we can apply Vaught's Theorem to conclude that $\mathcal{C}$ is isomorphic to $\mathcal{B}$. □

This completes the proof of Theorem 4.5. □

Remark. It is easy to see that in the only thing that we used about the sequence of potential Boolean algebras $\mathcal{D}_0, \mathcal{D}_1, \ldots$ is the that fact that the operations of $\mathcal{D}_k$ are the restriction of total computable functions and the universe of $\mathcal{D}_k$ is computable for each k. That is, we have proved the following.

Theorem 4.12. *Let $\mathcal{B}$ be any computable infinite atomic Boolean algebra and let $\mathcal{D}_0, \mathcal{D}_1, \ldots$ be any uniformly computable sequence of computable structures $\mathcal{D} - (D, \vee_D, \wedge_D, \neg_D, 0_D, 1_D)$ whose universe D is contained in $\{0, 1\}^*$. Then there is* *a computable Boolean algebra $\mathcal{C}$ which is isomorphic to $\mathcal{B}$ but is not computably isomorphic to any $\mathcal{D}_k$.*

Note that if $A_0, A_1, \ldots$ is any uniformly computable sequence of p-time subsets of $\{0, 1\}^*$, then the set of all structures $(A, \vee_A, \wedge_A, \neg_A, 0_A, 1_A)$ where $A = A_i$ for some i and $\vee_A$ and $\wedge_A$ are p-time functions from $\{0, 1\}^* \times \{0, 1\}^*$ into $\{0, 1\}^*$ and $\neg_A$ is a p-time function from $\{0, 1\}^*$ into $\{0, 1\}^*$ can be listed as a uniformly computable sequence of structures as Theorem 4.12. It follows that for any such sequence of p-time subsets $A_0, A_1, \ldots$ and any computable infinite atomic Boolean algebra $\mathcal{B}$, there exists a computable Boolean algebra $\mathcal{C}$ which is isomorphic to $\mathcal{B}$ but which is not computably isomorphic to any polynomial-time Boolean algebra with universe A_i.

We now turn to our intermediate case, namely computable Boolean algebras $\mathcal{B}$ where we assume that $\mathrm{At}(\mathcal{B})$ is computable but that $I(\mathrm{At}(\mathcal{B}))$ is not necessarily computable.

4.3. Computable Boolean algebras with computable set of atoms

Suppose that $\tilde{\mathcal{B}}$ is a countably infinite computable Boolean algebra such that $\mathrm{At}(\tilde{\mathcal{B}})$ is computable. Now if $\tilde{\mathcal{B}}$ is isomorphic to either $\tilde{\mathcal{N}} \times \mathcal{B}$ or $\tilde{\mathcal{Q}} \times \mathcal{B}$ where $\mathcal{B}$ is a computable Boolean algebra, then by part (b) and (c) of Theorem 4.3, $\tilde{\mathcal{B}}$ is computably isomorphic to a polynomial-time Boolean algebra with universe $\mathrm{Bin}(\omega)$ and is computably isomorphic to a polynomial-time Boolean algebra with universe $\mathrm{Tal}(\omega)$. It follows from Theorem 3.5 that the only other possibility is that $\tilde{\mathcal{B}}$ is isomorphic to $\tilde{\mathcal{C}}$. In this case we can not guarantee that $\tilde{\mathcal{B}}$ is computably isomorphic to a polynomial-time Boolean algebra with universe $\mathrm{Bin}(\omega)$ and is computably isomorphic to a polynomial-time Boolean algebra with universe $\mathrm{Tal}(\omega)$. That is, we have the following.

Theorem 4.13. *Let G be any infinite polynomial-time subset of $\{0, 1\}^*$. Then there exists a computable Boolean algebra $\tilde{\mathcal{B}}$ such that $\mathrm{At}(\tilde{\mathcal{B}})$ is computable and $\tilde{\mathcal{B}}$ is isomorphic to $\tilde{\mathcal{C}}$ which is not computably isomorphic to any polynomial-time Boolean algebra whose universe is G.*

Proof. Our proof of the theorem will be based on a modification of the proof of Theorem 4.5. That is, assume the notation of Theorem 4.5 and assume that $\mathcal{B}$ is a computable copy of the atomless Boolean algebra $\tilde{\mathcal{Q}}$. Now suppose we carry out the construction of the Theorem 4.5. It is not difficult to show that since at any given stage s, no atom of E_s^s can be an atom of E, it will be the case that the computable Boolean algebra $\mathcal{C}$ produced by the construction of Theorem 4.5 will be atomic. Thus $\mathcal{C}$ would have the property that it is a computable atomic Boolean algebra such that $\mathcal{C}/I(\mathrm{At}(\mathcal{C}))$ is isomorphic to $\mathcal{B}/I(\mathrm{At}(\mathcal{B}))$. Of course, since $\mathcal{B}$ is atomless, $\mathcal{B}/I(\mathrm{At}(\mathcal{B}))$ is just isomorphic to $\mathcal{B}$. Thus $\mathcal{C}$ would be an atomic Boolean algebra such that $\mathcal{C}/I(\mathrm{At}(\mathcal{C}))$ is isomorphic to $\tilde{\mathcal{Q}}$. However it is easy to see that $\tilde{\mathcal{C}}/I(\mathrm{At}(\tilde{\mathcal{C}}))$ is also isomorphic to $\tilde{\mathfrak{Q}}$ and hence we could once again apply Vaught's Theorem to conclude that $\mathfrak{C}$ is isomorphic to $\tilde{\mathfrak{C}}$. Moreover, $\mathfrak{C}$ would have the property that it is not computably isomorphic to any polynomial Boolean algebra with universe G. The only thing that the construction of Theorem 4.5 does not ensure is that $\text{At}(\mathfrak{C})$ is computable.

Our idea then is to modify the construction of $\mathfrak{C}$ to ensure that $\text{At}(\mathfrak{C})$ is computable. This is actually quite simple to do. The basic idea is that we will declare an element a of C_s to be an atom of $\mathfrak{C}$ if and only if a is an atom of C_s which is under some $f^s_{i,e,j}$. Now the construction of Theorem 4.5 does allow some elements of the form $g^s_{i,e,j}$ to be become a part of some $f^{s+1}_{x,y,z}$ and thus become an atom of $\mathfrak{C}$. We can not allow this to happen in our construction since we have declared that $g^s_{i,e,j}$ is not an atom of $\mathfrak{C}$. However all we have to do is introducc some extra splitting of $g^s_{i,e,j}$ at stage $s+1$ before we can incorporate it into $f^{s+1}_{x,y,z}$. Thus below we shall give the modification of the construction of Theorem 4.5 which is required. To make clear the modifications of the construction of Theorem 4.5 that we use, we shall put the modifications in italics.

Assume the notation of Theorem 4.5 where $\mathcal{B}$ is a computable copy of the atomless Boolean algebra $\tilde{\mathfrak{Q}}$.

Construction

Stage 0 is as in the proof of Theorem 4.5.

Stage $s+1$: First check the consistency of $\mathcal{D}_j$ being a Boolean algebra for each $j \leq s+1$ as before and then look for a requirement R_e where $e = \langle n, k \rangle \leq s$ such that R_e is not canceled, there does not exists a Δ_e marker which is currently resting on some element of C_s, and $\phi_n(e^s_{i,e,j}) \downarrow$ for all $i \leq e+1$ and $j \in \{0, 1\}$.

If there is no such e, extend the construction as before.

Next suppose that there is a requirement R_e where $e = [n, k \leq s$ such that R_e is not canceled, there does not exists a Δ_e marker which is currently resting on some element of C_s, and $\phi_n(e^s_{i,e,j}) \downarrow$ for all $i \leq e+1$ and $j \in \{0, 1\}$. Let $f = [u, w]$ be the least such requirement. Let $h^s_{i,e,j} = \phi_{u,s}(e^s_{i,f,j})$ for all $1 \leq i \leq f$ and $j \in \{0, 1\}$. Then begin to compute the subalgebras $\mathcal{D}_{w,t} = \text{sub}_{\mathcal{D}_w}(\{g_0, \ldots, g_t\})$ in order until either we find that $\mathcal{D}_w$ is not consistent with being a Boolean algebra with universe G, in which case we cancel all requirements of the form R_e where $e = \langle v, w \rangle$, or we find t, k and j such that the cardinality of $[h^s_{k,f,j}]_{\mathcal{D}_{w,t}}$ is strictly greater than $4^{|C_s|}$.

Here we require that $|[h^s_{k,f,j}]_{\mathcal{D}_{w,t}}|$ *is strictly greater than* $4^{|C_s|}$ *rather than just* $|[h^s_{k,f,j}]_{\mathcal{D}_{w,t}}|$ *is strictly greater than* $|C_s|^2$ *as in Theorem 4.5.. This is required since we will introduce a number of new splittings under* $e^s_{i,e,j}$ *at stage* $s+1$ *to ensure that all the* $g^s_{x,y,z}$ *which are under* $e^s_{k,f,j}$ *are not atoms of* $\mathfrak{C}$. *This will increase the size of* $|[e^s_{k,f,j}]_{\mathfrak{C}}|$ *but it will not get as big as* $4^{|C_s|}$.

If we find that R_f is canceled, then we extend C_s to C_{s+1} as described above. Otherwise assume that t, k are j are the least triple such that the cardinality of $[h^s_{k,f,j}]_{\mathcal{D}_{w,t}}$

is strictly greater than $4^{|C_s|}$. We then have the two cases, when $e^s_{k,f}$ is an atom of E^s_s or is not an atom of E^s_s. The construction proceeds as in Theorem 4.5 in each case.

In either case, we let C_{s+1} be the subalgebra of $\tilde{\mathcal{Q}}$ generated by C_s plus all elements of the form $e^{s+1}_{x,y,z}$, $f^{s+1}_{x,y,z}$, and $e^{s+1}_{x,y,z}$ where $y \leq s+1$ plus all elements of the form $d^s_{x,y,z,i}$ which were defined at stage $s+1$.

End of construction

It is easy to see that since $\mathcal{B}$ is an atomless Boolean algebra, it will be the case that for every atom of E^s_s of the form $e^s_{x,y}$, there will be a least $t > s$ such that $e^t_{x,y}$ is not an atom of E^t_t. This means that $e^{t-1}_{x,y}$ will be the atom that is split to extend our sequence of isomorphisms at stage t. Our construction then ensures that all of $e^{t-1}_{x,y,0}$, $g^{t-1}_{x,y,0}$, $e^{t-1}_{x,y,1}$, and $g^{t-1}_{x,y,1}$ will be $\geq_{C_t}$ some atom of C_t which is contained in some $f^t_{u,w,j}$. Thus since no $f^t_{u,w,j}$ can be split after stage t, it will follow that every nonzero element of $\mathcal{C}$ contains an atom of $\mathcal{C}$. Thus $\mathcal{C}$ is an atomic Boolean algebra. Moreover our construction ensures that a is an atom of $\mathcal{C}$ if and only if a is an atom of some C_s which is $\leq_{C_s}$ some $f^s_{x,y,z}$. It follows that $\mathrm{At}(\mathcal{C})$ is computable.

We can then prove the same sequence of lemmas as in the proof of the Theorem 4.5 to establish that $\mathcal{C}$ has the properties required by the theorem. In fact, the proofs of the first four lemmas below are almost exactly the same as the corresponding proofs in Theorem 4.5 so that we shall not repeat their proofs.

Lemma 4.14. *For any given requirement R_e, we act for R_e at most finitely often, that is, $\{s : e(s) = e\}$ is finite for all e.*

Lemma 4.15. *For all k, $\lim_s E^s_k = E_k$ and $\lim_s F^s_k = F_k$ exists and hence $F = \bigcup_k F_k$ is an isomorphism from $\mathcal{B}$ onto $E = \bigcup_k E_k$.*

Lemma 4.16. *For all e, $1 \leq i \leq e+1$, and $j \in \{0,1\}$, $\lim_s e^s_{i,e,j} = e_{i,e,j}$, $\lim_s f^s_{i,e,j} = f_{i,e,j}$, and $\lim_s g^s_{i,e,j} = g_{i,e,j}$ exist.*

Lemma 4.17. *For all e, requirement R_e is satisfied.*

Lemma 4.18. *$\mathcal{C}$ is isomorphic to $\tilde{\mathcal{C}}$.*

Proof. We can prove that exactly as the proof of Lemma 4.11 that every element of the form $e^s_{x,y,z}$, $f^s_{x,y,z}$, $g^s_{x,y,z}$ are either in the ideal generated by the atoms of $\mathcal{C}$ or $\equiv_{I(\mathrm{At}(\mathcal{C}))}$ to some element in $E = \bigcup_s E^s_s$. The only other elements that are introduced in the construction are the various $d^s_{x,y,z,j}$ which were used to ensure that the $g^s_{x,y,z}$ are not atoms of $\mathcal{C}$. However all such elements are immediately incorporated into some $f^{s+1}_{x',y',z'}$ at stage $s+1$ and thus will be atoms of $\mathcal{C}$. It follows that $\mathcal{C}/I(\mathrm{At}(\mathcal{C}))$ is isomorphic to $E/I(\mathrm{At}(E))$. However since $\mathcal{B}$ is atomless, E is atomless and hence $\mathcal{C}/I(\mathrm{At}(\mathcal{C}))$ is isomorphic to the atomless Boolean algebra $\tilde{\mathcal{Q}}$. However $\tilde{\mathcal{C}}$ is an atomic Boolean algebra such that $\tilde{\mathfrak{C}}/I(\mathrm{At}(\tilde{\mathfrak{C}}))$ is isomorphic to $\tilde{\mathfrak{Q}}$. Thus by Vaught's Theorem, $\tilde{\mathfrak{C}}$ is isomorphic to $\mathfrak{C}$. □

This completes the proof of Theorem 4.13. □

Remark. We can make the same remark as the remark following Theorem 4.5. That is, the only thing that we used about the sequence of potential Boolean algebras $\mathfrak{D}_0, \mathfrak{D}_1, \ldots$ is that fact that the operations of $\mathfrak{D}_k$ are the restriction of total computable functions and the universe of $\mathfrak{D}_k$ is computable for each k. Hence we have proved the following.

Theorem 4.19. *Let $\mathfrak{D}_0, \mathfrak{D}_1, \ldots$ be any uniformly computable sequence of computable structures $\mathfrak{D} - (D, \vee_D, \wedge_D, \neg_D, 0_D, 1_D)$ whose universe D is contained in $\{0, 1\}^*$. Then there is a computable Boolean algebra $\mathfrak{C}$ which is isomorphic to $\tilde{\mathfrak{C}}$ and has the property that* $\mathrm{At}(\mathfrak{C})$ *is computable but $\mathfrak{C}$ is not computably isomorphic to any $\mathfrak{D}_k$.*

Then once again if $A_0, A_1, \ldots$ is any uniformly computable sequence of polynomial-time subsets of $\{0, 1\}^*$, then the set of all structures $(A, \vee_A, \wedge_A, \neg_A, 0_A, 1_A)$ where $A = A_i$ for some i and $\vee_A$ and $\wedge_A$ are p-time functions from $\{0, 1\}^* \times \{0, 1\}^*$ into $\{0, 1\}^*$ and $\neg_A$ is a p-time function from $\{0, 1\}^*$ into $\{0, 1\}^*$ can be listed as a uniformly computable sequence of structures as in Theorem 4.19. It follows that for any such sequence of polynomial-time subsets $A_0, A_1, \ldots$, there exists a computable Boolean algebra $\mathfrak{C}$ which is isomorphic to $\tilde{\mathfrak{C}}$ and has the property that $\mathrm{At}(\mathfrak{C})$ is computable but $\mathfrak{C}$ is not computably isomorphic to any polynomial-time Boolean algebra with universe A_i.

We conclude by noting that Remmel followed the corresponding analysis of polynomial-time linear orderings with an examination of polynomial-time categoricity for linear orderings. The notion of polynomial-time categoricity was introduced in [5] and further developed in [6, 7]. The next task would to examine the polynomial-time categoricity of Boolean algebras.

References

[1] Cenzer, D. and Remmel, J. B., Polynomial-time versus recursive models, Ann. Pure Appl. Logic 54 (1991), 17–58.

[2] Cenzer, D. and Remmel, J. B., Polynomial-time Abelian groups, Ann. Pure Appl. Logic 56 (1992), 313–363.

[3] Cenzer, D. and Remmel, J. B., Recursively presented games and strategies, Math. Social Sci. 24 (1992), 117–139.

[4] Cenzer, D. and Remmel, J. B., Feasible graphs and colorings, Math. Logic Quart. 41 (1995), 327–352.

[5] Cenzer, D. and Remmel, J. B., Feasibly categorical abelian groups, in: Feasible Mathematics II (eds. P. Clote and J. Remmel), Prog. Comput. Sci. Appl. Logic 13, Birkhäuser, 1995, 91–154.

[6] Cenzer, D. and Remmel, J. B., Feasibly categorical models, in: Logic and Computational Complexity (ed. D. Leivant), Lecture Notes in Comput. Sci. 960, Springer-Verlag, 1995, 300–312.

[7] Cenzer, D. and Remmel, J. B., Complexity and categoricity, Inform. and Comput. 140 (1) (1998), 2–25.

[8] Cenzer, D. and Remmel, J. B., Feasible graphs with standard universe, in: Computability Theory, Oberwolfach 1996, Special issue of Ann. Pure Appl. Logic, to appear.

[9] Cenzer, D. and Remmel, J. B., Complexity Theoretic Model Theory and Algebra, in: Handbook of Recursive Mathematics (eds. Yu. L. Ershov, S. S. Goncharov, A. Nerode and J. B. Remmel), Stud. Logic Found. Math. 138, Elsevier, 1998, 381–513.

[10] Cenzer, D. and Downey, R. and Remmel, J. B., Feasible torsion-free Abelian groups, to appear.

[11] Grigorieff, S., Every Recursive linear ordering has a copy in DTIME(n), J. Symbolic Logic 55 (1990), 260–276.

[12] Goncharov, S. S., Some properties of the constructivization of Boolean algebras, , Sibirsk. Mat. Zh. 16 (1975), 203–214.

[13] Goncharov, S. S., Countable Boolean algebras and decidability, Plenum, 1997.

[14] Hopcroft, J. and Ullman, J., Formal Languages and Their Relations to Automata, Addison Wesley, 1969.

[15] Nerode, A., Remmel, J. B., Complexity theoretic algebra II, the free Boolean algebra, Ann. Pure Appl. Logic 44 (1989), 71–99.

[16] Nerode, A. and Remmel, J. B., Polynomial-time equivalence types, Logic and Computation: Proc. of a Workshop held at CMU, June 3–July 2, 1987 (ed. W. Sieg), Contemp. Math. 106, American Mathematical Society, Providence, R. I., 1990.

[17] P. Odifreddi, Classical Recursion Theory, North-Holland, Amsterdam, 1989.

[18] Remmel, J. B., Recursively Enumerable Boolean Algebras, Ann. Math. Logic 14 (1978), 75–107.

[19] Remmel, J. B., Recursive Boolean algebras with recursive sets of atoms, J. Symbolic Logic 46 (1981), 595–616.

[20] Remmel, J. B., Recursive Boolean Algebras, in: Handbook of Boolean Algebras (ed. J. Monk), North-Holland, 1989, 1099–1165.

[21] Remmel, J. B., When is every recursive linear ordering of type μ recursively isomorphic to a p-time linear order over the binary representation of the natural numbers?, Feasible Mathematics (eds. S. Buss and P. Scott), Prog. Comput. Sci. Appl. Logic 9, Birkhäuser, 1990, 321–341.

[22] Remmel, J. B., Polynomial-time categoricity and linear orderings, in: Logical Methods (eds. J. Crossley, J. Remmel, R. Shore, and M. Sweedler), Prog. Comput. Sci. Appl. Logic 12, Birkhäuser, 1993, 321–341.

[23] Vaught, R. L., Topics in the theory of arithmetical classes and Boolean algebras, Ph.D. Dissertation, University of California, Berkeley, 1954.

The proof-theoretic strength of the Dushnik–Miller Theorem for countable linear orders

*Rodney G. Downey and Steffen Lempp**

Abstract. We show that the Dushnik–Miller Theorem for countable linear orderings (stating that any countable linear ordering has a nontrivial self-embedding) is equivalent (over recursive comprehension (RCA_0)) to arithmetic comprehension (ACA_0).

1991 Mathematics Subject Classification:03F35, 06A05

This paper presents a result in reverse mathematics, a program initiated by H. Friedman and S. Simpson, trying to determine the weakest possible "set-theoretical" axiom (system) to prove a given theorem of "ordinary" mathematics by trying to prove the axiom from the theorem (over a weaker "base system").

The "set-theoretical" axiom systems we will be concerned with are weak subsystems of second-order arithmetic. (We refer to Simpson [Si99] for a detailed exposition of such systems.) In particular, we will use the axiom system RCA_0 of recursive comprehension (with Σ_1^0-induction) as a base system and exhibit a theorem which is equivalent (over the base system RCA_0) to the axiom system ACA_0 of arithmetic comprehension (with Σ_1^0-induction).

The area of "ordinary" mathematics we study in the context of reverse mathematics is that of linear orderings. In particular, we will characterize the proof-theoretic strength of the

Dushnik–Miller Theorem on countable linear orderings [DM40]. *Let $\mathcal{L}$ be a countably infinite linear ordering. Then there is a nontrivial* self-embedding *of $\mathcal{L}$, i.e., an order-preserving injection of $\mathcal{L}$ into itself which is not the identity.*

We will thus show the following

Theorem. *Over the base system RCA_0, the Dushnik–Miller Theorem on countable linear orderings is equivalent to the axiom system ACA_0.*

Proof. We first prove the easy direction, namely, that ACA_0 is strong enough to prove the Dushnik–Miller Theorem on countable linear orderings. Fix a countably infinite

*The first author was partially supported by the New Zealand Marsden Fund for Basic Science, and both authors by a U. S./N. Z. Binational Grant. The second author's research was partially supported by NSF grant DMS-9504474.

linear ordering $\mathcal{L}$. Call $C \subseteq L$ a *convex* subset of $\mathcal{L}$ if C contains any point z between any points $x, y \in C$.

First assume that L contains a convex subset C of order type ω. Then the map i which is the identity off C and moves every element of C to its immediate successor is a nontrivial self-embedding of $\mathcal{L}$ and can be defined by a first-order formula (in the language of arithmetic), thus can be shown to exist by ACA_0. The case where $\mathcal{L}$ contains a convex subset of order type ω^* (i.e., ω under the reverse ordering) is handled similarly.

So assume that $\mathcal{L}$ does not contain any convex subsets of order type ω or ω^*. Call a convex subset C of $\mathcal{L}$ discrete if every element of C (except the least, if any) has an immediate predecessor in C, and every element of C (except the greatest, if any) has an immediate successor in C. By our assumption, any discrete subset of $\mathcal{L}$ must be finite. By picking one element from each maximal discrete subset of $\mathcal{L}$ (except the first and last maximal discrete subset, if any), we see that there is an infinite subset of $\mathcal{L}$ which is densely ordered without endpoints. Since ACA_0 actually allows full first-order induction, we can now define the self-embedding as follows: List the points of $\mathcal{L}$ as $\{x_n\}_{n\in\omega}$. When picking an image $i(x_n)$ for x_n, simply ensure that $i(x_n)$ has infinitely many points to its left and right and is infinitely far apart from $i(x_0), \dots, i(x_{n-1})$. Since the choice of $i(x_n)$ can be made in an arithmetic way, i can be shown to exist by ACA_0. This concludes the proof of the easy direction.

As for the hard direction, we need to show that the second-order part of any model of "RCA_0 plus Dushnik–Miller Theorem on countable linear orderings" is closed under Σ^0_1-comprehension. So fix any set $A \in \mathcal{S}$ (where $\mathcal{S}$ is the collection of subsets included in the given second-order model of "RCA_0 plus Dushnik–Miller"). We need to show that its "Turing jump" A' is also in $\mathcal{S}$. We do so by defining a countable linear ordering $\mathcal{L}$ computable in A such that any nontrivial self-embedding i of $\mathcal{L}$ can compute A'.

Fix an A-computable enumeration $\{A'_s\}_{s\in\omega}$ of A', and let $c(x) = \mu s \geq x(A' \restriction (x+1) = A'_s \restriction (x+1))$ be the associated A-computable *computation function* of A'. Since $A' \leq_T A \oplus c$, it suffices to ensure the following

Claim 1. *Any nontrivial self-embedding i of $\mathcal{L}$ can compute the computation function c.*

We define the linear ordering $\mathcal{L}$ of order type $(M, <)$ with universe M in stages and start by letting $\mathcal{L}_0$ be the ordering $0 \leq_{\mathcal{L}} 2 \leq_{\mathcal{L}} 4 \leq_{\mathcal{L}} \cdots$ of all even integers in M. We establish Claim 1 by ensuring the existence of a function e satisfying the following

Claim 2. *There is a strictly increasing function $e : M \to L_0$ such that for all $x \in M$,*

(1) $\forall n_0, n_1 < c(x)\,(e(x) <_{\mathcal{L}} n_0 <_{\mathcal{L}} n_1 \to d(e(x), n_0) > d(n_0, n_1))$,

(2) $e(x+1) = \mu y \in L_0 \forall n\,(n < c(x) \to n <_{\mathcal{L}} y)$

where $d(n_0, n_1)$ is the ($\mathcal{M}$-finite) distance between n_0 and n_1 in $\mathcal{L}$.

We first establish Claim 1 from Claim 2: Fix any nontrivial self-embedding i of $\mathcal{L}$. By Σ_1^0-induction, e is monotonic, so the range of e is cofinal in $(M, <)$, and so there is $x_0 \in M$ such that for all $x \geq x_0$, we have $e(x) <_{\mathcal{L}} ie(x)$. Also, by Σ_1^0-induction and (1) of Claim 2, for all $x \geq x_0$, we have that one of $ie(x)$ and $i^2e(x)$ is $\geq c(x)$ (since $d(e(x), ie(x)) \leq d(ie(x), i^2e(x))$). So from $e(x)$ and i we can compute $c(x)$. Finally, by (2) of Claim 2, we can also compute $e(x+1)$. Thus i allows us to compute c as desired, establishing Claim 1 from Claim 2.

The proof of Claim 2 is a finite-injury priority argument (using Σ_1^0-induction). We have to maintain (1) and (2) of Claim 2 at any stage s for all $x \leq s$ (evaluating $c(x)$ for these x's at stage s). Note that the definition of the function e is fixed by (2) at any stage s (assuming $e(0) = 0$). The only problem arises if some number x enters A' at a stage $s > 0$, thus making (1) false. In that case, add all currently unused elements $x \leq s$ in $M - L_{s-1}$ into L_s just to the left of $e(x)$, and add sufficiently many unused elements $x > s$ in $M - L_{s-1}$ into L_s just to the right of x to make (1) true. Note that this action will not interfere with keeping (1) satisfied for any $x' < x$. It is now easy to verify that this construction will produce the desired linear ordering satisfying Claim 2. □

References

[DM40] Dushnik, B. and Miller, E. W., Concerning similarity transformations of linearly ordered sets, Bull. Amer. Math. Soc. 46 (1940), 322–326.

[Ro82] Rosenstein, J. G., Linear Orderings, Academic Press, New York, 1982.

[Si99] Simpson, S. G., Subsystems of second order arithmetic, Perspect. Math. Logic, Springer-Verlag, 1999.

Effectively nowhere simple relations on computable structures

Valentina S. Harizanov

Abstract. Let $\mathcal{A}$ be a computable structure and let R be an additional relation on its domain. The notion of "quasi-simplicity" of R on $\mathcal{A}$, first studied by G. Hird, is analogous to the computability-theoretic notion of simplicity, given the definability of various subrelations of $\neg R$. In the present paper, we define corresponding versions of the notions "nowhere simple" and "effectively nowhere simple." We establish a sufficient condition for existence of noncomputable effectively nowhere simple relations on a restricted class of computable structures.

1. Introduction and notation

We will denote structures by script letters, and their domains by the corresponding capital Latin letters. Let $\mathcal{A}$ be a structure for L. For $X \subseteq A$, let L_X be the language $L \cup \{\mathbf{a} : a \in X\}$, L expanded by adding a constant $\mathbf{a}$ for every $a \in X$. Let $\mathcal{A}_X = (\mathcal{A}, a)_{a \in X}$ be the expansion of $\mathcal{A}$ to the language L_X such that for every $a \in X$, $\mathbf{a}$ is interpreted by a. The *atomic diagram* of $\mathcal{A}$ is the set of all atomic and negated atomic sentences of L_A which are true in $\mathcal{A}_A$. A structure $\mathcal{A}$ is computable if its domain A is a computable subset of ω and its relations and operations are uniformly computable. That is, $\mathcal{A}$ is computable if its atomic diagram is computable.

A structure $\mathcal{B}$ isomorphic to a computable structure $\mathcal{A}$ is not necessarily computable. However, even if $\mathcal{B}$ is computable, it can still lose many of the computable properties of $\mathcal{A}$. One of the important and interesting questions in computable model theory is how a specific aspect of a computable structure may change if the structure is isomorphically transformed so that it remains computable.

Let $\mathcal{A}$ be a fixed computable structure. A computable property of $\mathcal{A}$ which Ash and Nerode considered is given by a new computable relation R on the domain A of $\mathcal{A}$. Here, we call R *new* if R is not named in the language of $\mathcal{A}$. By $\mathrm{Im}_{\mathcal{A}}(R)$ we denote the set of images of R under all isomorphisms from $\mathcal{A}$ to other computable structures. For example, Ash and Nerode [2] investigated the conditions under which every relation in $\mathrm{Im}_{\mathcal{A}}(R)$ must be c. e. Such R is called *intrinsically c. e.* on $\mathcal{A}$. In [7] we found conditions for $(\mathcal{A}, R)$ under which there is a c. e. relation in $\mathrm{Im}_{\mathcal{A}}(R)$ of an arbitrary c. e. degree.

A sequence of variables displayed after a formula contains a subsequence of all free variables occurring in the formula. An $L_{\omega_1\omega}$ formula $\alpha(\vec{x})$ is a Σ_1 formula if it is equivalent to a formula of the form

$$\bigvee_{i\in I} \exists\vec{y_i}\theta_i(\vec{x},\vec{y_i}),$$

where for every $i \in I$, $\theta_i(\vec{x},\vec{y_i})$ is a finitary quantifier-free formula. If the index set I is c. e., then $\alpha(\vec{x})$ is a *computable* Σ_1 formula. If I is finite, then $\alpha(\vec{x})$ is a finitary Σ_1 formula.

For simplicity, we shall assume throughout the text that all new relations on computable structures are unary. We fix a new computable relation R on the domain of a computable structure $\mathcal{A}$ for a language L. Ash and Nerode [2] introduced a computable syntactic condition for R on $\mathcal{A}$, called being a formally c. e. relation. The relation R is *formally c. e.* (on $\mathcal{A}$) if R can be defined by a computable Σ_1 formula with parameters. That is, there is a sequence $\vec{c} \in A^{<\omega}$ and a computable Σ_1 formula $\alpha(\vec{\mathbf{c}}, x)$ such that the following equivalence holds for every $a \in A$:

$$R(a) \Leftrightarrow \mathcal{A}_A \models \alpha(\vec{\mathbf{c}}, \mathbf{a}).$$

Clearly, if R is formally c. e. on $\mathcal{A}$, then R is intrinsically c. e. on $\mathcal{A}$. Ash and Nerode [2] also established the converse for a computable relation R, under an additional decidability condition on $(\mathcal{A}, R)$.

Let $\mathcal{B}$ be a computable structure for L, and let X be a new relation the domain B. The complement of X with respect to B is $\overline{X}$. Let $\mathbf{R}$ be a symbol for X. Since we are interested in the case when X is c. e., certain first-order formulae with positive occurrences of $\mathbf{R}$ in the expanded language $L \cup \{\mathbf{R}\}$ play a special role (see [3], [7] and [10]). A Σ_1 formula in $L \cup \{\mathbf{R}\}$, possibly with individual constants (parameters), in which $\mathbf{R}$ occurs only positively is also called a Σ_1^Γ formula. This notation was introduced in [1], where a hierarchy of infinitary formulae was defined in a general setting in which Γ is a function assigning computable ordinals to relation symbols. A unary relation F on B is definable by a computable Σ_1^Γ formula $\alpha(\vec{d}, x)$ (with parameters $\vec{d} \in B^{<\omega}$) if for every $b \in B$:

$$F(b) \Leftrightarrow (\mathcal{B}_B, X) \models \alpha(\vec{\mathbf{d}}, \mathbf{b}).$$

The subsets of $\overline{X}$ which are definable by computable Σ_1^Γ formulae, and their subsets, play the role of finite sets to some extent.

We say that the relation X is *quasi-simple* on $\mathcal{B}$ if X is c. e., $\overline{X}$ is not definable by a computable Σ_1^Γ formula, and for every c. e. $W \subseteq \overline{X}$, there is a unary relation F definable by a computable Σ_1^Γ formula such that

$$W \subseteq F \subseteq \overline{X}.$$

Quasi-simplicity was first investigated by Hird [8], [10]. He proved that, under certain decidability conditions on $(\mathcal{A}, R)$, $\mathrm{Im}_{\mathcal{A}}(R)$ contains a quasi-simple relation. Ash,

Knight and Remmel [3] gave conditions on $(\mathcal{A}, R)$ which are sufficient for obtaining a quasi-simple relation of an arbitrary c. e. degree in $\mathrm{Im}_{\mathcal{A}}(R)$.

For a formula θ, let $\theta^1 =_{\mathrm{def}} \theta$ and $\theta^0 =_{\mathrm{def}} \neg\theta$. If f is a partial function, then $\mathrm{dom}(f)$ is the domain of f, $\mathrm{range}(f)$ is the range of f, and $f(a)\downarrow$ denotes that $a \in \mathrm{dom}(f)$. The length of a sequence $\vec{x}$ is denoted by $\mathrm{lh}(\vec{x})$. If $\vec{x} = (x_0, \dots, x_{m-1})$ and f is a function, then $f(\vec{x}) =_{\mathrm{def}} (f(x_0), \dots, f(x_{m-1}))$. The concatenation of sequences is denoted by $^\frown$.

We fix $\langle\cdot,\cdot\rangle$ to be a computable bijection from ω^2 onto ω, which is strictly increasing with respect to both coordinates. Let $W_0, W_1, \dots$ be a standard computable enumeration of all c. e. sets. By $\equiv_T$ we denote Turing equivalence of sets.

2. Nowhere simple relations on structures

Shore [15] introduced the concepts of nowhere simple and effectively nowhere simple sets in computability theory. Let $X \subseteq \omega$. We say that X is *nowhere simple* if X is c. e., and for every c. e. set W such that $W - X$ infinite, there is an infinite c. e. set W' such that $W' \subseteq W - X$. Thus, nowhere simplicity is definable in the lattice of all c. e. sets. Similarly, X is *effectively nowhere simple* if X is c. e., and there is a unary computable function g such that for every $e \in \omega$, $W_{g(e)} \subseteq W_e - X$ and

$$W_e - X \text{ is infinite} \Rightarrow W_{g(e)} \text{ is infinite.}$$

The function g is called a *witness function* for X. Clearly, every computable set is effectively nowhere simple. Miller and Remmel [12] established that X is effectively nowhere simple if there is a c. e. set W such that $W \cap X = \emptyset$ and for every $e \in \omega$,

$$W_e - X \text{ is infinite} \Rightarrow W_e \cap W \text{ is infinite.}$$

The set W is called a *witness set* for X. Hence, effective nowhere simplicity is definable in the lattice of all c. e. sets. Shore [15], and Miller and Remmel [12] proved that every c. e. Turing degree contains an effectively nowhere simple set.

Let $\mathcal{V}_\infty$ be a computable infinite dimensional vector space over a computable field F. We also assume that $\mathcal{V}_\infty$ has a dependence algorithm. We consider only subspaces of $\mathcal{V}_\infty$. For any subset $S \subseteq V_\infty$, by $\mathcal{S}^*$ we denote the subspace (with the domain S^*) generated by S. Let $V_e = W_e^*$. Then $\mathcal{V}_0, \mathcal{V}_1, \mathcal{V}_2, \dots$ is a computable list of all c. e. subspaces of $\mathcal{V}_\infty$. A subspace $\mathcal{V}$ of $\mathcal{V}_\infty$ is *nowhere simple* (see [14], [4] and [5]) if $\mathcal{V}$ is c. e., and for every $e \in \omega$, there is a c. e. subspace $\mathcal{V}'_e$ such that $V'_e \subseteq V_e$, $V'_e \cap V = \{0\}$, and

$$\dim(\mathcal{V}_e/\mathcal{V}) = \infty \Rightarrow \dim(\mathcal{V}'_e) = \infty.$$

If, in addition, there is a unary computable function g such that for every $e \in \omega$, $V'_e = V_{g(e)}$ then $\mathcal{V}$ is called *effectively nowhere simple*. Nerode and Remmel [14] proved that a nowhere simple subset of a computable basis of $\mathcal{V}_\infty$ generates a nowhere simple subspace of $\mathcal{V}_\infty$. Similarly, an effectively nowhere simple subset of a computable basis generates an effectively nowhere simple subspace. Let **c** be a nonzero c. e. Turing degree. Downey and Remmel [5] established that **c** contains a simple subset of a given computable basis of $\mathcal{V}_\infty$, which generates an effectively nowhere simple subspace $\mathcal{V}$ of $\mathcal{V}_\infty$. Moreover, $\mathcal{V}$ has a *witness space* $\mathcal{W}$, where $\mathcal{W}$ is c. e., $W \cap V = \{0\}$ and for every $e \in \omega$,

$$\dim(\mathcal{V}_e/\mathcal{V}) = \infty \Rightarrow \dim([\mathcal{V}_e \cap (\mathcal{W} \cup \mathcal{V})^*]/\mathcal{V}) = \infty.$$

Downey and Remmel [5] proved that the existence of a witness space guarantees effective nowhere simplicity. It is not known whether every effectively nowhere simple subspace has a witness space. It is not even known for an effectively nowhere simple subspace $\mathcal{V}$ and $v \in V_\infty - V$, whether $V \cup \{v\}$ generates an effectively nowhere simple subspace.

We consider the following general definition of nowhere simplicity and effective nowhere simplicity of relations on computable structures.

Definition 2.1. Let $\mathcal{B}$ be a computable structure and let X be a new unary relation on B. Then X is *nowhere simple* on $\mathcal{B}$ if X is c. e., and for every c. e. set W, there is a c. e. set $W' \subseteq W - X$ such that if $W - X$ is not contained in any subset F of $\overline{X}$ definable by a computable Σ_1^Γ formula, then the same is true of W'.

The relation X is *effectively nowhere simple* on $\mathcal{B}$ if there is a computable unary function g such that for every $e \in \omega$, $W_{g(e)} \subseteq W_e - X$ and if $W_e - X$ is not contained in any subset F of $\overline{X}$ definable by a computable Σ_1^Γ formula, then the same is true of $W_{g(e)}$.

3. Main result

We fix a computable structure $\mathcal{A}$ for a language L, and a new computable unary relation R on A. Let $\vec{c} \in A^{<\omega}$ and $a \in A$. The property of $\vec{c}$ and a, termed a is free over $\vec{c}$, played an essential role in the results in [7] and [3]. We say that a is *free over* $\vec{c}$ if $a \in \overline{R}$ ($\overline{R} = A - R$) and for every finitary Σ_1^Γ formula $\alpha(\vec{c}, x)$, if

$$(\mathcal{A}_A, R) \models \alpha(\vec{c}, \mathbf{a})$$

then

$$(\exists a' \in R)[(\mathcal{A}_A, R) \models \alpha(\vec{c}, \mathbf{a}')].$$

Let the set of all free elements over $\vec{c}$ be denoted by $\mathrm{fr}(\vec{c})$. Note that $\mathrm{fr}(\vec{c}) \subseteq \overline{R}$. Clearly, if $a \in \mathrm{fr}(\vec{c})$ and $\vec{d}$ is a subsequence of $\vec{c}$, then $a \in \mathrm{fr}(\vec{d})$. As in [3], let

$$\mathrm{bd}(\vec{c}) =_{\mathrm{def}} \{a \in \overline{R}\colon a \text{ is not free over } \vec{c}\}.$$

Thus, if $a \in \mathrm{bd}(\vec{c})$ and $\vec{c}$ is a subsequence of $\vec{d}$, then $a \in \mathrm{bd}(\vec{d})$. A maximal relation on $\overline{R}$ (with respect to the set-theoretic inclusion) which is definable by a computable Σ_1^Γ formula with parameters $\vec{c}$ is of the form $\mathrm{bd}(\vec{c})$. Conversely, if $\mathrm{bd}(\vec{c})$ is definable by a computable Σ_1^Γ formula with parameters $\vec{c}$, then $\mathrm{bd}(\vec{c})$ is a maximal relation on $\overline{R}$ definable by such a formula.

Theorem 3.1. *Let the following conditions hold for a computable structure $\mathcal{A}$ and a new computable unary relation R on A:*

(1) *for every $\vec{c} \in A^{<\omega}$, there is an element a such that $a \in \mathrm{fr}(\vec{c})$;*

(2) *for every $\vec{c} \in A^{<\omega}$, $\mathrm{bd}(\vec{c})$ is definable by a computable Σ_1^Γ formula with parameters $\vec{c}$;*

(3) *there is an algorithm which for a given $\vec{c} \in A^{<\omega}$ and $a \in A$, decides whether $a \in \mathrm{fr}(\vec{c})$ (equivalently, whether $a \in \mathrm{bd}(\vec{c})$);*

(4) *for every $\vec{c} \in A^{<\omega}$ and $a, v \in \overline{R}$, if $a \in \mathrm{fr}(\vec{c})$, and $v \in \mathrm{bd}(\vec{c})$, then $a \in \mathrm{fr}(\vec{c}^\frown v)$;*

(5) *for every $\vec{c} \in A^{<\omega}$, $\vec{u} \in \overline{R}^{<\omega}$ and $a, v \in \overline{R}$, if $a \in \mathrm{fr}(\vec{c}^\frown \vec{u}), v \in \mathrm{fr}(\vec{c}^\frown a)$, and $v \in \mathrm{fr}(\vec{u})$, then $a \in \mathrm{fr}(\vec{c}^\frown \vec{u}^\frown v)$.*

Then there is a computable structure $\mathcal{B}$ isomorphic to $\mathcal{A}$ such that the image of R on B is noncomputable and effectively nowhere simple on $\mathcal{B}$.

Proof. For simplicity, we assume that $A = \omega$. We will construct a computable structure $\mathcal{B}$ with domain $B = \omega$ and an isomorphism f from $\mathcal{B}$ to $\mathcal{A}$ such that $f^{-1}(R)$ is noncomputable and effectively nowhere simple on $\mathcal{B}$. Let s be an arbitrary stage of the construction. We will define a finite set Ψ^s of formulae of the open diagram of $\mathcal{B}$ and certain formulae of the form $\mathbf{R}(\mathbf{b})$ for $b \in B$. We will also define a finite partial isomorphism f_s from $\mathcal{B}$ to $\mathcal{A}$. A finite isomorphism from $\mathcal{B}$ to $\mathcal{A}$ at stage s, is an injective function p with a finite domain such that for every $\theta \in \Psi^s$, if $\theta = \theta(\mathbf{b}_0, \dots, \mathbf{b}_{l-1})$ for some $b_0, \dots, b_{l-1} \in B$, then $p(b_0)\downarrow, \dots, p(b_{l-1})\downarrow$ and $\mathcal{A} \models \theta[p(b_0), \dots, p(b_{l-1})]$, and if $\mathbf{R}(\mathbf{b}) \in \Psi^s$ then $p(b) \in R$. Let $X_s = f_s^{-1}(R)$. Define $\Psi^{-1} = \emptyset$, $f_{-1} = \emptyset$ and $X_0 = \emptyset$. At the end of the construction, we will have that $f = \lim_{s \to \infty} f_s$ exists. Let $X =_{\mathrm{def}} f^{-1}(R)$ and $\Psi = \bigcup_{s \geq 0} \Psi^s$. The construction will ensure that X is c. e.

Let $(\theta_e)_{e \in \omega}$ be an effective list of all atomic formulae in L_A. The construction will meet the following requirements for every $e \geq 0$,

$P_e^0 : (\theta_e \in \Psi \text{ or } \neg\theta_e \in \Psi)$, and $(b \in X \Rightarrow \mathbf{R}(\mathbf{b}) \in \Psi)$;
$P_e^1 : e \in \mathrm{dom}(f)$;
$P_e^2 : e \in \mathrm{range}(f)$;
$Q_e : \overline{X} \neq W_e$;

$N_{\langle e,k\rangle}$: If $W_e - X$ is not contained in any subset of $\overline{X}$ definable by a computable Σ_1^Γ formula, we define $u_{e,k}$ such that $u_{e,k} \in W_e - X$ and

$$f(u_{e,k}) \in \mathrm{fr}\,\bigl(f(u_{e,0},{}^\frown \cdots {}^\frown u_{e,k-1})\bigr),$$

where for $i \in \{0, \ldots, k-1\}$, $u_{e,i}$ is previously defined by $N_{\langle e,i\rangle}$.

The priority ordering of the requirements is:

$$P_0^0,\ P_0^1,\ P_0^2,\ Q_0,\ N_0,\ P_1^0,\ P_1^1,\ P_1^2,\ Q_1,\ N_1,\ldots.$$

Let $e, i \in \omega$. We set that at stage -1, $u_{e,i}$ is undefined. To assure effective nowhere simplicity, we will have that the element $u_{e,i}$ once defined will remain unchanged, and $\{u_{e,j} : j \in \omega \wedge (u_{e,j} \text{ is defined})\}$ will be a required c. e. subset W_e' of $W_e - X$. Since the construction will be effective uniformly in e, it will follow from the $s-m-n$ theorem that there is a unary computable function g such that $W_{g(e)} = W_e'$.

The strategy for meeting a single requirement Q_e is to chose an element d_e such that

$$d_e \in W_e \ \Leftrightarrow\ d_e \in X.$$

We will have $d_e = \lim\limits_{s\to\infty} d_e^s$, where d_e^s is the approximation to d_e at stage s, and d_e^s may be undefined for finitely many s. If d_e^s is defined, then $d_0^s, \ldots, d_{e-1}^s$ are also defined. We assume that for every $e \in \omega$, d_e^{-1} is undefined. For $n \in \omega$, let $\vec{d^s_{<n}} =_{\mathrm{def}} (d_0^s, \ldots, d_e^s)$, where e is the largest number such that $e < n$ and d_e^s is defined. The construction will ensure that all elements of $\vec{d^s_{<n}}$ are in $\mathrm{dom}(f_s)$.

Let $\vec{u^s}$ be the vector consisting of all elements $u_{e,i}$, $e, i \in \omega$, which are defined at stage s (listed in the order in which they are first defined). The construction will ensure that all elements of $\vec{u^s}$ are contained in $\mathrm{dom}(f_s)$. For $k \in \omega$, if $u_{e,k}$ is defined at some stage, then $u_{e,o}, \ldots, u_{e,k-1}$ are also defined at that stage. Let $\vec{u_{e,<k}} =_{\mathrm{def}} (u_{e,o}, \ldots, u_{e,k-1})$. We have the following definitions which will be used in the construction.

Requirement P_e^0 requires attention and can be attacked at s if $\theta_e \notin \Psi^{s-1}$, $\neg\theta_e \notin \Psi^{s-1}$ and all elements of B occurring in θ_e are in the domain of f_{s-1}. Requirement P_e^0 also requires attention if θ is of the form $\mathbf{R}(\mathbf{b})$, $\theta \notin \Psi^{s-1}$, $f_{s-1}(b)\!\downarrow$ and $f_{s-1}(b) \in R$.

Requirement P_e^1 requires attention and can be attacked at s if $e \notin \mathrm{dom}(f_{s-1})$.

Requirement P_e^2 requires attention and can be attacked at s if $e \notin \mathrm{range}(f_{s-1})$.

Requirement Q_e requires attention and can be attacked at s if d_e^{s-1} is undefined at $s-1$, or $d_e^{s-1} \in W_{e,s} - W_{e,s-1}$.

Requirement $N_{\langle e,k\rangle}$ requires attention at s if $u_{e,k}$ is undefined at $s-1$. Such a requirement $N_{\langle e,k\rangle}$ can be attacked at s if $u_{e,o}, \ldots, u_{e,k-1}$ are defined at $s-1$, all elements of $\vec{c}$, where

$$\vec{c} = \bigl(0, \ldots, \langle e,k\rangle\bigr)^\frown\bigl(f_{s-1}^{-1}(0), \ldots, f_{s-1}^{-1}(\langle e,k\rangle)\bigr)^\frown \vec{d^{s-1}_{<\langle e,k\rangle+1}},$$

are in $\mathrm{dom}(f_{s-1})$, and there is $u \in W_{e,s}$ such that $u \in \mathrm{dom}(f_{s-1})$, and

$$f_{s-1}(u) \in \mathrm{fr}\left(f_{s-1}(\vec{c}^\frown \vec{u}_{e,<k})\right).$$

Notice that $f_{s-1}(u) \in \overline{R}$.

Requirement P_e^0 is never injured.

Requirement P_e^1 is injured at s if $f_s(e) \neq f_{s-1}(e)$.

Requirement P_e^2 is injured at s if $f_s^{-1}(e) \neq f_{s-1}^{-1}(e)$.

Requirement Q_e is injured at s if d_e^{s-1} is defined, and d_e^s is undefined; or if d_e^s is defined, $d_e^s \in \overline{R}$, and

$$f_s(d_e^s) \in \mathrm{bd}\left(f_s(0,\dots,e)^\frown(0,\dots,e)^\frown f_s(\vec{d}^s_{<e})^\frown f_s(\vec{u}^s)\right).$$

Requirement $N_{\langle e,k\rangle}$ is never injured.

Construction. *Stage* $s \geq 0$.

Let Req be the highest priority requirement which requires attention and which can be attacked at stage s.

(i) Req $= P_e^0$

Let $\theta_e = \theta_e(\mathbf{b}_0,\dots,\mathbf{b}_{l-1})$ for some $b_0,\dots,b_{l-1} \in \omega$. Define $\Psi^s = \Psi^{s-1} \cup \{\theta_e^k\}$, where $k \in \{0,1\}$ is such that $\mathcal{A} \models \theta_e^k[f_{s-1}(b_0),\dots,f_{s-1}(b_{l-1})]$. Let $f_s =_{\mathrm{def}} f_{s-1}$. The constants associated with Q-requirements are as at stage $s-1$.

(ii) Req $= P_e^1$

We define $f_s = f_{s-1} \cup \{(e,a)\}$, where a is the least unused element in A. If $a \in R$, we set $\Psi^s = \Psi^{s-1} \cup \{\mathbf{R}(\mathbf{e})\}$; otherwise, we set $\Psi^s = \Psi^{s-1}$. The constants associated with Q-requirements are as at stage $s-1$.

(iii) Req $= P_e^2$

We define $f_s = f_{s-1} \cup \{(b,e)\}$, where b is the least unused element in B. If $e \in R$, we set $\Psi^s = \Psi^{s-1} \cup \{\mathbf{R}(\mathbf{b})\}$; otherwise, we set $\Psi^s = \Psi^{s-1}$. The constants associated with Q-requirements are as at stage $s-1$.

(iv) Req $= Q_e$

Let $\vec{c} = (0,\dots,e)^\frown(f_{s-1}^{-1}(0),\dots,f_{s-1}^{-1}(e))^\frown \vec{d}^{s-1}_{<e}{}^\frown \vec{u}^{s-1}$. It follows by construction that all elements of $\vec{c}$ are in $\mathrm{dom}(f_{s-1})$.

Case (a): d_e^{s-1} is undefined.

Let a be the least element in $A - f_{s-1}(W_{e,s})$ such that a is free over $f_{s-1}(\vec{c})$. This a exists by condition (1) of the theorem. If $a \in \mathrm{range}(f_{s-1})$, we set $d_e^s =_{\mathrm{def}} f_{s-1}^{-1}(a)$ and $f_s =_{\mathrm{def}} f_{s-1}$. If $a \notin \mathrm{range}(f_{s-1})$, we choose b to be the least new element in $B - W_{e,s}$, and set $d_e^s =_{\mathrm{def}} b$ and $f_s =_{\mathrm{def}} f_{s-1} \cup \{(b,a)\}$. Notice that in both subcases, $d_e^s \notin X_s$ and $d_e^s \notin W_{e,s}$. For every $e' \neq e$, let $d_{e'}^s =_{\mathrm{def}} d_{e'}^{s-1}$. Let $\Psi^s =_{\mathrm{def}} \Psi^{s-1}$.

Case (b): $d_e^{s-1} \in W_{e,s} - W_{e,s-1}$.

It will follow by construction that $a = f_{s-1}(d_e^{s-1})$ is free over $f_{s-1}(\vec{c})$. Hence $a \in \overline{R}$. Let $\psi(\vec{\mathbf{c}}, x)$ be the Σ_1^Γ formula with parameters $\vec{\mathbf{c}}$ which captures the part of the diagram of $(\mathcal{B}, X)$, determined by stage $s-1$. That is, $\psi(\vec{\mathbf{c}}, x) = (\exists \vec{y})\delta(\vec{\mathbf{c}}, x, \vec{y})$, where $\delta(\vec{\mathbf{c}}, d_e^{s-1}, \vec{\mathbf{b}})$ is the conjunction of all formulas of Ψ^{s-1}, and $\mathrm{lh}(\vec{y}) = \mathrm{lh}(\vec{b})$. Clearly, $(\mathcal{B}, X) \models \psi[\vec{c}, d_e^{s-1}]$, so $(\mathcal{A}, R) \models \psi[f_{s-1}(\vec{c}), a]$. We choose a' to be the least element such that $a' \in R$ and $(\mathcal{A}, R) \models \psi[f_{s-1}(\vec{c}), a']$. Let $\vec{a} \in A^{\mathrm{lh}(\vec{b})}$ be the least sequence such that $(\mathcal{A}, R) \models \delta[f_{s-1}(\vec{c}), a', \vec{a}]$. We set

$$f_s =_{\text{def}} (\vec{c}^\frown d_e^{s-1\frown}\vec{b},\ f_{s-1}(\vec{c})^\frown a'^\frown \vec{a}).$$

For every $e' \leq e$, let $d_{e'}^s =_{\text{def}} d_{e'}^{s-1}$, and for every $e' > e$, let $d_{e'}^s$ be undefined. Let $\Psi^s =_{\text{def}} \Psi^{s-1} \cup \{\mathbf{R}(\mathbf{b}) : b \in X_s - X_{s-1}\}$.

(v) $\mathrm{Req} = N_j$, where $j = \langle e, k\rangle$.

Let $\vec{c_j} = (0, \dots, j)^\frown(f_{s-1}^{-1}(0), \dots, f_{s-1}^{-1}(j))^\frown \vec{d}_{<j+1}^{s-1\frown} \vec{u}_{e,<k}$. It follows by construction that all elements of $\vec{c_j}$ are in $\mathrm{dom}(f_{s-1})$. We choose the least element $u \in W_{e,s}$ such that $u \in \mathrm{dom}(f_{s-1})$ and $f_{s-1}(u) \in \mathrm{fr}(f_{s-1}(\vec{c_j}))$. Hence $u \notin X_{s-1}$. We set

$$u_{e,k} =_{\text{def}} u.$$

For every $e' \leq j$, let $d_{e'}^s =_{\text{def}} d_{e'}^{s-1}$, and for every $e' > j$, let $d_{e'}^s$ be undefined. Let $f_s =_{\text{def}} f_{s-1}$ and $\Psi^s =_{\text{def}} \Psi^{s-1}$.

End of the construction.

Lemma 3.2. *No Q-requirement is injured by an action of a lower priority N-requirement.*

Proof. Let us consider N_j for some $j = \langle e, k\rangle$, and $Q_{e'}$ for some $e' \leq j$. Let s be a stage at which the requirement N_j is being satisfied. We will use the same notation as in (v) of the construction. We assume that $f_{s-1}(d_{e'}^{s-1})$ is free over

$$f_{s-1}(0, \dots, e')^\frown(0, \dots, e')^\frown f_{s-1}(\vec{d}_{<e'}^{s-1})^\frown f_{s-1}(\vec{u}^{s-1}).$$

We want to show that $f_s(d_{e'}^s)$ remains free over

$$f_s(0, \dots, e')^\frown(0, \dots, e')^\frown f_s(\vec{d}_{<e'}^s)^\frown f_s(\vec{u}^s).$$

Since $d_{e'}^s = d_{e'}^{s-1}$, $f_s(d_{e'}^s) = f_{s-1}(d_{e'}^{s-1})$, $f_s(0, \dots, e') = f_{s-1}(0, \dots, e')$, $\vec{d}_{<e'}^s = \vec{d}_{<e'}^{s-1}$, $f_s(\vec{d}_{<e'}^s) = f_{s-1}(\vec{d}_{<e'}^{s-1})$, $\vec{u}^s = \vec{u}^{s-1\frown} u_{e,k}$, $f_s(\vec{u}^{s-1}) = f_{s-1}(\vec{u}^{s-1})$, and $f_s(u_{e,k}) = f_{s-1}(u_{e,k})$, it is sufficient to conclude that $f_{s-1}(d_{e'}^{s-1})$ is free over

$$f_{s-1}(0, \dots, e')^\frown(0, \dots, e')^\frown f_{s-1}(\vec{d}_{<e'}^{s-1})^\frown f_{s-1}(\vec{u}^{s-1})^\frown f_{s-1}(u_{e,k}).$$

First, we assume that $f_{s-1}(u_{e,k}) \in \mathrm{bd}(f_{s-1}(\vec{u}^{s-1}))$. Then the required conclusion follows from condition (4) of the theorem if we set

$$\vec{c} = f_{s-1}(0, \dots, e')^\frown(0, \dots, e')^\frown f_{s-1}(\vec{d}_{<e'}^{s-1})^\frown f_{s-1}(\vec{u}^{s-1}),$$

$$a = f_{s-1}(d^{s-1}_{e'}) \text{ and } v = f_{s-1}(u_{e,k}).$$

That is, $a \in \mathrm{fr}(\vec{c})$ by our first assumption, and $v \in \mathrm{bd}(\vec{c})$ since, by our second assumption, v is bounded over a subsequence of $\vec{c}$.

Now, we assume that $f_{s-1}(u_{e,k}) \in \mathrm{fr}(f_{s-1}(\overrightarrow{u^{s-1}}))$. Then the required conclusion follows from condition (5) of the theorem if we set

$$\vec{c} = f_{s-1}(0, \ldots, e')^\frown(0, \ldots, e')^\frown f_{s-1}(\overrightarrow{d^{s-1}_{<e'}}),$$

$$\vec{u} = f_{s-1}(\overrightarrow{u^{s-1}}),$$

$$a = f_{s-1}(d^{s-1}_{e'}) \text{ and } v = f_{s-1}(u_{e,k}).$$

That is, $a \in \mathrm{fr}(\vec{c}^\frown\vec{u})$ by our first assumption, $v \in \mathrm{fr}(\vec{c}^\frown a)$ by a condition in (v) of the construction, and $v \in \mathrm{fr}(\vec{u})$ by our second assumption. □

It follows from the construction that a Q-requirement can be injured at most once by every higher priority N-requirement. Since P^1-requirements, P^2-requirements and Q-requirements may be injured only by higher priority Q-requirements, it is easy to show that each requirement is attacked and injured only finitely often. Thus, all P-requirements are met. Hence there is an isomorphism f from $\mathcal{B}$ to $\mathcal{A}$ such that $f = \lim_{s\to\infty} f_s$. The construction is effective because of the condition (3) of the theorem. Therefore, $\mathcal{B}$ is a computable structure. Let $X = \bigcup_{s\geq 0} X_s$. Clearly, X is c. e. It follows by construction that $X = f^{-1}(R)$. Since every Q-requirement is satisfied, X is noncomputable.

Lemma 3.3. *Every N-requirement is satisfied.*

Proof. We fix $e \in \omega$. If $u_{e,k}$ is defined for every $k \in \omega$, then every $N_{\langle e,k\rangle}$ is satisfied. Therefore, let us assume that k is the least number such that $u_{e,k}$ is not defined. Let s_0 be a stage by which all requirements of higher priority than $N_{\langle e,k\rangle}$ have been attacked for the last time. Hence, at s_0, the sequence of numbers in B coming from the higher priority P-requirements and Q-requirements has reached its final value $\vec{b}$. Then for $s > s_0$, for every $u \in W_{e,s} - X_s$ such that $u \in \mathrm{dom}(f_{s-1})$, we have that $f_{s-1}(u) \in \mathrm{bd}(f(\vec{b})^\frown f(u_{e,<k}))$. It then follows from condition (2) of the theorem that $W_e - X$ is contained in a subset of $\overline{X}$ which is definable by a computable Σ^Γ_1 formula. Hence $N_{\langle e,k\rangle}$ is satisfied for every k. □

Finally, we will prove that X is effectively nowhere simple on $\mathcal{B}$. Let $e \in \omega$. If for some k, $u_{e,k}$ is not defined, then $W_e - X$ is contained in some $F \subseteq \overline{X}$ such that F is definable by a computable Σ^Γ_1 formula. We now assume that $u_{e,k}$ is defined for every k. We set $W'_e = \{u_{e,0}, u_{e,1}, u_{e,2}, \ldots\}$. It is enough to show that W'_e is not contained in any subset of $\overline{X}$ which is definable by a computable Σ^Γ_1 formula. Let us assume otherwise. That is, let W'_e be contained in some subset of $\overline{X}$ which is definable by a computable Σ_1^Γ formula with parameters $\vec{d}$. Then $f(u_{e,k}) \in \mathrm{bd}(f(\vec{d}))$ for every k. This is not possible since, by (v) of the construction, for every k such that $\langle e, k\rangle \geq \max(\mathrm{range}(\vec{d}))$, we have that $f(u_{e,k}) \in \mathrm{fr}(f(\vec{d}))$. □

Theorem 3.1 can be modified, by using coding and permitting, as described in Theorem 2.5 in [7], to obtain an effectively nowhere simple set in $\mathrm{Im}_{\mathcal{A}}(R)$ of an arbitrary c. e. degree.

Theorem 3.4. *Let conditions* (1)–(5) *from Theorem* 3.1 *hold for a unary computable relation R on the domain of a computable structure $\mathcal{A}$. Then, for every c. e. degree* **c**, *there is a computable structure $\mathcal{B}$ isomorphic to $\mathcal{A}$ such that the image X of R in B is of degree* **c** *and effectively nowhere simple on $\mathcal{B}$.*

Proof. We just sketch the main idea. Let $C \subseteq \omega$ be an arbitrary noncomputable c. e. set such that at every stage s, C receives at most one new element c, and $c \leq s$. We replace the Q-requirements of Theorem 3.1 by the requirement $X \equiv_T C$. At every stage s of the construction, we have a sequence of movable markers on the elements in $\overline{X}_s$, where $\overline{X}_s = \{d_0^s < d_1^s < d_2^s < \cdots\}$. More precisely, we effectively define a partial binary function γ with the intention that if $c \in C_s - C_{s-1}$, then $\gamma(c, s-1)\downarrow$ and $d_{\gamma(c,s-1)}^{s-1} \in X_s$. □

Example 3.1. Let $\mathcal{A}$ be $(\omega, =)$ and let $R \subseteq A$ be a computable infinite co-infinite set. It is shown in [3] that $\mathcal{A}$ and R satisfy conditions (1)–(4). The condition (5) is also satisfied because for $\vec{c} \in A^{<\omega}$ and $a \in \overline{R}$, we have

$$a \in \mathrm{fr}(\vec{c}) \Leftrightarrow a \notin \mathrm{range}(\vec{c}).$$

Hence, we obtain Shore's and Miller–Remmel's theorem that every c. e. Turing degree contains an effectively nowhere simple set.

Example 3.2. Let $\mathcal{A}$ be a computable linear order of type η and let $R \subseteq A$ be a computable dense co-dense set. It is shown in [3] that $\mathcal{A}$ and R satisfy conditions (1)–(4). The condition (5) is also satisfied because for $\vec{c} \in A^{<\omega}$ and $a \in \overline{R}$, we have

$$a \in \mathrm{fr}(\vec{c}) \Leftrightarrow a \notin \mathrm{range}(\vec{c}).$$

Thus, we have the following corollary.

Corollary 3.5. *For every c. e. set C, there is a computable linear order of type η with a dense co-dense effectively nowhere simple set X such that $X \equiv_T C$.*

Example 3.3. Let $\mathcal{B}$ be a computable linear order of type $\omega + \omega^*$ and let X be the ω-part of B. Then it is easy to see that X can not be nowhere simple unless it is computable. That is, if X is c. e. and if W is an infinite c. e. set such that $W \subseteq \overline{X}$, then X is computable. On the other hand, Ash, Knight and Remmel [3] have shown that for any noncomputable c. e. set C, there is a computable linear order of type $\omega + \omega^*$

with the ω-part X such that X is a simple set and $X \equiv_T C$. Indeed, if $\mathcal{A} = (\omega, \prec)$ is a computable linear order of type $\omega + \omega^*$ with the computable ω-part R, it is shown in [3] that conditions (1)-(4) are satisfied. Condition (5) is not satisfied because for $\vec{c} \in A^{<\omega}$ and $a \in \overline{R}$, we have

$$a \in \mathrm{fr}(\vec{c}) \Leftrightarrow a \prec c_0,$$

where $\mathrm{range}(\vec{c}) \cap \overline{R} = \{c_0 \prec \cdots \prec c_l\}$.

Acknowledgement. I thank Rumen Dimitrov for proofreading. I also thank the referee for suggestions.

References

[1] C. J. Ash and J. F. Knight, Ramified systems, Ann. Pure Appl. Logic 70 (1994), 205–221.

[2] C. J. Ash and A. Nerode, Intrinsically recursive relations, in: Aspects of Effective Algebra (ed. J. N. Crossley), U.D.A. Book Co., Steel's Creek, Australia, 1981, 26–41.

[3] C. J. Ash, J. F. Knight, and J. B. Remmel, Quasi-simple relations in copies of a given recursive structure, Ann. Pure Appl. Logic 86 (1997), 203–218.

[4] R. Downey, Nowhere simplicity in matroids, J. Austral. Math. Soc. Ser. A 35 (1983), 28–45.

[5] R. G. Downey and J. B. Remmel, Effectively and noneffectively nowhere simple subspaces, in: Logical Methods (eds. J. N. Crossley, J. B. Remmel, R. A. Shore, and M. E. Sweedler), Birkhäuser, Boston, 1993, 314–351.

[6] V. S. Harizanov, Effectively and noneffectively nowhere simple sets, Math. Logic Quart. 42 (1996), 241–248.

[7] V. S. Harizanov, Some effects of Ash–Nerode and other decidability conditions on degree spectra, Ann. Pure Appl. Logic 55 (1991), 51–65.

[8] G. R. Hird, Recursive Properties of Relations on Models, Ph.D. Thesis, Monash University, Melbourne, Australia, 1984.

[9] G. Hird, Recursive properties of intervals of recursive linear orders, in: Logical Methods (eds. J. N. Crossley, J. B. Remmel, R. A. Shore, and M. E. Sweedler), Birkhäuser, Boston, 1993, 422–437.

[10] G. R. Hird, Recursive properties of relations on models, Ann. Pure Appl. Logic 63 (1993), 241–269.

[11] G. Metakides and A. Nerode, Recursion theory on fields and abstract dependence, J. Algebra 65 (1980), 36–59.

[12] D. Miller and J. B. Remmel, Effectively nowhere simple sets, J. Symbolic Logic 49 (1984), 129–136.

[13] A. Nerode and J. B. Remmel, A survey of lattices of r. e. substructures, in: Recursion Theory (eds. A. Nerode and R. Shore), Proc. Sympos. Pure Math. 42, American Mathematical Society, Providence, 1985, 323–375.

[14] A. Nerode and J. B. Remmel, Recursion theory on matroids, in: Patras Logic Symposium (ed. G. Metakides,), Stud. Logic Found. Math. 109, North-Holland, Amsterdam, 1982, 41–65.

[15] R. A. Shore, Nowhere simple sets and the lattice of recursively enumerable sets, J. Symbolic Logic 43 (1978), 322–330.

[16] R. I. Soare, Recursively Enumerable Sets and Degrees. A Study of Computable Functions and Computably Generated Sets, Springer-Verlag, Berlin, 1987.

Jump traces with large gaps

Peter G. Hinman

Abstract. We settle the main question left open in [Hi-Sl] by establishing that for all n there exists a Turing degree $\mathbf{a}$ with jump trace of the form $(\overrightarrow{0}\,;\overleftarrow{n})$ — that is for all $k \in \omega$, $\mathbf{0}^{(k)} \leq_T \mathbf{a}^{(k)} \leq_T \mathbf{0}^{(n+k)}$ and $\mathbf{a}^{(k)}$ is incomparable with $\mathbf{0}^{(k+1)}, \ldots, \mathbf{0}^{(n+k-1)}$. It follows that every consistent jump trace is realized.

1. Introduction and summary

In [Hi-Sl; 1.3] the *jump trace* of an arithmetical Turing degree $\mathbf{a}$ is defined as

$$\mathrm{JTr}(\mathbf{a}) = (h_0, h_1, \ldots, h_k, \ldots; \ldots, \ell_k, \ldots, \ell_1, \ell_0),$$

where for all k, $\mathbf{0}^{(h_k+k)} \leq_T \mathbf{a}^{(k)} \leq_T \mathbf{0}^{(\ell_k+k)}$ and these bounds are best possible — that is, $\mathbf{a}^{(k)}$ is incomparable with all of $\mathbf{0}^{(h_k+k+1)}, \ldots, \mathbf{0}^{(\ell_k+k-1)}$ or equivalently $\mathbf{0}^{(h_k+k+1)} \not\leq_T \mathbf{a}^{(k)} \not\leq_T \mathbf{0}^{(\ell_k+k-1)}$. This notion is a generalization of the familiar notions of highness and lowness — a degree $\mathbf{a}$ is properly high$_k$ iff its jump trace is $(0, \ldots, 0, \overrightarrow{1}\,; \overleftarrow{1})$, where the initial sequence of 0's has length k and $\overrightarrow{1}$ and $\overleftarrow{1}$ denote constant infinite sequences, low$_k$ iff its jump trace is $(\overrightarrow{0}\,; \overleftarrow{0}, 1, \ldots, 1)$, where the terminal sequence of 1's has length k, and intermediate iff its jump trace is $(\overrightarrow{0}\,; \overleftarrow{1})$. (These facts account for the choice of letters h and ℓ.) Intuitively, h_k is a lower bound on the information content of $\mathbf{a}$ in an k-jump context, and ℓ_k is an upper bound.

The elementary properties of the jump operator imply

1.1 Lemma. *For any jump trace* $(h_0, h_1, \ldots, h_k, \ldots; \ldots, \ell_k, \ldots, \ell_1, \ell_0)$, *for all* k,

(i) $h_k \leq h_{k+1} \leq \ell_{k+1} \leq \ell_k$;

(ii) $\ell_k \leq \ell_{k+1} + 1$.

It follows that any jump trace is of the form

$$(h_0, h_1, \ldots, h_{k-1}, \overrightarrow{h}\,; \overleftarrow{\ell}, \ell_{k-1}, \ldots, \ell_1, \ell_0),$$

with

$$h_0 \leq h_1 \leq \cdots \leq h_{k-1} \leq h \leq \ell \leq \ell_{k-1} \leq \cdots \leq \ell_1 \leq \ell_0 \quad \text{and} \quad \ell_i \leq \ell_{i+1} + 1.$$

We call such a sequence a *gap-*$(\ell - h)$ *trace*. A trace is *realized* iff it is the jump trace of some degree. We proved as [Hi-Sl; Theorem 1.5] the

1.2 Theorem. *For* $n \leq 2$, *every gap-n trace is realized. For* $n > 2$, *if the simplest gap-n trace* $(\vec{0}\,;\,\overleftarrow{n})$ *is realized, then every gap-n trace is realized.*

Our main result here is that the hypothesis $n \leq 2$ is unnecessary:

1.3 Theorem. *For all n, there is a Turing degree* $\mathbf{a}$ *with* $\mathrm{JTr}(\mathbf{a}) = (\vec{0}\,;\,\overleftarrow{n})$*—that is for all* $k \in \omega$, $\mathbf{0}^{(k)} \leq_T \mathbf{a}^{(k)} \leq_T \mathbf{0}^{(n+k)}$ *and* $\mathbf{a}^{(k)}$ *is incomparable with* $\mathbf{0}^{(k+1)}, \ldots, \mathbf{0}^{(n+k-1)}$.

1.4 Corollary. *For all n, every gap-n trace is realized.*

The key to the proof of Theorem 1.2 was a generalization of an unpublished result of Harrington. Interestingly the main tools here are two different generalizations of the same result. The other ingredients are a generalization of a result of Jockusch and Shore [JoSh; Corollary 4.2] together with some wizardry using the Recursion Theorem inspired by the methods of [LaYi]. Of course, the first use of these techniques is M. Simpson's finite-injury proof of the Sacks Jump Theorem [So; Remark VII.1.8].

2. Jump inversion

Our notation generally follows that of [So], and we expect the reader to be somewhat familiar with the principal results of Chapters I–VIII of that book. Our main notational deviation is the following. Fix an enumeration $\langle W_e e \in \omega\rangle$ of the r. e. sets and set $X^{\langle e\rangle} = X \oplus W_e^X = X \oplus \{x : \exists y\, \langle x, X \upharpoonright y\rangle \in W_e\}$, $X^{\langle\,\rangle} = X$, and

$$X^{\langle e_0,\ldots,e_m\rangle} = \left(X^{\langle e_0,\ldots,e_{m-1}\rangle}\right)^{\langle e_m\rangle}.$$

Thus always $X \leq_T X^{\langle e_0,\ldots,e_m\rangle} \leq_T \mathbf{0}^{(m+1)}$. Sets of the form $X^{\langle e_0,\ldots,e_m\rangle}$ and their degrees are called $(m+1)$-REA$[X]$. m is generally omitted when it is 1 and X when it is $\emptyset$.

For example, in this notation, the uniform relativized version of the Sacks Jump Theorem asserts that there exists a primitive recursive function f such that for all sets X and all indices e, $X^{\langle f(e)\rangle\prime} \equiv_T X'^{\langle e\rangle}$ and $X <_T X^{\langle f(e)\rangle}$.

A basic tool in all that we do is the following result of Jockusch and Shore.

2.5 Lemma ([Jo-Sh; Corollary 4.2]). *For any index d and any low set B, there exists an r.e. set C such that* $B \leq_T C$ *and* $C^{\langle d\rangle} \equiv_T \emptyset'$. *More generally, there exists a primitive recursive function f such that for all sets X and indices b and d, if* $X^{\langle b\rangle}$ *is low*$[X]$, *then* $X^{\langle b\rangle} \leq_T X^{\langle f(b,d)\rangle}$ *and* $X^{\langle f(b,d),d\rangle} \equiv_T X'$. *Furthermore, indices for all of these Turing reductions are primitive recursively calculable from b and d.*

The other basic tool is the following generalization of an unpublished theorem of Harrington — the generalization consists in the relativization to the low set C.

2.6 Theorem. *For any set S which is REA$[\emptyset']$ and any low r.e. set C, there exists an r.e. set A and an REA$[A]$ set B such that $C \leq_T A$ and*

$$A' \equiv_T B' \equiv_T S \equiv_T B \oplus \emptyset'.$$

More generally, there exist primitive recursive functions f_0 and f_1 such that for all sets X, and all natural numbers c and z, if $X^{\langle c \rangle}$ is low$[X]$, then $X^{\langle c \rangle} \leq_T X^{\langle f_0(c,z) \rangle}$ and

$$X^{\langle f_0(c,z) \rangle \prime} \equiv_T X^{\langle f_0(c,z), f_1(c,z) \rangle \prime} \equiv_T X'^{z} \equiv_T X^{\langle f_0(c,z), f_1(c,z) \rangle} \oplus X'.$$

Furthermore, indices for all of these Turing reductions are primitive recursively calculable from c and z.

Proof. We prove the first version; the relativized and uniform version follows by the usual observation that the proof is effective and independent of a relativizing parameter. The proof is a blend of the original proof of Harrington with the technique of Robinson for working above a low degree, which is explained as part of [So; Theorem XI.3.2]. Although there is no proof of Harrington's Theorem in print, a generalization of it including upper-cone avoidance appears as [Hi-Sl; Theorem 2.5]. Therefore, rather than presenting here a complete proof, we shall explain how to modify a simplified version of this proof and will expect the reader to have access to it.

As in [Hi-Sl; Theorem 2.5], given S and C as specified, we shall actually construct an r.e. set A and an r.e.$[A]$ set B such that

$$(A \oplus C)' \equiv_T (A \oplus B \oplus C)' \equiv_T S \equiv_T A \oplus B \oplus C \oplus \emptyset'.$$

We fix a set E such that for all e,

$$e \in S \implies E^{[e]} \text{ is finite}, \qquad e \notin S \implies E^{[e]} = \omega^{[e]},$$

Our strategy is to include all but finitely much of each column of E in A to guarantee that $S \leq_T A'$ while controlling computations to ensure that $(A \oplus B \oplus C)'$ is computable from both S and $A \oplus B \oplus C \oplus \emptyset'$. The role of B is to provide enough information about the limiting behavior of the columns of A to enable $C \oplus \emptyset'$ to compute moduli for them.

The construction of A is necessarily an infinite injury one, but that of $B = B[A]$ is finite injury with restraints designed to protect computations of the form $\{e\}_t^{A \oplus B_t \oplus C_t}(e)$ from injury by subsequent enumerations into B below the use u; finiteness of the positive requirements ensures that this will happen at most finitely often. Of course, such a computation may also be injured by changes in C, and in general this may occur infinitely often. The lowness of C provides a way to make an informed guess whether or not C will change below u; the guess may be wrong, but only finitely often. Thus if we guess that C will not change, we C-*certify* the computation and proceed as before; in the opposite case we ignore the computation and wait until it is certified at a later stage before protecting it with a restraint.

Certification works as follows. Let $\langle F_n : n \in \omega \rangle$ be the canonical enumeration of the finite sets. Since C is low, the set

$$C^* := \{ e : (\exists n \in W_e)\ F_n \cap C = \emptyset \}$$

is recursive in $\emptyset'$ and hence is the limit of a recursive sequence $\langle C_s^* : s \in \omega \rangle$. When we encounter the computation $\{e\}_t^{A \oplus B_t \oplus C_t}(e)$ with use u, we first determine n_t such that $F_{n_t} = \{ y < u : y \notin C_t \}$ and enumerate n_t into an auxiliary r.e. set W_c that we build as part of the construction. If no such y will subsequently be enumerated into C, then $c \in C^*$ and hence for all sufficiently large v, $c \in C_v^*$. Hence a search for the least v such that either $c \in C_v^*$ or some $y < u$ enters C_v is well-defined; the computation is certified in the first case. Of course, certification may occur in cases when a C-change does occur later, but this can happen only finitely often.

The Recursion Theorem is used to enable us to use the index of the set W_c in the course of building this very set. This is slightly more involved than it might seem, because we are simultaneously using infinitely many such c, one for each potential computation. One way to formalize this procedure is the following. View the construction as dependent on a parameter a which ranges over indices for primitive recursive functions — we temporarily denote the a-th such function by $[a]$. In determining whether or not to certify the computation $\{e\}_t^{A \oplus B_t \oplus C_t}(e)$, we search for v such that $[a](e, t) \in C_v^*$. Then there exists a primitive recursive function f such that $f(a, e, t)$ is an index of the auxiliary r.e. set used at this stage. The (Primitive) Recursion Theorem now provides an index $\bar{a}$ such that for all e and t, $[\bar{a}](e, t) = f(\bar{a}, e, t)$, and the successful construction is the one based on the parameter value $\bar{a}$. We leave the remaining details, which may now be seen as a precise definition of the function f, for the reader to glean from a study of [So; Theorem XI.3.2].

We turn now to the discussion of how to modify the proof of [Hi-Sl; Theorem 2.5]. Of course, we have no need of the length functions l^B and $\hat{l}^A$. The construction of $B[Z]$ proceeds as given there (page 568) with now

$$r^B(e, t) = \begin{cases} u(Z \oplus B_t[Z] \oplus C_t; e, e, t), & \text{if this is } C\text{-certified;} \\ 0 & \text{otherwise.} \end{cases}$$

Clearly Remark 2.6 is still valid.

In the construction of A we need to make the following changes. First we include C in the computations:

$$\{\hat{e}\}_k^{A_s \oplus B_t[A_s] \oplus C_t}(x) \simeq \begin{cases} \{e\}_k^{A_s \oplus B_t[A_s] \oplus C_t}(x), & \text{if this converges with use } u < t \leq a_s; \\ \uparrow, & \text{otherwise;} \end{cases}$$

$$\hat{k}^A(e, s, t) = \text{least } k \leq t \left[\{\hat{e}\}_t^{A_s \oplus B_t[A_s] \oplus C_t}(e)\downarrow \implies \{\hat{e}\}_k^{A_s \oplus B_t[A_s] \oplus C_t}(e)\downarrow \right];$$

$$\hat{u}^A(e,s,t) = \begin{cases} u(A_s \oplus B_t[A_s] \oplus C_t; e, e, k), & \text{if } \{\hat{e}\}_k^{A_s \oplus B_t[A_s] \oplus C_t}(e)\downarrow; \\ 0, & \text{otherwise.} \end{cases}$$

Now let $\Gamma_e(s,t)$ denote the condition: $\{\hat{e}\}_t^{A_s \oplus B_t[A_s] \oplus C_t}(e)\downarrow$ and for all w

$$\hat{k}^A(e,s,t) \le w < t \implies B_w[A_s] \upharpoonright \hat{u}^A(e,s,t) \ne B_t[A_s] \upharpoonright \hat{u}^A(e,s,t)$$
$$\text{and } C_t \upharpoonright \hat{u}^A(e,s,t) = C \upharpoonright \hat{u}^A(e,s,t),$$

and define $\hat{r}^A(e,s)$, $\hat{R}_s^A(e)$ and A_{s+1} as before. Finally, the set of *e-true* stages is defined by

$$T^e = \left\{ s : A_s^{[<e]} \oplus C_s \upharpoonright a_s^e = A^{[<e]} \oplus C \upharpoonright a_s^e \right\},$$

where a_s^e is the smallest new element of $A_s^{[<e]} \oplus C_s$, and $\Delta_e(s,t)$ denotes the condition

$$s \in T^e \wedge t \le s \wedge \Gamma_e(s,t) \wedge B_t^{[<e]} \upharpoonright \hat{u}^A(e,s,t) = B^{[<e]} \upharpoonright \hat{u}^A(e,s,t).$$

The statements of most of the lemmas 2.7–2.14 of [Hi-Sl; Theorem 2.5] require modifications which are generally clear from the above changes in the definitions. We shall assume that the reader can make many of these and will direct our comments for the most part to the changes in the proofs. The main idea of the proof of Lemma 2.7 is unchanged: $s \in T^e$ and $\Gamma_e(s,t)$ guarantee that the relevant computations will not suffer any further injuries due to enumerations into A. In place of 2.7 (i)(d) and (ii) we have

$$\hat{u}^A(e,\bar{s},\bar{t}) = \hat{u}^A(e,s,\bar{t}) = u(A \oplus B_t \oplus C_t; e, e, t) = r^B(e,\bar{t}).$$

Note that because of the inclusion of a C-correctness condition in the definition of Γ_e, all of these computations will be certified. Similarly, this condition makes the modified versions of 2.7 (iii) work as before, since there will be no further C-injuries.

Lemma 2.8 holds and is proved exactly as given. Lemma 2.9 now reads

$$\{e\}^{A \oplus B \oplus C}(e)\downarrow \iff \exists s\, \exists t\, \Delta_e(s,t).$$

The implication ($\Longleftarrow$) is immediate from (the modified) 2.7 and ($\Longrightarrow$) is proved much as before. The only relevant part of Definition 2.10 is $t_e :=$ least t $(B_t^{[<e]} = B^{[<e]})$. The argument for Lemma 2.11 is much the same (and simpler) because of the inclusion of C-correctness in T^e. The modified statement of Lemma 2.12 is the same except that part (i) asserts only that t_e exists, which is immediate from the induction hypothesis, and the proofs of parts (ii) and (iii) are as before.

The main new point of the proof is that still $R^B(e) = \lim_t R_t^B(e)$ exists. This is guaranteed by certification as follows. Suppose that e is minimal such that $\lim_t r^B(e,t)$ does not exist. Clearly this can happen only if $\{e\}^{A \oplus B \oplus C}(e)\uparrow$ but for infinitely many t, $\{e\}_t^{A \oplus B_t \oplus C_t}(e)\downarrow$ and is C-certified. By the nature of the finite-injury priority construction, for sufficiently large t, these computations will suffer no B-injuries and for infinitely many t numbers n_t will be enumerated into the set W_c described above, and the computation will be certified by the discovery of a $v > t$ such that $c \in C^*_v$. Since $C^* = \lim_{t\to\infty} C^*_t$, it follows that $c \in C^*$ so that for some $n \in W_c$, $F_n \cap C = \emptyset$. But then $n = n_t$ for some t, and by the definition of n_t it follows that the computation $\{e\}_t^{A\oplus B_t \oplus C_t}$ uses correct C information. Since it already uses correct A and B information, we have $\{e\}^{A\oplus B\oplus C}(e)\downarrow$, contrary to hypothesis. The rest of 2.12 (iv) now follows as before.

For Lemma 2.13, we note that now T^e, Γ_e, and Δ_e are C-recursive, and we may proceed much as before to compute S-recursively indices for them as well as for $A^{[e]}$ and $B^{[e]}$ relative to C. Of course, we use here again the hypothesis that C is low both to compute $\bar{x}$ as the answer to a one quantifier question relative to C and hence $\emptyset'$-recursively, and to complete the proof via the equivalences in the middle of page 572. The proof of the modified Lemma 2.14 is again nearly identical. □

3. Iterated jump inversion

The preceding result can be iterated as follows.

3.7 Theorem. *For any $n > 0$ and any set S which is n-REA$[\emptyset']$, there exist an n-REA set A and an REA$[A]$ set B such that*

$$A' \equiv_T B' \equiv_T S \equiv_T B \oplus \emptyset'.$$

More generally, there exist primitive recursive functions $f_0, \ldots, f_n$ such that for all sets X, and all n-tuples of natural numbers $\mathbf{z} = (z_0, \ldots, z_{n-1})$,

$$\begin{aligned} X^{\langle f_0(c,\mathbf{z}),\ldots,f_{n-1}(c,\mathbf{z})\rangle'} &\equiv_T X^{\langle f_0(c,\mathbf{z}),\ldots,f_{n-1}(c,\mathbf{z}),f_n(c,\mathbf{z})\rangle'} \\ &\equiv_T X'^{z_0,\ldots,z_{n-1}} \\ &\equiv_T X^{\langle f_0(c,\mathbf{z}),\ldots,f_{n-1}(c,\mathbf{z}),f_n(c,\mathbf{z})\rangle} \oplus X'. \end{aligned}$$

Furthermore, indices for all of these Turing reductions are primitive recursively calculable from c and $z_0, \ldots, z_{n-1}$.

Proof. We prove the first version; the relativized and uniform version follows by the usual observation that the proof is effective and independent of a relativizing parameter. We proceed by induction. The case $n = 1$ is Harrington's Theorem (2.6 without C); assume as induction hypothesis that the result holds for n and fix a set T which is $(n+1)$-REA$[\emptyset']$. For some n-REA$[\emptyset']$ set S, T is REA$[S]$, and by the induction hypothesis, there exist an n-REA set A and an REA$[A]$ set B such that

$$A' \equiv_T B' \equiv_T S \equiv_T B \oplus \emptyset'.$$

Now B is low[A] and T is REA[A'], so by the preceding theorem, there exist an REA[A] set C and an REA[C] set D such that $B \leq_T C$ and

$$C' \equiv_T D' \equiv_T T \equiv_T D \oplus A'.$$

Then C is $(n+1)$-REA and since $B \leq_T C \leq_T D$,

$$D \oplus A' \equiv_T D \oplus B \oplus \emptyset' \equiv_T D \oplus \emptyset'.$$

Hence C and D are as required.

Next we combine this result with the technique of 2.5.

3.8 Theorem. *For any $n > 0$, any set S which is n-REA[$\emptyset'$], and any index d, there exists an $(n+1)$-REA set C such that*

$$C^{\langle d \rangle} \equiv_T S \equiv_T C \oplus \emptyset'.$$

More generally, there exist primitive recursive functions $f_0, \ldots, f_n$ such that for all sets X, and all natural numbers $\mathbf{z} = (z_0, \ldots, z_{n-1})$,

$$X^{\langle f_0(d,\mathbf{z}),\ldots,f_n(d,\mathbf{z}),d \rangle} \equiv_T X^{\prime z_0,\ldots,z_{n-1}} \equiv_T X^{\langle f_0(d,\mathbf{z}),\ldots,f_n(d,\mathbf{z}) \rangle} \oplus X'.$$

Furthermore, indices for all of these Turing reductions are primitive recursively calculable from c and $z_0, \ldots, z_{n-1}$.

Proof. Given S and d, by the preceding theorem, there exist an n-REA set A and an REA[A] set B such that

$$A' \equiv_T B' \equiv_T S \equiv_T B \oplus \emptyset'.$$

Now B is low[A], so by Theorem 2.5, there exists a set C which is REA[A] such that $B \leq_T C$ and $C^{\langle d \rangle} \equiv_T A' \equiv_T S$. Then also

$$S \leq_T B \oplus \emptyset' \leq_T C \oplus \emptyset' \leq_T S,$$

so C is as required. □

3.9 Theorem. *For all $n > 0$, there exist primitive recursive functions $g_0, \ldots, g_{n-1}$ such that for all sets X and all n-tuples of indices $\mathbf{d} = (d_0, \ldots, d_{n-1})$,*

(i) $X^{\langle g_0(\mathbf{d}),\ldots,g_{n-1}(\mathbf{d}),\mathbf{d} \rangle} \equiv_T X^{(n)}$;

(ii) *if $n > 1$, then* $X^{\langle g_0(\mathbf{d}),\ldots,g_{n-1}(\mathbf{d}),d_0 \rangle} \equiv_T X^{\langle g_0(\mathbf{d}),\ldots,g_{n-1}(\mathbf{d}) \rangle} \oplus X'$;

(iii) $X^{\langle g_0(\mathbf{d}),\ldots,g_{n-1}(\mathbf{d}) \rangle} \not\leq_T X^{(n-1)}$;

(iv) $X <_T X^{\langle g_0(\mathbf{d}) \rangle}$.

Furthermore, indices for all of these Turing reductions are primitive recursively calculable from $d_0, \ldots, d_{n-1}$.

Proof. We proceed by induction. The case $n = 1$ is just Theorem 2.5, so assume as induction hypothesis that the result holds for n. Let $\gamma_0, \ldots, \gamma_{n-1}$ denote the functions guaranteed by this hypothesis and $f_0, \ldots, f_n$ the functions guaranteed by case n of Theorem 3.8. Fix a set X and indices $\mathbf{d} = (d_0, \ldots, d_n)$. Then for any indices $\mathbf{z} = (z_0, \ldots, z_{n-1})$,

$$X^{\langle f_0(d_0,\mathbf{z}),\ldots,f_n(d_0,\mathbf{z}),d_0\rangle} \equiv_T X'^{z_0,\ldots,z_{n-1}} \equiv_T X^{\langle f_0(d_0,\mathbf{z}),\ldots,f_n(d_0,\mathbf{z})\rangle} \oplus X'.$$

By the effectivenes of indices for the reductions, there exist primitive recursive functions $\alpha_0, \ldots, \alpha_{n-1}$ such that

$$X^{\langle f_0(d_0,\mathbf{z}),\ldots,f_n(d_0,\mathbf{z}),\mathbf{d}\rangle} \equiv_T X'^{z_0,\ldots,z_{n-1},\alpha_0(\mathbf{d},\mathbf{z}),\ldots,\alpha_{n-1}(\mathbf{d},\mathbf{z})}.$$

Now by the Recursion Theorem, there exist primitive recursive functions $\bar{z}_0, \ldots, \bar{z}_{n-1}$, such that — writing simply $\bar{z}_i$ for $\bar{z}_i(\mathbf{d})$ and $\bar{\alpha}_j$ for $\alpha_j(\mathbf{d}, \bar{z}_0, \ldots, \bar{z}_{n-1})$ — for any set Y and $i < n$,

$$Y^{\langle \bar{z}_i\rangle} = Y^{\langle \gamma_i(\bar{\alpha}_0,\ldots,\bar{\alpha}_{n-1})\rangle}.$$

For $i \le n$, set $g_i(\mathbf{d}) := f_i(d_0, \bar{z}_0, \ldots, \bar{z}_{n-1})$; recall that the $\bar{z}_i$ depend on all of $d_0, \ldots, d_n$. Then we have for part (i),

$$\begin{aligned}
X^{\langle g_0(\mathbf{d}),\ldots,g_n(\mathbf{d}),\mathbf{d}\rangle} &\equiv_T X^{\langle f_0(d_0,\bar{z}_0,\ldots,\bar{z}_{n-1}),\ldots,f_n(d_0,\bar{z}_0,\ldots,\bar{z}_{n-1}),\mathbf{d}\rangle} \\
&\equiv_T X'^{\bar{z}_0,\ldots,\bar{z}_{n-1},\bar{\alpha}_0,\ldots,\bar{\alpha}_{n-1}} \\
&\equiv_T X'^{\gamma_0(\bar{\alpha}_0,\ldots,\bar{\alpha}_{n-1}),\ldots,\gamma_{n-1}(\bar{\alpha}_0,\ldots,\bar{\alpha}_{n-1}),\bar{\alpha}_0,\ldots,\bar{\alpha}_{n-1}} \\
&\equiv_T (X')^{(n)} \equiv_T X^{(n+1)}.
\end{aligned}$$

(ii) is immediate from the properties of $f_0, \ldots, f_n$. For (iii), suppose towards a contradiction that $X^{\langle g_0(\mathbf{d}),\ldots,g_n(\mathbf{d})\rangle} \le_T X^{(n)}$. Then

$$\begin{aligned}
X'^{\gamma_0(\bar{\alpha}_0,\ldots,\bar{\alpha}_{n-1}),\ldots,\gamma_{n-1}(\bar{\alpha}_0,\ldots,\bar{\alpha}_{n-1})} &\equiv_T X'^{\bar{z}_0,\ldots,\bar{z}_{n-1}} \\
&\equiv_T X^{\langle f_0(d_0,\bar{z}_0,\ldots,\bar{z}_{n-1}),\ldots,f_n(d_0,\bar{z}_0,\ldots,\bar{z}_{n-1})\rangle} \oplus X' \\
&\equiv_T X^{\langle g_0(\mathbf{d}),\ldots,g_n(\mathbf{d})\rangle} \oplus X' \\
&\le_T X^{(n)} \equiv_T (X')^{(n-1)},
\end{aligned}$$

contrary to (iii) of the induction hypothesis. Finally, (iv) is immediate, since otherwise, for some $b_1, \ldots, b_n$,

$$X^{\langle g_0(\mathbf{d}),\ldots,g_n(\mathbf{d})\rangle} \equiv_T X^{\langle b_1,\ldots,b_n\rangle} \le_T X^{(n)},$$

contrary to (iii). □

Finally, we have

Proof (of Theorem 1.3). Fix $n > 1$ and let $g_0, \ldots, g_{n-1}$ be the functions guaranteed by the preceding theorem. By the Recursion Theorem, there exist indices $\bar{d}_0, \ldots, \bar{d}_{n-1}$

such that for $i < n$ and any set Y, $Y^{\langle \bar{d}_i \rangle} = Y^{\langle g_i(\bar{d}_0,\dots,\bar{d}_{n-1})\rangle}$. Then $Y <_T Y^{\langle \bar{d}_0 \rangle}$,

$$X^{\langle \bar{d}_0,\dots,\bar{d}_{n-1},\bar{d}_0,\dots,\bar{d}_{n-1}\rangle} \equiv_T X^{(n)} \qquad \text{and} \qquad X^{\langle \bar{d}_0,\dots,\bar{d}_{n-1}\rangle} \not\leq_T X^{(n-1)}.$$

Also $X' \not\leq_T X^{\langle \bar{d}_0,\dots,\bar{d}_{n-1}\rangle}$, since otherwise

$$X^{\langle \bar{d}_0,\dots,\bar{d}_{n-1},\bar{d}_0\rangle} \equiv_T X^{\langle \bar{d}_0,\dots,\bar{d}_{n-1}\rangle} \oplus X' \equiv_T X^{\langle \bar{d}_0,\dots,\bar{d}_{n-1}\rangle},$$

contrary to the mentioned property of $\bar{d}_0$. Now set $A_0 = \emptyset$, $A_{l+1} = A_l^{\langle \bar{d}_0,\dots,\bar{d}_{n-1}\rangle}$, and $A = A_1$. Then it follows from the preceding that for all l and n,

$$A_{l+2} \equiv_T A_l^{(n)} \qquad \text{and} \qquad A_l' \not\leq_T A_{l+1} \not\leq_T A_l^{(n-1)}.$$

Hence by induction we have for all l,

$$A_{2l} \equiv_T \emptyset^{(ln)}, \quad A_{2l+1} \equiv_T A^{(ln)}, \quad \text{and} \quad \emptyset^{(ln+1)} \not\leq_T A_{2l+1} \not\leq_T \emptyset^{(n+ln-1)}.$$

Since A is n-REA, we have $\emptyset \leq_T A \leq_T \emptyset^{(n)}$, whence for all k, $\emptyset^{(k)} \leq_T A^{(k)} \leq_T \emptyset^{(n+k)}$. The preceding ensures that for $k = ln$,

$$\emptyset^{(k+1)} \not\leq_T A^{(k)} \not\leq_T \emptyset^{(n+k-1)},$$

and it follows from the monotonicity of the jump operator that these inequalities hold also for all k.

The following generalization of the Sacks Jump theorem leads to a slightly different proof of Theorem 1.3.

3.10 Theorem. *For all $n > 0$, there exist primitive recursive functions $h_0, \dots, h_{n-1}$ such that for all sets X and all n-tuples of indices $\mathbf{e} = (e_0, \dots, e_{n-1})$,*

(i) $X^{\langle h_0(\mathbf{e}),\dots,h_{n-1}(\mathbf{e})\rangle(n)} \equiv_T X^{(n)\mathbf{e}}$;

(ii) *if $n > 1$, then $X^{\langle h_0(\mathbf{e}),\dots,h_{n-1}(\mathbf{e})\rangle}$ is Turing incomparable with $X', \dots, X^{(n-1)}$.*

Furthermore, indices for all of these Turing reductions are primitive recursively calculable from $e_0, \dots, e_{n-1}$.

Proof. Let $g_0, \dots, g_{n-1}$ be as in Theorem 3.9. By the effectiveness of indices for the reductions, there exist primitive recursive functions $\alpha_0, \dots, \alpha_{n-1}$ such that for any $\mathbf{d} = (d_0, \dots, d_{n-1})$ and $\mathbf{e} = (e_0, \dots, e_{n-1})$,

$$X^{(n)\mathbf{e}} \equiv_T X^{\langle g_0(\mathbf{d}),\dots,g_{n-1}(\mathbf{d}),\mathbf{d},\alpha_0(\mathbf{d},\mathbf{e}),\dots,\alpha_{n-1}(\mathbf{d},\mathbf{e})\rangle}.$$

In particular, if we substitute $g_i(\mathbf{d})$ for d_i and write $\mathbf{g(d)}$ for $g_0(\mathbf{d}), \dots, g_{n-1}(\mathbf{d})$, then for all $\mathbf{d}$ and $\mathbf{e}$,

$$X^{(n)\mathbf{e}} \equiv_T X^{\langle g_0(\mathbf{g(d)}),\dots,g_{n-1}(\mathbf{g(d)}),\mathbf{g(d)},\alpha_0(\mathbf{g(d)},\mathbf{e}),\dots,\alpha_{n-1}(\mathbf{g(d)},\mathbf{e})\rangle}.$$

Now, by the Recursion Theorem, choose primitive recursive functions $\bar{d}_i(\mathbf{e})$ such that for all $i < n$ and all Y,

$$Y^{\langle \bar{d}_i(\mathbf{e})\rangle} = Y^{\langle \alpha_i(\mathbf{g}(\bar{d}_0(\mathbf{e}),\dots,\bar{d}_{n-1}(\mathbf{e})),\mathbf{e})\rangle}.$$

Then setting, for $i < n$, $\bar{d}_i := \bar{d}_i(\mathbf{e})$, $\bar{\mathbf{d}} := (\bar{d}_0, \ldots, \bar{d}_{n-1})$, and $h_i(\mathbf{e}) := g_i(\mathbf{g}(\bar{\mathbf{d}}))$, we have

$$\begin{aligned} X^{(n)^{\mathbf{e}}} &\equiv_T X^{\langle h_0(\mathbf{e}),\ldots,h_{n-1}(\mathbf{e}),\mathbf{g}(\bar{\mathbf{d}}),\alpha_0(\mathbf{g}(\bar{\mathbf{d}}),\mathbf{e}),\ldots,\alpha_{n-1}(\mathbf{g}(\bar{\mathbf{d}}),\mathbf{e})\rangle} \\ &\equiv_T X^{\langle h_0(\mathbf{e}),\ldots,h_{n-1}(\mathbf{e}),\mathbf{g}(\bar{\mathbf{d}}),\bar{\mathbf{d}}\rangle} \\ &\equiv_T X^{\langle h_0(\mathbf{e}),\ldots,h_{n-1}(\mathbf{e})\rangle^{(n)}}. \end{aligned}$$

It is immediate from Theorem 3.9 and the definition of the functions h_i that

$$X^{\langle h_0(\mathbf{e}),\ldots,h_{n-1}(\mathbf{e})\rangle} \not\leq_T X^{(n-1)}.$$

Suppose contrary to (ii) that $X' \leq_T X^{\langle h_0(\mathbf{e}),\ldots,h_{n-1}(\mathbf{e})\rangle}$. Then by 3.9(ii),

$$X^{\langle h_0(\mathbf{e}),\ldots,h_{n-1}(\mathbf{e}),g_0(\bar{\mathbf{d}})\rangle} \equiv_T X^{\langle h_0(\mathbf{e}),\ldots,h_{n-1}(\mathbf{e})\rangle} \oplus X' \equiv_T X^{\langle h_0(\mathbf{e}),\ldots,h_{n-1}(\mathbf{e})\rangle},$$

contrary to 3.9(iv).

3.11 Corollary. *For all $n > 0$, there exist indices $\bar{e}_0, \ldots, \bar{e}_{n-1}$ such that for all sets X,*

(i) $X^{\langle \bar{e}_0,\ldots,\bar{e}_{n-1}\rangle^{(n)}} \equiv_T X^{(n)^{\bar{e}_0,\ldots,\bar{e}_{n-1}}}$;

(ii) *if $n > 1$, $X^{\bar{e}_0,\ldots,\bar{e}_{n-1}}$ is Turing incomparable with $X', \ldots, X^{(n-1)}$.*

Proof. This follows easily from the preceding theorem via the Recursion Theorem: it suffices to choose $\bar{e}_0, \ldots, \bar{e}_{n-1}$ such that for all $i < n$ and all sets Y, $Y^{\langle h_i(\bar{e}_0,\ldots,\bar{e}_{n-1})\rangle} = Y^{\langle \bar{e}_i \rangle}$. □

This now provides a second

Proof (of Theorem 1.3). With $\bar{e}_0, \ldots, \bar{e}_{n-1}$ as in the preceding corollary (and $n > 1$), set $A = \emptyset^{\langle \bar{e}_0,\ldots,\bar{e}_{n-1}\rangle}$. Clearly $\emptyset \leq_T A \leq_T \emptyset^{(n)}$ and A is incomparable with $\emptyset', \ldots, \emptyset^{(n-1)}$. It follows by induction that for all $l \in \omega$ and $k = ln$ that $\emptyset^{(k)} \leq_T A^{(k)} \leq_T \emptyset^{(k)}$ and A is incomparable with $\emptyset^{(k+1)}, \ldots, \emptyset^{(k+n-1)}$. But then these inequalities hold also for all values of k. □

References

[HiSl] P. G. Hinman and T. A. Slaman, Jump Embeddings in the Turing Degrees, J. Symbolic Log. 56 (1991), 563–591.

[JoSh] C. G. Jockusch and R. A. Shore, Pseudo-jump operators. I: the r.e. case, Trans. Amer. Math. Soc. 275 (1983), 599–609.

[LaYi] A. H. Lachlan and X. Yi, Jump Operators for REA Operators, Math. Logic Quart. 39 (1993), 1–6.

[So] R. I. Soare, Recursively enumerable sets and degrees, Springer-Verlag, Berlin, Heidelberg, etc., xviii + 437 pp., 1987.

Weak recursive degrees and a problem of Spector

Shamil Ishmukhametov

Abstract. We introduce a concept of weak recursive degrees and show that each weak recursive degree possesses a strong minimal cover (s.m.c.). Since the class of r.e. array nonrecursive (a.n.r.) degrees defined by Downey, Jockusch and Stob [1990] is complementary to the class of the weak recursive degrees in the r.e. degrees **R** and no a.n.r. degree can possess a s.m.c., we obtain as a corollary that a r.e. degree possesses a s.m.c. iff it is weak recursive.

A degree **m** is a strong minimal cover for a degree **a**, if $\mathbf{a} < \mathbf{m}$ and for each degree **c**, $\mathbf{c} < \mathbf{m} \rightarrow \mathbf{c} \leq \mathbf{a}$. The problem of describing degrees possessing strong minimal covers was first raised by Spector [1956], but there was little progress since that time. Spector's construction of a minimal degree shows that the degree **0** possesses a s.m.c. Titgemeyer [1965] built a minimal degree possessing a s.m.c. and thus proved that the 3-element chain is embeddable as an initial segment in the degrees. In [1970], Yates raised the problem of the existence of s.m.c. for every minimal degree, and this problem has not been solved yet. Cooper [1972], [1995] constructed a r.e. degree possessing a s.m.c., and Kumabe [ta] showed the existence of such a 1-generic degree. Ishmukhametov [ta1] proved that each level of α-r.e. degrees contains degrees possessing s.m.c. and showed [ta2] that there exists a high r.e. degree **a** such that in the lower cone $\mathbf{D}(\leq \mathbf{a})$ some minimal degree has no a s.m.c.

On the other hand, more progress has been made in the description of the degrees that cannot be s.m.c. Clearly, all r.e. degrees and all n-r.e. degrees for $n \in \omega$ are such. In general, all degrees relatively enumerable in a lesser degree are splittable, so they cannot be s.m.c. For example, the class of all degrees greater than or equal to $\mathbf{0}'$ and the class of 1-generic degrees are such. In [1990] and [1996], Downey, Jockusch and Stob defined a notion of an array nonrecursive degree as a degree **a** such that for any function $f \leq_{tt} K$, where K is a creative set, there is a function g recursive in **a** such that $g(n) \geq f(n)$ for infinitely many n. They showed that any degree **a** from $\overline{\mathrm{GL}_2}$ (i. e. satisfying $\mathbf{a}'' > (\mathbf{a} \cup 0')')$ is a.n.r., but that there are also low and low_2 a.n.r. degrees. The most important fact is that the a.n.r. degrees are closed upwards, splittable and cuppable to all higher degrees (more generally, any recursively presented lattice with distinct least and greatest elements can be embedded in the lower cone $\mathbf{D}$ $(\leq \mathbf{a})$ bounded by an a.n.r. degree **a** preserving least and greatest elements and lattice operations, Fejer [1989], Downey, Jockusch and Stob [1996]).

So no a.n.r. degree **a** can be a s.m.c. or possess a s.m.c.

Decreasing the gap between a.n.r. degrees and degrees possessing a s.m.c, we introduce a notion of weak recursive degrees and show that any weak recursive degree has a s.m.c.

Definition. A degree **a** is weak recursive if there is a recursive function $p(x)$ such that for every function f recursive in **a**, there is a weak array $\{W_{h(n)}\}_{n=0}^{\infty}$ such that for every $n \in \omega$:

1. $|W_{h(n)}| \leq p(n)$,
2. $f(n) \in W_{h(n)}$,
3. $x \neq y \to W_{h(x)} \cap W_{h(y)} = \emptyset$.

It follows immediately from the definition that the class of weak recursive degrees WR is closed downwards, i. e. for any degrees **a**, **b**, if $\mathbf{a} < \mathbf{b}$ and $\mathbf{b} \in$ WR, then $\mathbf{a} \in$ WR.

If a weak recursive degree **a** is located below $\mathbf{0}'$, then every function f recursive in **a** is p-r.e. i. e. has a recursive approximation $f(s, n)$ such that for every n, $|\{f(s, n) : s \in \omega\}| \leq p(n)$.

Since a r.e. degree **a** is p-r.e. for some increasing recursive function p iff it is not array nonrecursive (Downey, Jockusch and Stob [1990]), we obtain that the class of r.e. weak recursive degrees is complementary to the class of array nonrecursive degrees in **R**.

Thus, our theorem gives a decision to Spector's problem mentioned above for the case of the r.e. degrees. We list definable properties of weak recursive r.e. degrees in the proposition below:

Proposition. *Let* **a** *be a r.e. degree. Then the following conditions are equivalent:*

(1) **a** *is weak recursive,*

(2) **a** *is not array nonrecursive,*

(3) *Any* **b**, $\mathbf{b} \leq \mathbf{a}$ *has a strong minimal cover,*

(4) *There is a degree* $\mathbf{b} \geq \mathbf{a}$ *which is not the sup of a minimal pair (in all degrees) (Downey* [1993], *Theorem* 1.1*),*

(5) *There is a non-splittable degree* $\mathbf{b} \geq \mathbf{a}$,

Since the weak recursive degrees are closed downwards, we obtain a definable subclass of minimal degrees possessing a s.m.c., namely, the class of minimal degrees bounded by weak recursive r.e. degrees.

At the same time, we give a partial answer to the question of Downey, Jockusch and Stob [1996] about the definability of the class of a.n.r. degrees by showing that the class of r.e. a.n.r. degrees is definable (in all degrees) by any property listed in items (3)-(5) of the Proposition using the fact that **R** is definable in all degrees (Cooper [ta]).

Our notation is standard and follows Odifreddi [1989].

1. Weak recursive degrees

We shall prove the theorem below using a combination of non-effective and effective methods.

Theorem 1. *Given a weak recursive degree* **a**, *there exists a strong minimal cover* **m** *for* **a**.

Proof. Let $A \in \mathbf{a}$. We shall construct a set M such that the following conditions hold:

$$N_e : B_e = \Phi_e^{M \oplus A} \to (M \oplus A = \Gamma_e^{B_e} \vee B_e = \Delta_e^A),$$
$$P_e : M \neq \Theta_e^A, \quad e \in \omega,$$

where $\{\Phi_e, \Theta_e\}_{e \in \omega}$ is a recursive enumeration of all possible pairs of partial recursive functionals, Γ and Δ are p.r.functionals built during the construction, and $C \oplus D = \{2x+1 : x \in C\} \cup \{2x : x \in D\}$. □

We begin with some definitions.

Definition 1. Given a set X, a string $\sigma \in 2^{<\omega}$ is called X-true if for each even $n = 2m < |\sigma|, \sigma(n) = X(m)$.

Definition 2. Given a partial recursive functional Φ and a set X, an X-true string σ is called X-splittable on Φ, if there exist X-true strings σ_1 and σ_2, extending σ and a natural number y such that $\Phi^{\sigma_0} \upharpoonright y \downarrow \neq \Phi^{\sigma_1} \upharpoonright y \downarrow$.

Definition 3. Given a p.r.functional Φ, a set X, and an X-true string σ, Φ is called X-splittable above σ, if any X-true string extending σ is X-splittable. Φ is called X-splittable if it is splittable above the empty string. Notice that the question of whether a p.r.f. Φ is X-splittable can be answered recursively in X''.

Consider the module for a requirement N.

The basic module

1. First check if the functional Φ is A-splittable. If yes, then go to 2, otherwise, find a least A-true node τ such that there are no A-splittings above τ. Notice that for any

set D such that $D \oplus A$ extends τ, $\Phi^{D \oplus A}$ is either partial or recursive in A. Define $M \oplus A \upharpoonright |\tau| = \tau$.

2. We shall construct a p.r.f. Γ reducing $M \oplus A$ to $\Phi^{M,A}$. This Γ is defined by stages. At stage n, we try to find an A-splitting $\langle \tau_0, \tau_1 \rangle$ above each A-true string τ defined at stage $n-1$. In general, we build an infinite A-splitting tree, at every level n of which we have 2^n A-splittings.

But the main problem is that we cannot choose an A-splitting extending a given initial segment τ of A recursively. Therefore, we arrange a construction in which we have in store several A-true strings τ, and if the search of an A-splitting $\langle \tau_0, \tau_1 \rangle$ extending the current τ fails, then we replace τ and begin a new search. The property of a degree **a** being weak recursive allows us to bound the number of such searches.

Fix some recursive enumeration of all finite sets of finite strings (elements of $\{0, 1\}^{<\omega}$). We define by induction a function f such that:

1. Each value $f(n)$ is a pair $\langle k, y_n \rangle$, where k is the code for a set F_n of A-true nodes.

2. F_0 is the empty set, $y_0 = 0$.

   F_n for $n > 0$ is by definition a set F such that:

2.1. Each string $\beta \in F$ is an A-true extension of some α from F_{n-1}.

2.2. For any $\alpha \in F_{n-1}$ there are strings τ_0 and τ_1 in F extending α such that $\langle \tau_0, \tau_1 \rangle$ forms an A-splitting on Φ and for each τ_j, $j \in \{0, 1\}$, there are $p(n+1)$ strings β_i in F, $1 \le i \le p(n+1)$ extending τ_j such that the strings $\rho_i = \Phi^{\beta_i} \upharpoonright y_{n+1}$ are defined and pairwise incomparable.

It is not difficult, given a set F_{n-1}, to find a set F_n satisfying conditions 2.1–2.2 A-recursively. Indeed, in order to do so, enumerate all possible pairs $\langle \sigma, l \rangle$ such that $\Phi^{\sigma} \upharpoonright l \downarrow$, and select pairs with σ A-true until a set F satisfying 2.1–2.2 appears. Define F_n equal to this F. The A-splittability of Φ ensures the appearance of some F.

Therefore, the function f can be chosen being recursive in A, and there exists a disjoint weak array $\{W_{h(n)}\}_{n=0}^{\infty}$ satisfying $|W_{h(n)}| \le p(n)$ and $f(n) \in W_{h(n)}$ for every $n \in \omega$. Assume that such $\{W_{h(n)}\}_{n=0}^{\infty}$ has been found.

Construction

I. 1. Compute $f(0) = \langle k, y \rangle$. This computation is the only non-effective action in this procedure. All other instructions will be recursive.

Restore by $f(0)$ the set F_0 and find the A-splitting $\langle \tau_0, \tau_1 \rangle$ and A-true nodes β_i^j, $1 \le i \le p(1)$, $j \in \{0, 1\}$ from F_0 extending appropriate τ_j.

2. Enumerate the set $W_{h(1)}$. Let $a_1, a_2, a_3, \ldots,$ be its elements in order of their appearance. By the choice of h, the length of this sequence does not exceed $p(1)$, and at least one of the a_i coincides with $f(1)$. At stage s_i, when a new a_i appears in $W_{h(1)}$, carry out the following instructions:

Let $a_i = \langle k_i, y_i \rangle$. Restore by k_i the set F_i and find for each i, j, $1 \le i \le p(1)$, $j \in \{0, 1\}$ a splitting $\langle \tau^j_{0,i}, \tau^j_{1,i} \rangle$ extending β^j_i, and sequences of $p(2)$ incomparable nodes α extending each $\tau^j_{0,i}$, $\tau^j_{1,i}$. Wait for a stage s when every $\rho = \Phi^\alpha_s \restriction y_i \downarrow$. Check, if they are pairwise incomparable and agree at even elements. If not, then the hypothesis $a_i = f(1)$ is incorrect. Drop this a_i. Otherwise, define $\Gamma^\rho \restriction |\alpha| = \alpha$ for all α extending $\tau^j_{0,i}$ and $\tau^j_{1,i}$, $j \in \{0, 1\}$.

Notice that there are no conflicts between the definitions of Γ for different i, $1 \le i \le p(1)$, since appropriate $\rho = \Phi^\alpha_s \restriction y_i$ extend incomparable nodes $\Phi^{\beta^j_i} \restriction y$, $j \in \{0, 1\}$ (where y is as in I.1) and therefore are themselves pairwise incomparable.

Since there is an i such that $a_i = f(1)$, there are at least two A-true nodes α which form an A-splitting on Φ.

II. Fix some $n > 0$ and consider the instructions for this n:

Enumerate set $W_{h(n)}$. Let $a_1, a_2, \ldots$ be its elements in order of appearance. At stage s_i, when a new a_i appears in $W_{h(n)}$, carry out the following instructions:

2.1. Let $a_i = \langle k_i, y_i \rangle$. Restore by k_i the set F_i and find an element $a = \langle k', y' \rangle$ of $W_{h(n-1)}$ such that nodes of F_i extend nodes of the set F' numbered by k'. If there is no such a, wait for its appearance in $W_{h(n-1)}$. If the hypothesis $a_i = f(n)$ is correct, then such a will eventually be found.

2.2. By definition, the set F_n coded by $f(n)$ contains nodes of two kinds. The nodes of the first kind form 2^n A-splittings on Φ, the nodes of the second kind are extensions of the former. For each node τ of the first kind, there are $p(n+1)$ nodes of the second kind extending τ. We assume that sets F coded by elements of $W_{h(e)}$, $e \in \omega$ have the same configuration.

For each node τ of the first kind, from F' find $p(n)$ nodes β_i, $1 \le i \le p(n)$ of the second kind extending τ. Connect each element $a_i = \langle k, y \rangle$ enumerated in $W_{h(n)}$ with an appropriate node β_i. Denote by F_i the set decoded by a_i. Choose from this set nodes β extending β_i. Wait for a stage s when $\rho = \Phi^\beta \restriction y \downarrow$ for all chosen β. Check if they are pairwise incomparable, then define $\Gamma^\rho \restriction |\beta| = \beta$.

The following lemmas can be proved by induction.

Lemma 1. *For each n, there exist 2^n incomparable nodes $\rho = \Phi^\beta \restriction y_n$ for some A-true β, at which Γ is defined and equal to β.*

Lemma 2. *There exists an infinite A-recursive set T of A-true nodes β such that:*

1. *For any $\beta \in T$, there is a string ρ at which Γ is defined equal to β.*
2. *Any $\beta \in T$ has at least two incomparable extensions τ_0 and τ_1 in T such that the pair $\langle \tau_0, \tau_1 \rangle$ forms an A-splitting on Φ.*

Now we are ready to describe the whole construction.

At stage $e \in \omega$, we satisfy the requirements N_e and P_e and define the set of A-true nodes T_e such that T_e contains infinitely many A-true splittings and for each e, $T_e \supset T_{e+1}$.

Stage 0. 1. Check whether Φ_0 is A-splittable. If yes, go to 2, otherwise, find a least A-true string τ of odd length $2k-1$ such that there are no A-splittings on Φ_0 extending τ. Define T_0 equal to the set of all A-true nodes extending $\tau * i$, where $i \in \{0, 1\}$ is chosen different from $\Theta_0^A(k)$. Then go to the next stage.

2. Apply the instructions of the basic module to the functional Φ_0. Construct the functional Γ_0 and choose the set T recursive in A as in Lemma 2 of the basic module.

Choose a least A-splitting $\langle \tau_0, \tau_1 \rangle$ in T. Let $y = 2z+1$ be a least number such that $\tau_0(y) \neq \tau_1(y)$ (some y exists since both τ_0 and τ_1 are A-true and therefore coincide at even natural numbers).

Choose $i \in \{0, 1\}$ in such a way that $\tau_i(y)$ differs from $\Theta_0^A(z)$, and define T_0 equal to the set of all nodes $\tau \in T$ extending τ_i.

Assume that for all $k \leq e$, the sets T_k are already defined, and each A-true node in T_e contains at least two incomparable A-true extensions in T_e.

Stage $e+1$. 1. Check whether the set T_e contains a node τ such that there is no A-splitting $\langle \tau_0, \tau_1 \rangle$ above τ on Φ_{e+1} such that both τ_j are chosen from T_e.

If yes, choose the least such τ of odd length and find two incomparable nodes $\langle \tau_0, \tau_1 \rangle$ in T_e extending τ. Find a least $y = 2z+1$ such that $\tau_0(y) \neq \tau_1(y)$ and choose $i \in \{0, 1\}$ such that $\tau_i(y)$ differs from $\Theta_{e+1}^A(z)$. Define T_{e+1} equal to the set of all nodes $\tau \in T_e$ extending τ_i.

If not, then go to 2.

2. Define a function f for the functional Φ_{e+1} as in the basic module but with the restriction that all nodes for required A-splittings are chosen from the set T_e. All other instructions do not differ from the instructions of stage 0 where the functional Φ_0 is replaced by Φ_{e+1}. This ends the construction.

By the construction, each T_{e+1}, $e \in \omega$, is a proper subset of T_e. There exists just one set B any initial segment of which has an extension in every set T_n, $n \in \omega$. Define $M = \{x : B(2x+1) = 1\}$. By the construction, for each e, there exists a number z_e such that $M(z_e) \neq \Theta_e^A(z_e)$. Therefore $M \not\leq_T A$.

Now we prove that all requirements N_e, $e \in \omega$, are satisfied. Indeed, if Φ_e is not A-splittable in T_{e-1}, then $M \oplus A$ contains an initial segment τ above which there are no A-splittings from T_{e-1}, and $\Phi_e^{M \oplus A}$ is either partial or recursive in A. Otherwise, by the construction, T_e contains infinitely many initial segments of $M \oplus A$, and for any node β from T_e, there exists $\rho = \Phi_e^\beta \restriction y$ for some y such that $\Gamma_e^\rho \restriction |\beta| = \beta$. This ensures $\Gamma_e^{B_e} = M \oplus A$, where $B_e = \Phi^{M \oplus A}$.

2. Open Problems

There are a number of open problems connected with the Theorem. The main problem is whether one can decrease the oracle $\mathbf{a}''$ used in the construction in order to obtain a strong minimal cover below $\mathbf{a}'$ or $\mathbf{0}'$.

Another problem is to solve the problem of Yates for minimal degrees below $\mathbf{0}'$. We do not know whether all minimal degrees below $\mathbf{0}'$ are weak recursive, or whether our theorem gives a solution to this problem.

References

[1] S. B. Cooper [1996], Strong minimal covers for recursively enumerable degrees, Math. Logic Quart. 42, 191–196.

[2] S. B. Cooper [ta], On a conjecture of Kleene and Post, Math. Logic Quart., to appear.

[3] R. Downey [1993], Array nonrecursive degrees and lattice embeddings of the diamond, Ill. J. Math. 37, 349–374.

[4] R. Downey, C. Jockusch, and M. Stob [1990], Array nonrecursive sets and multiple permitting arguments, in: Recursion Theory Week (eds. K. Ambos-Spies, G. H. Muller, and G. Sacks), Lecture Notes in Math. 1432, Springer-Verlag, Berlin, 141–173.

[5] R. Downey, C. Jockusch, and M. Stob [1996], Array nonrecursive sets and genericity, in: Computability, Enumerability, Unsolvability: Directions in Recursion Theory (eds S. B. Cooper, T. A. Slaman, S. S. Wainer), London Math. Soc. Lecture Notes Ser. 224, Cambridge University Press, 93–104.

[6] P. Fejer [1989], Embedding lattices with top preserved below non-GL_2 degrees, Z. Math. Logik Grundlag. Math. 35, 527–539.

[7] Sh. Ishmukhametov [ta1], Minimal covers for recursively enumerable degrees, preprint.

[8] Sh. T. Ishmukhametov [ta2], On a new method for constructing minimal Turing degrees, Izv. Vyssh. Ucheb. Zaved., to appear (in Russian).

[9] M. Kumabe [ta], A 1-generic degree with a strong minimal cover, J. Symbolic Logic, to appear.

[10] P. Odifreddi [1989], Classical Recursion Theory, Stud. Logic Found. Math. 125, North-Holland, Amsterdam.

[11] C. Spector [1956], On degrees of recursive unsolvability, Ann. of Math. (2) 64, 581–592.

[12] D. Titgemeyer [1965], Untersuchungen über die Struktur des Kleene–Postschen Halbverbandes der Grade der rekursiven Unlösbarkeit, Arch. Math. Logik Grundlagenforschung 8, 45–62.

[13] C. E. M. Yates [1970], Initial segments of the degrees of unsolvability, Part II: Minimal degrees, J. Symbolic Logic 35, 243–266.

Compositions of permutations and algorithmic reducibilities

K. V. Korovin

Abstract. For wtt-reducibility or any other stronger reducibility, there exists a set $F \leq_T \emptyset'$ such that the set $\{f \mid f$ is a permutation of ω and the graph of f is reducible to $F\}$ is not closed under compositions. Let A be a set such that $A \leq_T \emptyset'$. Then there is a permutation f such that the graph of f is 2-c.e. and the graph of f^2 is not wtt-reducible to A.

All necessary definitions can be found in [3, 4]. Algorithmic properties of permutations on natural numbers are studied in many papers. The authors studied groups

$$G_d = \{f \mid f \text{ is a permutation of } \omega \text{ and } \deg f \leq \mathbf{d}\},$$

where $\mathbf{d}$ is a Turing degree, and $\deg f$ means the Turing degree of f. It turns out that any group G_d characterizes the degree $\mathbf{d}$, i. e. two such groups are isomorphic iff their corresponding degrees coincide and the embedding on such groups is equivalent to Turing reducibility on degrees (see [2]). In [2] A. S. Morozov stated the following problem: is it possible to extend these results to other reducibilities?

In this paper we give the negative answer to this question for a broad class of algorithmic reducibilities.

Let $\langle x, y\rangle$ denotes the standard pairing function from $\omega \times \omega$ onto ω. Let $(\cdot)_1$ and $(\cdot)_2$ be computable functions $\omega \to \omega$ such that $\langle (x)_1, (x)_2\rangle = x$, $(\langle x_1, x_2\rangle)_1 = x_1$, $(\langle x_1, x_2\rangle)_2 = x_2$.

As usual for $n > 2$ $\langle x_1, \dots, x_n\rangle$ denotes $\langle \dots \langle\langle x_1, x_2\rangle, x_3\rangle, \dots, x_n\rangle$. D_n denotes the finite set with the canonical index n, i.e. $n = \sum_{j \in D_n} 2^j$.

For $f : \omega \to \omega$ Γ_f denotes the graph of f, i.e. the set $\{\langle x, y\rangle \mid f(x) = y\}$. A set A is *weak truth-table reducible* to B ($A \leq_{\mathrm{wtt}} B$) if there exist a natural number z and a computable function g such that

1) $\chi_A = \{z\}^B$;

2) the set $D_{g(x)}$ contains all questions to the oracle B during the computation of $\{z\}^B(x)$. We will say that the pair $(\{z\}, g)$ reduces A to B.

Definition. A finite function $f : \omega \to \omega$ is a *path* if

$$\Gamma_f = \{\langle m_0, m_1\rangle, \langle m_1, m_2\rangle, \dots \langle m_{k-1}, m_k\rangle\},$$

where $m_i \neq m_j$ for $i \neq j$. Let $b(f) = m_0$ denotes the first element of the path and $e(f) = m_k$ denotes the last element of the path.

Definition. For $D \subseteq \omega$, we define $\text{field}(D) = \{(m)_1 \mid m \in D\} \cup \{(m)_2 \mid m \in D\}$. For a function f, $\text{field}(f) = \text{field}(\Gamma_f) = \text{dom}(f) \cup \text{range}(f)$.

We will consider a reducibility of a function f to a set A as a reducibility of the graph of f to A. In this case the following theorem holds.

Theorem 1. *Let $\preceq$ be the* wtt- *or any other stronger reducibility (i.e. $A \preceq B$ implies $A \leq_{\text{wtt}} B$). There exists a set $F \leq_T \emptyset'$ such that the set*

$$\{f \mid f \text{ is a permutation of } \omega \text{ and } \Gamma_f \preceq F\}$$

is not closed under the composition.

Proof. We will construct a permutation f with the following properties: $f \leq_T \emptyset'$ and $\Gamma_{f^2} \not\leq_{\text{wtt}} \Gamma_f$. Obviously, the graph of f will be the required set F. We construct the permutation f by stages by a finite extension $0'$-oracle construction, so that

$$\Gamma_{f_0} \subseteq \Gamma_{f_1} \subseteq \cdots \subseteq \bigcup_{i \in \omega} \Gamma_{f_i} = \Gamma_f.$$

Here each f_i will be a path. It suffices to meet, for each s, the following requirement: the pair of functions $(\{(s)_1\}, \{(s)_2\})$ does not reduce Γ_{f^2} to Γ_f.

Stage $s = 0$. Define $\Gamma_{f_0} = \{\langle 0, 1\rangle\}$.

Stage $s > 0$. Define $p_s = \mu n(n \notin \text{field}(f_s))$, $q_s = \mu n(n \notin \text{field}(f_s) \cup \{p_s\})$, $\Gamma_{f_s'} = \Gamma_{f_s} \cup \{\langle p_s, b(f_s)\rangle, \langle e(f_s), q_s\rangle\}$ and $\beta_s = \mu n(n \notin \text{field}(f_s'))$. We denote $(s)_1$ by z and $\{(s)_2\}$ by g. Let us consider all possible cases for $g(\langle e(f_s'), \beta_s\rangle)$.

Case 1. $g(\langle e(f_s'), \beta_s\rangle)$ is undefined. Then we define $f_{s+1} = f_s$ and go to the next stage.

Case 2. $g(\langle e(f_s'), \beta_s\rangle)$ is defined. Then we define $l_s = g(\langle e(f_s'), \beta_s\rangle)$ and $D^+_{l_s} = D_{l_s} \cap \Gamma_{f_s'}$, $D^-_{l_s} = D_{l_s} \setminus D^+_{l_s}$.

We extend the path f_s' to the path f_s'' with the following condition: the graph of any extension of f_s'' contains no element of $D^-_{l_s}$. For any $m = \langle u, v\rangle \in D^-_{l_s}$, we proceed the following procedure.

A number c we call a *new element* if $c \notin \text{field}(D^-_{l_s})$, c does not belong to the field of the constructed path and $c \neq \beta_s$.

Let c be a new element, and a be the first element of the constructed path. If $u \neq \beta_s$ and u does not belong to the field of the path, then we add pairs (u, c) and (c, a) to the path. If $u = \beta_s$, v does not belong to the field of the path and $v \neq u$, then we add pairs (v, c) and (c, a) to the constructed path. In other cases we do nothing. We denote the constructed path by f_s''.

Let us consider the function $\{z\}$.If $\{z\}^{\Gamma_{f_s''}}(\langle e(f_s'', \beta_s\rangle)$ is defined and equal to 1, then we add pairs $\langle e(f_s''), c\rangle$, $\langle c, d\rangle$, $\langle d, \beta_s\rangle$ to the path, where c and d are sequentially chosen new elements. In other cases we add to the path pairs $\langle e(f_s''), c\rangle$, $\langle c, \beta_s\rangle$, where c is a new element. We denote the constructed path by f_{s+1}.

It is clear that$\Gamma_f = \bigcup_{i\in\omega}\Gamma_{fi}$ is the graph of a permutation, and $\Gamma_f \leq_T \emptyset'$. Now the theorem follows from Lemmas 1 and 2. □

Lemma 1. *If at stage s Case 2 holds, then for all m, $m \in D_{l_s}^-$ implies $m \notin \Gamma_f$.*

Proof. Let $m \in D_{l_s}^-$ and $m = \langle u, v\rangle$. By the construction, the permutation f is an infinite cycle, so $u = v$ implies $m \notin \Gamma_f$. Consider the case $u \in \mathrm{dom}(f_s'')$. If $u \in \mathrm{dom}(f_s')$, then $\langle u, v\rangle \notin \Gamma_f$ by the definition of $D_{l_s}^-$. If constructing f_s'' we include u into the field of f_s'', then $\langle u, c\rangle \in \Gamma_{f_s''}$ and c is a new element, i.e. $c \neq v$. If $u = e(f_s'')$, then $\langle u, c\rangle \in \Gamma_{f_{s+1}} \subset \Gamma_f$, where c is a new element, i.e. $\langle u, v\rangle \notin \Gamma_f$.

If $u = \beta_s$ and $u \neq v$, then $v \in \mathrm{dom}(f_{s+1})$ and $\beta_s = e(f_{s+1})$, so f is an infinite cycle. Therefore $\langle u, v\rangle \notin \Gamma_f$. □

Lemma 2. $\Gamma_{f^2} \not\leq_{\mathrm{wtt}} \Gamma_f$.

Proof. Assume that $\Gamma_{f^2} \leq_{\mathrm{wtt}} \Gamma_f$, i. e. there exists a pair of functions $(\{z\}, \{p\})$ which reduces Γ_{f^2} to Γ_f. Let us consider the stage $s = \langle z, p\rangle$.

Note that $\{z\}^{\Gamma_f}(\langle e(f_s''), \beta_s\rangle) = \{z\}^{\Gamma_{f_s''}}(\langle e(f_s''), \beta_s\rangle)$. It follows from the construction that $e(f_s'') = e(f_s')$, i.e. $\{p\}(\langle e(f_s''), \beta_s\rangle) = l_s$. If $m \in D_{l_s}$ and $m \in \Gamma_{f_s''}$, then $m \in \Gamma_f$, since $\Gamma_{f_s''} \subset \Gamma_f$. If $m \in D_{l_s}$ and $m \notin \Gamma_{f_s''}$, then $m \in D_{l_s}^-$, and by Lemma 1 we have $m \notin \Gamma_f$. Therefore $D_{l_s} \cap \Gamma_{f_s''} = D_{l_s} \cap \Gamma_f$, i.e.

$$\{z\}^{\Gamma_f}(\langle e(f_s''), \beta_s\rangle) = \{z\}^{\Gamma_{f_s''}}(\langle e(f_s''), \beta_s\rangle). \tag{1}$$

By the assumption, $\chi_{\Gamma_{f^2}} = \{z\}^{\Gamma_f}$. There are two possible cases.

Case 1. $\{z\}^{\Gamma_f}(\langle e(f_s''), \beta_s\rangle) = 1$. The equality (1) implies

$$\{z\}^{\Gamma_{f_s''}}(\langle e(f_s''), \beta_s\rangle) = 1.$$

By the construction of f_{s+1}, the pairs $\langle e(f_s''), c\rangle >$ and $\langle c, d\rangle$ belong to the graph of f, where d is a new element. We obtain that $f^2(e(f_s'')) = d \neq \beta_s$, $\langle e(f_s''), \beta_s\rangle \notin \Gamma_{f^2}$, i.e.,

$$0 = \chi_{\Gamma_{f^2}}(\langle e(f_s''), \beta_s\rangle) = \{z\}^{\Gamma_f}(\langle e(f_s''), \beta_s\rangle) = 1.$$

This is a contradiction.

Case 2. $\{z\}^{\Gamma_f}(\langle e(f_s''), \beta_s\rangle) = 0$. This is case is similar to Case 1. □

Remark. The set $\{f \mid f \text{ is a permutation and } \Gamma_f \in \Sigma_n^0\}$ is the group of all $\emptyset^{(n-1)}$-computable permutations. We denote $\mathrm{dom}(\{i\}_s^A)$ by $W_{i,s}^A$.

Theorem 2. *The set*

$$\{f \mid f \text{ is a permutation and } \Gamma_f \in \Pi^0_n\}$$

is not closed under compositions for all $n \geq 1$.

Proof. We will by stages construct a permutation f with the following properties: $\Gamma_f \in \Pi^0_n$ and $\Gamma_{f^2} \notin \Pi^0_n$. We $\emptyset^{(n-1)}$-computable enumerate the complement of Γ_f and for any i meet the requirement

$$R_i : \text{the complement of } \Gamma_{f^2} \text{ does not coincide with } W_i^{\emptyset^{(n-1)}}.$$

The requirement R_i will be satisfied by a fixed witness $\langle a_i, c_i \rangle$. At every stage s we construct a finite path f_s, and a finite set B_s that contains all pairs of elements of the field f_s, which are not in Γ_{f_s}. The required permutation f will be the limit of the sequence $\{f_s\}_{s \in \omega}$, and $B_0 \subset B_1 \subset \cdots \subset \bigcup_{i \in \omega} B_s = \omega \setminus \Gamma_f$.

Stage $s = 0$. Define $\Gamma_{f_0} = \{\langle 0, 1 \rangle\}$, $B_0 = \{\langle 1, 0 \rangle\}$.

Stage $s > 0$. Let $s = \langle i, j \rangle$. A number c is called a *new element* if c does not belong to the field of the constructed path. Let p_s and q_s be distinct new elements. Define $\Gamma_{f'_s} = \Gamma_{f_s} \cup \{\langle p_s, b(f_s) \rangle, \langle e(f_s), q_s \rangle\}$.

Now let us consider the function $\{i\}$.

Case 1. The function $\{i\}$ is not considered at previous stages. This means that $\forall s'(s' < s \to (s')_1 \neq i)$. Let a_i, b_i, c_i be distinct new elements. Define $\Gamma_{f_{s+1}} = \Gamma_{f'_s} \cup \{\langle e(f'_s), a_i \rangle, \langle a_i, b_i \rangle, \langle b_i, c_i \rangle\}$.

Case 2. The function $\{i\}$ at a previous stage was considered but the requirement R_i was not satisfied, and $\langle c_i, a_i \rangle \in W_{i,s}^{\emptyset^{(n-1)}}$. Let $n_{i_1}, n_{i_2}, n_{i_3}, n_{i_4}$ be distinct new elements. Define

$$\begin{aligned}\Gamma_{f_{s+1}} = {} & \left(\Gamma_{f'_s} \setminus \left\{\langle {f'}_s^{-1}(a_i), a_i \rangle, \langle a_i, b_i \rangle, \langle b_i, c_i \rangle, \langle c_i, f'_s(c_i) \rangle\right\}\right) \\ & \cup \left\{\left\langle {f'}_s^{-1}(a_i), n_{i_1} \right\rangle, \langle n_{i_1}, c_i \rangle, \langle c_i, n_{i_2} \rangle, \langle n_{i_2}, a_i \rangle,\right. \\ & \qquad \left. \langle a_i, n_{i_3} \rangle, \langle n_{i_3}, b_i \rangle, \langle b_i, n_{i_4} \rangle, \langle n_{i_4}, f'_s(c_i) \rangle\right\}.\end{aligned}$$

We changed f'_s:

$$\ldots \to {f'}_s^{-1}(a_i) \to a_i \to b_i \to c_i \to f'_s(c_i) \ldots$$

to the path f_{s+1}:

$$\ldots \to {f'}_s^{-1}(a_i) \to n_{i_1} \to c_i \to n_{i_2} \to a_i \to n_{i_3} \to b_i \to n_{i_4} \to f'_s(c_i) \ldots$$

We obtained $f^2_{s+1}(c_i) = a_i$, and $\langle c_i, a_i \rangle \in W_{i,s}^{\emptyset^{(n-1)}}$, i.e. the requirement R_i is satisfied.

In other cases we define $f_{s+1} = f'_s$.

Now let

$$B_{s+1} = \{\langle c, d\rangle \mid (c \in \text{field } f_{s+1}) \;\&\; (d \in \text{field } f_{s+1}) \;\&\; (\langle c, d\rangle \notin \Gamma_{f_{s+1}})\}.$$

It follows from the construction that the limit of the sequence $\{f_s\}_{s\in\omega}$ exists, we denote it as f. It is clear that f is a permutation.

Our construction is computable in $\emptyset^{(n-1)}$ and $B_0 \subset B_1 \subset \cdots \subset \bigcup_{i\in\omega} B_s = \omega\backslash\Gamma_f$, hence $\omega\backslash\Gamma_f$ is $\emptyset^{(n-1)}$-computably enumerable (c. e.) set. We satisfied all requirements $R_i, i \in \omega$.

The theorem is proved. □

Now we will consider permutations with limit computable graphs.

A function f is *computable in limit* if $f(x) = \lim_s g(s, x)$ for a total computable function g. Let $f(x) = \lim_s g(s, x)$ and g is a total computable function.

Define $k_{f,g} = |\{s \mid g(s+1, x) \neq g(s, x)\}|$. If above $\forall x k_{f,g}(x) \leq n$ then f is *n-computable*, and f is *ω-computable* if $\forall x k_{f,g}(x) \leq h(x)$ for a total computable function h. A set A is *computable in limit*, *n-c. e.*, *ω-c. e.*, if its characteristic function is computable in limit, n-computable, ω-computable accordingly (we suppose that $\chi_A(x) = \lim_s A(s, x)$, and $A(0, x) = 0$).

Theorem 3. *Let A be arbitrary computable in limit set. There exists a permutation f such that Γ_f is 2-c.e. and $\Gamma_{f^2} \not\leq_{\mathrm{wtt}} A$.*

Proof. We construct the permutation f in stages. It suffices to meet, for each $s = \langle i, i\rangle$ for some i, j), the following requirement

$$R_i: \text{ the pair } (\{(i)_1\}, \{(i)_2\}) \text{ does not reduce } \Gamma_{f^2} \text{ to } A.$$

At each stage s we construct a finite path f_s. The required permutation will be the limit of the sequence $\{f_s\}_{s\in\omega}$.

Again the requirement R_i we satisfy by a fixed witness $\langle a_i, c_i\rangle$. We ensure that if the computation of $\{(i)_1\}^A(\langle a_i, c_i\rangle)$ asks to the oracle A only questions on elements of $D_{\{(i)_2\}(\langle a_i, c_i\rangle)}$, then $\{(i)_1\}^A(\langle a_i, c_i\rangle) \neq \chi_{\Gamma_{f^2}}(\langle a_i, c_i\rangle)$. The set A is computable in limit, and let $A(s, x)$ be a total computable function such that $\chi_A = \lim_s A(s, x)$. Define $A_s = \{x \mid A(s, x) = 1\}$.

Stage $s = 0$. Define $\Gamma_{f_0} = \{\langle 0, 1\rangle\}$.

Stage $s > 0$. Let $s = \langle i, j\rangle$. We call a number c a *new element* if c does not belong to the field of the constructed path. Let p_s and q_s be distinct new elements. Define $\Gamma_{f'_s} = \Gamma_{f_s} \cup \{\langle p_s, b(f_s)\rangle, \langle e(f_s), q_s\rangle\}$.

Let us consider the pair of the functions $(\{(i)_1\}, \{(i)_2\})$.

Case 1. The pair $(\{(i)_1\}, \{(i)_2\})$ was not considered at a previous stage. This means that $\forall s'(s' < s \to (s')_1 \neq i)$. Let a_i, b_i, c_i be distinct new elements. Define $\Gamma_{f_{s+1}} = \Gamma_{f'_s} \cup \{\langle e(f'_s), a_i\rangle, \langle a_i, b_i\rangle, \langle b_i, c_i\rangle\}$.

Case 2. The pair $(\{(i)_1\}, \{(i)_2\})$ was considered at a previous stage, $\{(i)_2\}(\langle a_i, c_i\rangle)$ converges, $\{(i)_1\}_s^{A_s}(\langle a_i, c_i\rangle) = \chi_{\Gamma_{f_s^2}}(\langle a_i, c_i\rangle)$, and the computation of $\{(i)_1\}_s^{A_s}(\langle a_i, c_i\rangle)$ asks A_s only questions from $D_{\{(i)_2\}(\langle a_i, c_i\rangle)}$.

There are two possible cases for $\{(i)_1\}^{A_s}(\langle a_i, c_i\rangle)$.

Case 2.1. $\{(i)_1\}^{A_s}(\langle a_i, c_i\rangle) = 1 = \chi_{\Gamma_{f_s^2}}(\langle a_i, c_i\rangle)$. Let n_s be a new element. Then we define

$$\begin{aligned}\Gamma_{f_{s+1}} = {} & \big(\Gamma_{f'_s} \setminus \{\langle a_i, f'_s(a_i)\rangle, \langle f'_s(a_i), c_i\rangle, \langle c_i, f'_s(c_i)\rangle\}\big) \\ & \cup \{\langle a_i, f'_s(c_i)\rangle, \langle e(f'_s), f'_s(a_i)\rangle, \langle f'_s(a_i), n_s\rangle, \langle n_s, c_i\rangle\}.\end{aligned}$$

We changed the path f'_s:

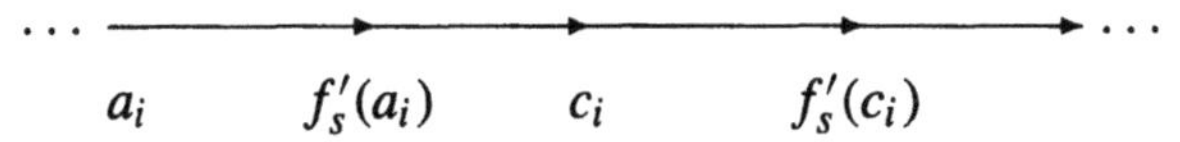

to the path f_{s+1}:

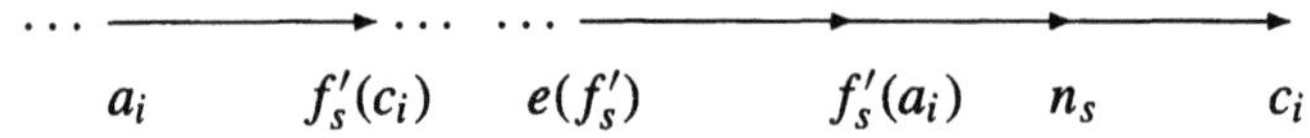

Case 2.2. $\{(i)_1\}^{A_s}(\langle a_i, c_i\rangle) = 0 = \chi_{\Gamma_{f_s^2}}(\langle a_i, c_i\rangle)$. Let n_{s_1}, n_{s_2} be distinct new elements. Then we define

$$\begin{aligned}\Gamma_{f_{s+1}} = {} & \big(\Gamma_{f'_s} \setminus \{\langle a_i, f'_s(a_i)\rangle, \langle {f'}_s^{-1}(c_i), c_i\rangle, \langle c_i, f'_s(c_i)\rangle\}\big) \\ & \cup \{\langle a_i, n_{s_1}\rangle, \langle n_{s_1}, c_i\rangle, \langle c_i, n_{s_2}\rangle, \langle n_{s_2}, f'_s(a_i)\rangle, \langle {f'}_s^{-1}(c_i), f'_s(c_i)\rangle\}.\end{aligned}$$

We changed the path f'_s:

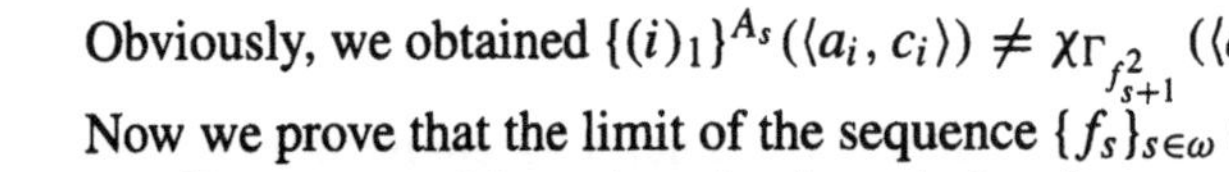

to the path f_{s+1}:

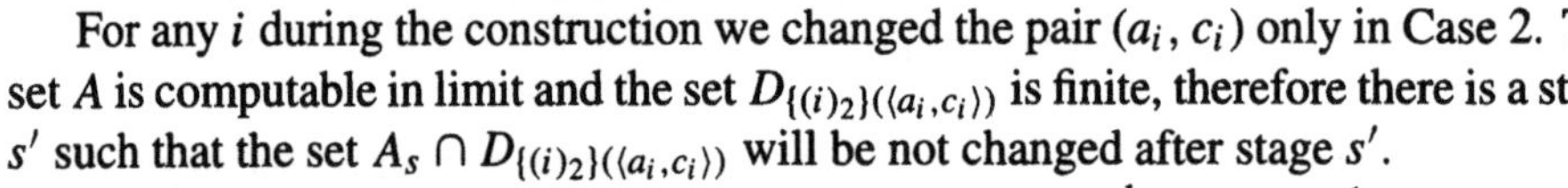

Obviously, we obtained $\{(i)_1\}^{A_s}(\langle a_i, c_i\rangle) \neq \chi_{\Gamma_{f_{s+1}^2}}(\langle a_i, c_i\rangle)$.

Now we prove that the limit of the sequence $\{f_s\}_{s\in\omega}$ exists.

It suffices to consider only pairs (a_i, c_i), $i \geq 0$.

For any i during the construction we changed the pair (a_i, c_i) only in Case 2. The set A is computable in limit and the set $D_{\{(i)_2\}(\langle a_i, c_i\rangle)}$ is finite, therefore there is a stage s' such that the set $A_s \cap D_{\{(i)_2\}(\langle a_i, c_i\rangle)}$ will be not changed after stage s'.

Obviously for any $t > s'$ we have $f_t(a_i) = f(a_i)$, $f_t^{-1}(a_i) = f^{-1}(a_i)$, $f_t(c_i) = f(c_i)$, $f_t^{-1}(c_i) = f^{-1}(c_i)$.

Therefore, the limit of the sequence $\{f_s\}_{s\in\omega}$ exists and the requirement R_i is satisfied at stage s'. This means that $\Gamma_{f^2} \not\leq_{\mathrm{wtt}} A$.

It follows from the construction that Γ_f is 2-c.e. The theorem is proved. □

Corollary 1. *Let A be arbitrary set such that $A \leq_T \emptyset'$. There exists a permutation f such that Γ_f is 2-c.e., but $\Gamma_{f^2} \not\leq_{\mathrm{wtt}} A$.*

Proof. A is computable in limit iff $A \leq_T 0'$, by the well-known Limit Lemma [4]. □

Since for any a set A is 2-c.e. iff $A \leq_{2tt} \emptyset'$, then

Corollary 2. *There exists a permutation f such that $\Gamma_f \leq_{2tt} \emptyset'$ and $\Gamma_{f^2} \not\leq_{\mathrm{wtt}} \emptyset'$.*

Corollary 3. *The set $\{f \mid f$ is a permutation and $\Gamma_f \leq_{tt} \emptyset'\}$ is not closed under composition.*

Corollary 4. *The set $\{f \mid f$ is a permutation and $\Gamma_f \leq_{\mathrm{wtt}} \emptyset'\}$ is not closed under compositions.*

It is known that a set A is ω-limit computable iff $A \leq_{tt} \emptyset'$ ([1]). Therefore, the set

$$\{f \mid f \text{ is a permutation and } \Gamma_f \text{ is } \omega\text{-limit computable}\}$$

is not closed under compositions.

I would like to acknowledge my thanks to Prof. A. S. Morozov for helpful suggestions and discussions.

References

[1] H. G. Carstens, Δ_2^0-Mengen. Arch. Math. Log. Grundlagenforsch. 18 (1976).

[2] A. S. Morozov. Groups of computable automorphisms, in: Handbook of recursive mathematics, Vol. 1, Stud. Logic Found. Math. 138, North-Holland, Amsterdam, 1998, 311–345.

[3] H. Rogers, Jr., Theory of Recursive Functions and Effective Computability, McGraw-Hill, New York, 1967.

[4] R. I. Soare Recursively Enumerable Sets and Degrees, A Study of Computable Functions and Computability Generated Sets, Springer-Verlag, Berlin, 1987.

Some properties of majorant-computability

M. V. Korovina and O. V. Kudinov

Abstract. We introduce the concept of majorant-computability over the reals which integrates methods and ideas of continuous mathematics and the modern mathematical logic, in particular, definability theory. Properties of majorant-computable real-valued functions and functionals are studied. In this framework we investigate definability of majorant-computable real-valued functions and functionals. A structure theorem is proved which connects the graph of a majorant-computable function with validity of a finite formula on the set of hereditarily finite sets on $\bar{\mathbf{R}}$, $\mathbf{HF}(\bar{\mathbf{R}})$ (where $\bar{\mathbf{R}}$ is an elementary proper extension of the standard reals). We establish a criterion of functionals to be majorant-computable. and find interesting examples of real-valued functions, functionals and subsets of the reals that are majorant-computable.

Keywords. Majorant-computable function, majorant-computable functionals, computational process, approximation, definability.

1. Introduction

The semiring of natural numbers is a classical structure for which the concept of computability has been defined and studied rather well. Although several definitions were independently proposed (Gödel, Church, or Turing), they turned out to be equivalent. In the case of computability over the reals the situation is essentially different from the classical one, since some nonequivalent notions have also been proposed by Moschovakis [10], Blum [3], Freedman [6], Ershov [5], Pour-El [14], and others. Apparently, it was caused by the fact that different authors took notice of different aspects of computational processes over abstract structures, in particularly, over the reals. So, in this case, it is desirable to find some more general and more exact formalization of the notion of computable real-valued function which would contain the formalizations considered so far as special cases. Such functions should be computable by appropriate computers, at least potentially, and have a strong mathematical description. In this article, we investigate such a notion of computability, majorant-computability. A characteristic property of this approach is that we do not use terminating algorithms in the definitions. The result of computation is defined by a nonterminating process computing approximations closer and closer to the result. The approach proposed in the article does not depend on the way of representing the reals. The use of nonstandard models of the first-order theory of the reals enables us to investigate the property of computability of partial real-valued functions. Definability of majorant-computable real-valued functions, functionals and subsets of the reals is studied. A structure theorem is proved which connects the graph of a majorant-computable function with validity of a finite formula in the set of hereditarily finite sets $\mathbf{HF}(\bar{\mathbf{R}})$ (where $\bar{\mathbf{R}}$ is an elementary proper extension of the standard real numbers). This property can be considered as a denotation semantic of computational processes. In our approach, the majorant-computable functions include an interesting class of real-valued total functions that possess meromorphic extensions onto **C**. This class, in particular, contains functions that are solutions of known differential equations. We introduce a notion of majorant-computable functionals over the class of majorant-computable real-valued total functions. and establish a criterion of majorant-computability of functionals. Some interesting examples of real-valued functions, functionals and subsets of the reals that are majorant-computable are given.

2. Basic notions

Throughout the article, $\langle \mathbf{R}, 0, 1, +, \cdot, \leq \rangle$ is the standard model of the reals, denoted also by **R**, where $+$ and $\cdot$ are regarded as predicate symbols. Let Th(**R**) be the first order theory of **R**. Let **N** denote the set of natural numbers and **Q** the set of rational numbers.

We use definability as one of the basic conceptions. Montague [9] proposed to consider computability from the point of view definability. Later, many authors among them Ershov [5], Gorgon [7] paid attention to properties of this approach applied to various basic models. We use the following notion of definability in a structure $\mathbf{M} = \langle M, \sigma_0 \rangle$, where σ_0 is a finite first-order language.

Bold face indicates sequences, in particular, $\mathbf{x} = x_1, \dots, x_n, \mathbf{y} = y_1, \dots, y_k$.

A formula $\Phi(\bar{x})$ is said to *define* the set $\{\mathbf{r} \in M^n \mid \mathbf{M} \models \Phi(\mathbf{r})\}$ (in **M**), and a set of this form is called a *definable* subset of M^n. A function $F : \mathbf{M}^m \to \mathbf{M}$ is said to be *definable* (in **M**) if its graph is definable (in **M**).

Recall some properties of definable sets and functions in **R**. The following statements are proved in [4, 11, 12]. Denote by $\mathbf{R}_0$ the real closure of the rational numbers. We use standard notations for real intervals (a, b), $[a, b]$, $(a, b]$, $[a, b)$. In addition, $\langle a, b \rangle$ denotes any of those intervals.

Proposition 1 (0-minimality property). 1. *A set $B \subseteq \mathbf{R}$ is definable if and only if there exist $n \in \omega$ and $a_1, b_1, \dots, a_n, b_n \in \mathbf{R}_0 \cup \{+\infty, -\infty\}$ such that*

$$B = \cup_{i \leq n} \langle a_i, b_i \rangle .$$

2. *Each definable subset of $\mathbf{R}^n$ is the disjoint union of finitely many cells, each of which is also definable. (A cell is a space that homeomorphic an $\mathbf{R}^k$, where $k \leq n$.)*

Proof. See [4]. □

Since the theory Th(**R**) admits elimination of quantifiers, it follows that 0-minimality property holds on every model of Th(**R**).

For a function f let us denote

$$\begin{aligned}\mathrm{dom}(f) &= \{x \mid \exists y\,(f(x) = y)\},\\ \mathrm{im}(f) &= \{y \mid \exists x\,(f(x) = y)\},\\ \Gamma_f &= \{(x, y) \mid [(x, y) \in \mathrm{dom}(f) \times \mathrm{im}(f)] \wedge [f(x) = y]\}.\end{aligned}$$

Definition 1. A partial real-valued function $f : \mathbf{R} \to \mathbf{R}$ is said to be *algebraic* if for some $a, b, c, d \in \mathbf{R}_0 \cup \{-\infty, +\infty\}$ and a polynomial $p \in \mathbf{Q}[x, y]$, the following conditions hold:

1. $\mathrm{dom}(f) = \langle a, b\rangle$;
2. $\mathrm{im}(f) = \langle c, d\rangle$;
3. $\Gamma_f = \{(x, y) \mid [(x, y) \in \langle a, b\rangle \times \langle c, d\rangle] \wedge [p(x, y) = 0]\}$.

Proposition 2. *A real-valued function $f : \mathbf{R} \to \mathbf{R}$ is definable if and only if the following conditions hold.*

1. *There exist $n \in \omega$ and $a_1, b_1, \ldots, a_n, b_n \in \mathbf{R}_0 \cup \{+\infty, -\infty\}$, such that $\mathrm{dom}(f) = \bigcup_{i\le n} \langle a_i, b_i\rangle$ and $\langle a_j, b_j\rangle \cap \langle a_i, b_i\rangle = \emptyset$ for $i \neq j$,*
2. *f is an algebraic function on an interval $\langle a_i, b_i\rangle$, $i \le n$.*

Proof. See [12] □

The previous proposition implies that a real-valued function f is definable if and only if f is the finite union of algebraic functions defined on disjoint intervals with algebraic endpoints. So, to study more reach class of real-valued functions then one above we need an extension of a basic model.

Let us construct the set of hereditarily finite sets HF(M) over a model **M**. This structure is rather well studied in the theory of admissible sets [1] and permits us to define the natural numbers, code and store information via formulas.

Let **M** be a model whose language σ_0 contains no function symbols and whose carrier set is M. We construct the set of hereditarily finite sets, HF(**M**), as follows:

1. $\mathrm{S}_0(M) \rightleftharpoons M$, $\mathrm{S}_{n+1}(M) \rightleftharpoons \mathcal{P}_\omega(\mathrm{S}_n(M)) \cup \mathrm{S}_n(M)$, where $n \in \omega$ and for every set B, $\mathcal{P}_\omega(B)$ is the set of all finite subsets of B.
2. $\mathrm{HF}(M) = \bigcup_{n\in\omega} \mathrm{S}_n(M)$.

We define **HF**(M) as the following model: $\mathbf{HF}(M) \rightleftharpoons \langle \mathrm{HF}(M), \sigma_0, \emptyset, \in\rangle$, where the constant $\emptyset \rightleftharpoons \{\,\}$ (the empty set) and the binary predicate symbol $\in$ have standard set-theoretic interpretation. Denote $\sigma = \sigma_0 \cup \{\in, \emptyset\}$. The notions of terms and atomic formulas are given in a standard way.

The set of Δ_0-*formulas* is the closure of the set of atomic formulas in the language σ under $\wedge, \vee, \neg, \exists x \in t$ and $\forall x \in t$, where $\exists x \in t\ \varphi$ denotes $\exists x(x \in t \wedge \varphi)$ and $\forall x \in t\ \varphi$ denotes $\forall x(x \in t \to \varphi)$.

The set of Σ-*formulas* is the closure of the set of Δ_0 formulas under $\wedge, \vee, \exists x \in t, \forall x \in t$, and $\exists$.

We define Π-*formulas* as negations of a Σ-formulas.

Definition 2. 1. A set $B \subseteq \mathbf{HF}(M)$ is Σ-*definable* if there exists a Σ-formula $\Phi(x)$ such that

$$x \in B \leftrightarrow \mathbf{HF}(M) \models \Phi(x).$$

2. A function $f : \mathbf{HF}(M) \to \mathbf{HF}(M)$ is Σ-*definable* if there exists a Σ-formula $\Phi(x, y)$ such that

$$f(x) = y \leftrightarrow \mathbf{HF}(M) \models \Phi(x, y).$$

In a similar way, we define the notions of Π-*definable functions* and *sets*.

The class of Δ-*definable functions* (*sets*) is the intersection of the class Σ-definable functions (respectively sets) and the class of Π-definable functions (sets).

Note that the sets M and M^n are Δ_0-definable. This fact makes $\mathbf{HF}(M)$ a suitable domain for studying functions from M^k to M and subsets of M^n.

To introduce the definition of majorant-computability we use a class of Σ-definable real-valued functions as a basic class. So, we recall properties of Σ-definable real-valued functions and subsets of $\mathbf{R^n}$. The following propositions give important properties of Σ-, Δ-, Π- definable sets and functions.

Proposition 3. *Let* $\tilde{\mathbf{R}}$ *be a model of* Th($\mathbf{R}$). *A set* $B \subseteq \tilde{\mathbf{R}}^n$ *is* Σ-*definable if and only if there exists an effective sequence of quantifier free formulas in the language of* Th($\mathbf{R}$) (*denote* σ_0) $\{\Phi_s(x)\}_{s\in\omega}$ *such that*

$$x \in B \leftrightarrow \mathbf{HF}(\tilde{\mathbf{R}}) \models \bigvee_{s\in\omega} \Phi_s(x).$$

Proof. The claim is immediate from the properties of the set of hereditarily finite sets and the fact that Th($\mathbf{R}$) admits elimination of quantifiers. □

Proposition 4. *Let* $\mathbf{R}$ *be the standard reals. The following sentences hold in* $\mathbf{HF}(\mathbf{R})$.

1. *The set* $\mathbf{N}$ *is* Δ-*definable.*
2. *The class of* Σ-*definable real-valued total functions is closed under composition.*
3. *The class of* Σ-*definable real-valued total functions coincides with the class of* Δ-*definable real-valued total functions.*

4. *Let* $f : \mathbf{R} \to \mathbf{R}$ *such that* $\mathrm{dom}(f) \subset \mathbf{N}$ *and* $\mathrm{im}(f) \subset \mathbf{N}$. *The function* f *is* Σ*-definable function if and only if* f *is a partial recursive function.*

5. *There is an universal* Σ*-definable real-valued function for the class of algebraic functions.*

6. *There is an universal real-valued* Σ*-definable function in the class of* Σ*-definable real-valued functions.*

Proof. See [11] □

Proposition 5. *A real-valued function* $f : \mathbf{R} \to \mathbf{R}$ *is* Σ*-definable if and only if the following conditions hold.*

1. *There exist effective sequences* $\{a_i\}_{i\in\omega}$, $\{b_i\}_{i\in\omega}$, *where* $a_i, b_i \in \mathbf{R}_0 \cup \{+\infty, -\infty\}$ *for* $i \in \omega$, *such that*

$$\mathrm{dom}(f) = \bigcup_{i\in\omega} \langle a_i, b_i \rangle$$

and

$$\langle a_j, b_j \rangle \cap \langle a_i, b_i \rangle = \emptyset \quad \textit{for } i \neq j.$$

2. *There exists effective sequences of algebraic functions* $\{g_i\}_{i\in\omega}$ *such that* f *coincides with* g_i *on the interval* $\langle a_i, b_i \rangle$.

Proof. See [12]. □

To introduce the notion of majorant-computability for partial real-valued functions we need a notion of prime extension of **R**.

Definition 3. A model $\bar{\mathbf{R}}$ of Th(**R**) is called a *prime extension* of **R** if there exists $t \in \bar{\mathbf{R}}$ such that $t > n$ for every natural n (we write $t > \mathbf{N}$) and $\bar{\mathbf{R}}$ is the real closure of the ordered field $\mathbf{R}(t)$.

Definition 4. Let $\bar{\mathbf{R}}$ be an elementary proper extension of **R** and let $t > \mathbf{N}$.

$$\mathrm{sp}(x) = \begin{cases} x^* \in \mathbf{R} \text{ if } |x - x^*| \leq \varepsilon, & \text{where } 0 < \varepsilon < \mathbf{R}^+,\ \epsilon = \frac{1}{t}, \\ +\infty & \text{if } x > \mathbf{R}, \\ -\infty & \text{if } x < \mathbf{R}. \end{cases}$$

Since $\bar{\mathbf{R}} \models \mathrm{Th}(\mathbf{R})$, $\bar{\mathbf{R}}$ is an elementary proper extension of **R** with the same true first-order formulas. Recall the properties of a prime extension.

Proposition 6. 1. *Every two prime extensions of* $\mathbf{R}$ *are isomorphic.*

2. *There exists an nonstandard element* $t > \mathbf{N}$ *in an elementary proper extension of* $\mathbf{R}$.

Proof. See [8]. □

Lemma 1. *Let* $p(x, \mathbf{y})$ *be a countable set of formulas with one variable* x *and real parameters* $\mathbf{y}$. *Given elementary proper extensions* $\mathbf{R}_1$, $\mathbf{R}_2$ *of* $\mathbf{R}$, $p(x, \mathbf{y})$ *holds in* $\mathbf{R}_1$ *if and only if* $p(x, \mathbf{y})$ *holds in* $\mathbf{R}_2$.

Proof. The claim is immediate from 0-minimality of $\mathbf{R}_1$ and $\mathbf{R}_2$. □

Lemma 2. *Let* $\bar{\mathbf{R}}$ *be a prime extension of* $\mathbf{R}$. *For every* Σ*-formula* $\varphi(z, \mathbf{x})$ *there exists a* Σ*-formula* $\varphi^*(\mathbf{x})$ *such that for all* $\mathbf{x} \in \mathbf{R}^n$ *and* $t > \mathbf{N}$,

$$\mathbf{HF}(\bar{\mathbf{R}}) \models \varphi(t, \mathbf{x}) \leftrightarrow \mathbf{HF}(\bar{\mathbf{R}}) \models \varphi^*(\mathbf{x}).$$

Proof. Let $\varphi(z, \mathbf{x})$ be a Σ-formula. Proposition 3 implies that there exists an effective sequence of quantifier free formulas in the language σ_0 such that

$$\mathbf{HF}(\bar{\mathbf{R}}) \models \varphi(t, \mathbf{x}) \leftrightarrow \mathbf{HF}(\bar{\mathbf{R}}) \models \bigvee_{i \in \omega} \varphi_i(t, \mathbf{x}).$$

By 0-minimality of $\bar{\mathbf{R}}$, $\mathbf{HF}(\bar{\mathbf{R}}) \models \varphi_i(t, \mathbf{x})$ for $t > \mathbf{N}$ if and only if there exists $c \in \mathbf{R}$ such that

$$\mathbf{HF}(\bar{\mathbf{R}}) \models \left(\forall t^* \geq c\right) \varphi_i(t^*, \mathbf{x}).$$

Put

$$\varphi^*(\mathbf{x}) \rightleftharpoons \bigvee_{i \in \omega} \exists t \forall t^* \geq t \varphi_i(t^*, \mathbf{x}).$$

By Proposition 3, the set $\{\mathbf{x} \mid \mathbf{HF}(\bar{\mathbf{R}}) \models \varphi^*(\mathbf{x})\}$ is Σ-definable by a Σ-formula that is a sought one. □

Lemma 3. *Let* $\bar{\mathbf{R}}$ *be a model of* $\mathrm{Th}(\mathbf{R})$ *and* B *be a countable* Σ*-definable in* $\mathbf{HF}(\bar{\mathbf{R}})$ *set. Then there exists a* Σ*-definable function numbering* B.

Proof. The claim is immediate from Proposition 1 and Proposition 3. □

3. Majorant-computable functions

Let us introduce a notion of majorant-computability for real-valued functions. We would use as a basic class the class of Σ-definable total function of type $f : \bar{\mathbf{R}}^n \to \bar{\mathbf{R}}$, where $\bar{\mathbf{R}}$ is an elementary proper extension of $\mathbf{R}$. A real-valued function is said to be majorant-computable if we can construct a special kind nonterminating process computing approximations closer and closer to the result.

Definition 5. A function $f : \mathbf{R}^n \to \mathbf{R}$ is called *majorant-computable* if there exist effective sequences of Σ-formulas $\{\Phi_s(a, \mathbf{x}, y)\}_{s\in\omega}$ and $\{G_s(a, \mathbf{x}, y)\}_{s\in\omega}$, with a parameter a, an elementary proper extension $\bar{\mathbf{R}}$ of $\mathbf{R}$ such that the following conditions hold.

1. There exists $t \in \bar{\mathbf{R}}$ such that $t > \mathbf{N}$.
2. For all $s \in \omega$, the formulas $\Phi_s(a, \mathbf{x}, y)$ and $G_s(a, \mathbf{x}, y)$ define total functions f_s and g_s as follows:
   a. $f_s : \bar{\mathbf{R}}^n \to \bar{\mathbf{R}}$ and $g_s : \bar{\mathbf{R}}^n \to \bar{\mathbf{R}}$;
   b. $f_s(\mathbf{x}) = y \leftrightarrow \mathbf{HF}(\bar{\mathbf{R}}) \models \Phi_s(t, \mathbf{x}, y)$, $g_s(\mathbf{x}) = y \leftrightarrow \mathbf{HF}(\bar{\mathbf{R}}) \models G_s(t, \mathbf{x}, y)$.
3. For all $\mathbf{x} \in \bar{\mathbf{R}}^n$, the sequence $\{f_s(\mathbf{x})\}_{s\in\omega}$ of the functions defined in 2) monotonically increases; the sequence $\{g_s(\mathbf{x})\}_{s\in\omega}$ of the functions defined in 2) monotonically decreases;
4. For all $s \in \omega$, $\mathbf{x} \in \mathrm{dom}(f)$, $f_s(\mathbf{x}) \le f(\mathbf{x}) \le g_s(\mathbf{x})$ and for all $\mathbf{x} \in \bar{\mathbf{R}}^n$, $f_s(\mathbf{x}) \le g_s(\mathbf{x})$;
5. $f(\mathbf{x}) = y \leftrightarrow \lim_{s\to\infty} \mathrm{sp}(f_s(\mathbf{x})) = y$ and $\lim_{s\to\infty} \mathrm{sp}(g_s(\mathbf{x})) = y$.

The sequence $\{f_s\}_{s\in\omega}$ defined in Definition 5 is called a *sequence of lower Σ-approximations* for f. The sequence $\{g_s\}_{s\in\omega}$ is called a *sequence of upper Σ-approximations* for f.

The following theorem connects the graph of a majorant-computable function with validity of a finite formula in the set of hereditarily finite sets $\mathbf{HF}(\bar{\mathbf{R}})$ (where $\bar{\mathbf{R}}$ is an elementary proper extension of the standard real numbers).

Definition 6. Let $\bar{\mathbf{R}}$ be an elementary proper extension of $\mathbf{R}$. A formula Φ is said to *determine* the function $f : \bar{\mathbf{R}}^n \to \bar{\mathbf{R}}$ in the model $\mathbf{HF}(\bar{\mathbf{R}})$ if the following sentence holds

$$f(\mathbf{x}) = y \leftrightarrow \big(\mathbf{HF}(\bar{\mathbf{R}}) \models \Phi(\mathbf{x}, y) \text{ and } \{\mathrm{sp}(z) \mid \Phi(\mathbf{x}, z)\} = \{y\}\big).$$

Theorem 1. *For all functions $f : \mathbf{R}^n \to \mathbf{R}$, the following assertion are equivalent:*

1. *The function f is majorant-computable.*
2. *There exist a prime extension $\bar{\mathbf{R}} \succ \mathbf{R}$ and a Π-formula that determines the function F in the model $\mathbf{HF}(\bar{\mathbf{R}})$ with the property $F|_{\mathbf{R}} = f$.*
3. *There exists a Π-formula that in any elementary proper extension $\hat{\mathbf{R}} \succ \mathbf{R}$ determines the function F with the property $F|_{\mathbf{R}} = f$.*

Proof. 1 → 2) Let f be majorant-computable. By Proposition 6, without loss of generality, we may assume that there exist a prime extension $\bar{\mathbf{R}}$ of $\mathbf{R}$, a sequence $\{f_s\}_{s\in\omega}$ of lower Σ-approximations for f, and a sequence $\{g_s\}_{s\in\omega}$ of upper Σ-approximations for f. Let $t > \mathbf{N}$. Put

$$B = \mathbf{Q} \cup \{\pm t^n \mid n \in \mathbf{Q}\}.$$

Using $\{f_s\}_{s\in\omega}$, $\{g_s\}_{s\in\omega}$, and cofinality of B in $\bar{\mathbf{R}}$, we construct a sequence $\{f_s^*\}_{s\in\omega}$ of new lower Σ-approximations for f and a sequence $\{g_s^*\}_{s\in\omega}$ of new upper Σ-approximations for f such that for all $s \in \omega$, the ranges of f_s and g_s are subsets of B.

Denote:

$$D_1(\mathbf{x}) = \{z \in B \mid \exists s\, (f_s(\mathbf{x}) \geq z)\},\ D_2(\mathbf{x}) = \{z \in B \mid \exists s\, (g_s(\mathbf{x}) \leq z)\}.$$

The sets $D_1(\mathbf{x})$ and $D_2(\mathbf{x})$ are Σ-definable and countable, so, by Lemma 4, there exist a function $h_1(n, \mathbf{x})$ numbering $D_1(\mathbf{x})$ and a function $h_2(n, \mathbf{x})$ numbering $D_2(\mathbf{x})$.

Next, put

$$f_s^*(\mathbf{x}) = \max_{n\leq s} h_1(n, \mathbf{x}),\ \ g_s^*(\mathbf{x}) = \min_{n\leq s} h_2(n, \mathbf{x}).$$

By construction, the sequences $\{f_s^*\}_{s\in\omega}$ and $\{g_s^*\}_{s\in\omega}$ are sought ones.

Let

$$f_s^*(\mathbf{x}) = y \rightleftharpoons \mathbf{HF}(\bar{\mathbf{R}}) \models \Phi_s^*(t, \mathbf{x}, y),$$
$$g_s^*(\mathbf{x}) = y \rightleftharpoons \mathbf{HF}(\bar{\mathbf{R}}) \models G_s^*(t, \mathbf{x}, y),$$

where Φ_s^*, G_s^* are Σ-formulas.

From Lemma 3 it follows that for all $s \in \omega$, there exist Σ-formulas Φ_s^{**} and G_s^{**} such that, for every $\mathbf{x} \in \mathbf{R}^n$:

$$\mathbf{HF}(\bar{\mathbf{R}}) \models G_s^*(t, \mathbf{x}, y) \leftrightarrow G_s^{**}(\mathbf{x}, y),$$

$$\mathbf{HF}(\bar{\mathbf{R}}) \models \Phi_s^*(t, \mathbf{x}, y) \leftrightarrow \Phi_s^{**}(\mathbf{x}, y).$$

The following formula is a sought one.

$$\Phi(\mathbf{x}, y) \rightleftharpoons \forall s \forall y_1 \forall y_2 \left[\left(\Phi_s^{**}(\mathbf{x}, y_1) \wedge G_s^{**}(\mathbf{x}, y_2)\right) \to (y_1 \leq y \leq y_2)\right].$$

2 → 3) This is obvious.

3 → 1) Let $\bar{\mathbf{R}}$ be a prime extension of $\mathbf{R}$ and let a Π-formula $\Phi(\mathbf{x}, y)$ determine the function F such that $F|_{\mathbf{R}} = f$. From Proposition 3 it follows that there exists an effective sequence of quantifier-free formulas $\{\varphi_i(\mathbf{x}, y)\}_{i\in\omega}$ such that

$$\mathbf{HF}(\bar{\mathbf{R}}) \models \Phi(\mathbf{x}, y) \leftrightarrow \mathbf{HF}(\bar{\mathbf{R}}) \models \bigwedge_{i\in\omega} \varphi_i(\mathbf{x}, y).$$

Denote

$$\Phi_s(\mathbf{x}, y) \rightleftharpoons \bigwedge_{i \le s} \varphi_i(\mathbf{x}, y).$$

Put

$$f_s(\mathbf{x}) = \begin{cases} \inf\{y \mid \Phi_s(\mathbf{x}, y)\} & \text{if } \inf\{y \mid \Phi_s(\mathbf{x}, y)\} \text{ exists,} \\ -t \text{ otherwise,} \end{cases}$$

$$g_s(\mathbf{x}) = \begin{cases} \sup\{y \mid \Phi_s(\mathbf{x}, y)\} & \text{if } \sup\{y \mid \Phi_s(\mathbf{x}, y)\} \text{ exists,} \\ t \text{ otherwise,} \end{cases}$$

where $t > \mathbf{N}$.

Then $\{f_s\}_{s\in\omega}$ and $\{g_s\}_{s\in\omega}$ are sought sequences. □

Remark 1. Whenever we want to prove some sentence for a majorant-computable function, without loss of generality, we can assume that the proper elementary extension of **R** in Definition 5 is a prime extension of **R**.

Remark 2. If f is a total function then the sequences $\{f_s\}_{s\in\omega}$ and $\{g_s\}_{s\in\omega}$ constructed in the proof of Theorem 1 uniformly converge to f on **R**.

Corollary 1. *Let* $f : \mathbf{R}^n \to \mathbf{R}$. *The total function* f *is majorant-computable if and only if there exist effective sequences* $\{f_s\}_{s\in\omega}$ *and* $\{g_s\}_{s\in\omega}$ *with the following properties*:

1. *For all* $s \in \omega$, *the functions* $f_s : \mathbf{R}^n \to \mathbf{R}$ *and* $g_s : \mathbf{R}^n \to \mathbf{R}$ *are majorant-computable total ones.*

2. *For all* $\mathbf{x} \in \mathbf{R}^n$, *the sequence* $\{f_s(\mathbf{x})\}_{s\in\omega}$ *monotonically increases, the sequence* $\{g_s(\mathbf{x})\}_{s\in\omega}$ *monotonically decreases.*

3. *For all* $s \in \omega$, $\mathbf{x} \in \mathrm{dom}(f)$ $f_s(\mathbf{x}) \le f(\mathbf{x}) \le g_s(\mathbf{x})$ *and for all* $\mathbf{x} \in \mathbf{R}^n$, $f_s(\mathbf{x}) \le g_s(\mathbf{x})$.

4. $f(\mathbf{x}) = y \leftrightarrow \lim_{s\to\infty} f_s(\mathbf{x}) = y$ *and* $\lim_{s\to\infty} g_s(\mathbf{x}) = y$.

Proof. The claim is immediate from Remark 2 and Theorem 1. □

Proposition 7. *If a function* $f : \mathbf{R}^n \to \mathbf{R}$ *is majorant-computable, then the* $\mathrm{dom}(f)$ *is* Π_2*-definable by a formula of type* $\big((\forall z \in \mathbf{R})\, \Phi(z, \mathbf{x})\big)$, *where* Φ *is a* Σ*-formula.*

Proof. Let $\{f_s\}_{s\in\omega}$ be a sequence of lower Σ-approximations for f, and let a sequence $\{g_s\}_{s\in\omega}$ be a sequence of upper Σ-approximations for f.

We obtain:

$$x \in \mathrm{dom}(f) \leftrightarrow \mathrm{HF}(\mathbf{R}) \models (\forall \varepsilon \in \mathbf{R})(\exists s \in \mathbf{N})\big|f_s(\mathbf{x}) - g_s(\mathbf{x})\big| \le \varepsilon.$$

By Σ-definability of the naturals, the set $\mathrm{dom}(f)$ is Π_2-definable. □

Definition 7. Let a function $f : \mathbf{R}^n \to \mathbf{R}$ be total. The *epigraph* of f is defined to be the set $U = \{(\mathbf{x}, y) \mid f(\mathbf{x}) < y\}$. The *ordinate* set of f is defined to be the set $D = \{(\mathbf{x}, y) \mid f(\mathbf{x}) > y\}$.

Corollary 2. *Let a function $f : \mathbf{R}^n \to \mathbf{R}$ be total. The function f is majorant-computable if and only if the epigraph of f and the ordinate set of f are Σ-definable sets.*

Proof. It is clear that

$$(\mathbf{x}, z) \in U \leftrightarrow \mathbf{HF}(\mathbf{R}) \models \exists s \exists y \left[(s \in \mathbf{N}) \wedge G^{**}(\mathbf{x}, y) \wedge (z > y) \right],$$

$$(\mathbf{x}, z) \in D \leftrightarrow \mathbf{HF}(\mathbf{R}) \models \exists s \exists y \left[(s \in \mathbf{N}) \wedge F^{**}(\mathbf{x}, y) \wedge (z < y) \right],$$

where the formulas G^{**} and F^{**} are those in the proof of Theorem 1.

Corollary 3. *Let a function $f : \mathbf{R}^n \to \mathbf{R}$ be total. The function f is majorant-computable if and only if there exists a Σ-formula $\Phi(m, \mathbf{x}, y)$ with the following conditions*:

1. *For all $m \in \omega$, $\Phi(m, \mathbf{x}, y)$ defines a total function $f_m : \mathbf{R}^n \to \mathbf{R}$,*
2. *The sequence $\{f_s\}_{s\in\omega}$ uniformly converge to f.*

Proof. The claim is immediate from Corollary 2. □

Corollary 4. *The class of continuous majorant-computable real-valued total functions is closed under composition.*

Proof. Consider some $f : \mathbf{R} \to \mathbf{R}$ and $h : \mathbf{R} \to \mathbf{R}$ which are majorant-computable continuous real-valued total functions. Let $\{f_s^1\}_{s\in\omega}$ be a sequence of lower Σ-approximations uniformly converging to f and let $\{f_s^2\}_{s\in\omega}$ be a sequence of upper Σ-approximations uniformly converging to f. Let $\{h_s^1\}_{s\in\omega}$ be a sequence of lower Σ-approximations uniformly converging to h and let $\{h_s^2\}_{s\in\omega}$ be a sequence of upper Σ-approximations uniformly converging to h.

Let us construct corresponding approximations g_s^1 and g_s^2 for $g = f \circ h$. Let $\nu : \mathbf{N} \to \mathbf{R}_0 \cup \{-\infty, +\infty\}$ be a bijection between $\mathbf{N}$ and $\mathbf{R}_0 \cup \{-\infty, +\infty\}$. From Corollary 1 and Proposition 5 it follows that for all $s \in \omega$, there exist recursive functions p_s^1 and p_s^2 such that, for all $n \in \omega$ and for all $x \in \langle \nu p_s^1(2n), \nu p_s^1(2n+1) \rangle$, the function $f_s^1(x)$ coincides with a continuous algebraic function, and for all $x \in \langle \nu p_s^2(2n), \nu p_s^2(2n+1) \rangle$, the function $f_s^2(x)$ coincides with a continuous algebraic function.

For $j = 1, 2$, put:

$$t_j(x, s) = \mu z\Bigg(\bigvee_{i \in \omega} \Big(\big[h_z^1(x),\ h_z^2(x)\big] \subseteq \big[\nu p_s(2i),\ \nu p_s(2i+1)\big] \Big) \Bigg)$$

and

$$g_s^1(x) = \inf_{y \in \left[h^1_{t_1(x,s)}(x),\ h^2_{t_1(x,s)}(x)\right]} f_s^1(y),$$

$$g_s^2(x) = \sup_{y \in \left[h^1_{t_2(x,s)}(x),\ h^2_{t_2(x,s)}(x)\right]} f_s^2(y).$$

By uniform convergence $f_s^i \rightrightarrows f$ and continuity of f, h, the sequences g_s^1, g_s^2 are the sought ones. So, h is a majorant-computable function. □

Corollary 5.

1. *The class of majorant-computable real-valued functions is countable.*
2. *There exists an universal majorant-computable function for the class of real majorant-computable functions.*

Proof. The claims are immediate from Theorem 1. □

Corollary 6. *Let* $f : \mathbf{R} \to \mathbf{R}$.

1. *If f is a Σ-definable real function, then f is a majorant-computable function.*
2. *If* $\operatorname{dom} f \subset \mathbf{N}$ *and* $\operatorname{im} f \subset \mathbf{N}$, *then f is a majorant-computable iff f is a partial recursive function.*
3. *If f is a majorant-computable total function, then f is a piecewise continuous function.*

Proof. The claims follow from the definition of majorant-computability. □

Consider an interesting class of real-valued total functions that possess meromorphic extensions onto **C**. This subclass contains, for example, solutions to known differential equations.

Let **C** be the set of complex numbers.

Definition 8. A function $f : \mathbf{R} \to \mathbf{R}$ is said to possess meromorphic extension onto **C** if there exists a meromorphic function $f^* : \mathbf{C} \to \mathbf{C}$ such that $f^*|_{\mathbf{R}} = f$

We recall the fact concerning meromorphic functions in the theory of complex numbers [2].

Proposition 8. *Let $f : \mathbf{R} \to \mathbf{R}$ be a total function that possesses meromorphic extension onto C. There exist entire functions $a(z)$ and $b(z)$ such that*

1. $f(z) = \frac{a(z)}{b(z)}$,
2. *$a(z)$ and $b(z)$ have Taylor expansions with coefficients in* $\mathbf{R}$.

Proof. The claim is follows from the Schwartz principle [2]. □

Theorem 2. *Let $f : \mathbf{R} \to \mathbf{R}$ be a total function and let f^* be its meromorphic extension onto* $\mathbf{C}$. *Suppose that f^* can be written as $f^*(z) = \frac{a(z)}{b(z)}$, where functions $a(z)$ and $b(z)$ are those as in Proposition 7. The function f is majorant-computable if the following conditions on f hold:*

1. *There exist Σ-definable real-valued functions A and B such that*
$$\max_{|z| \leq w} |a(z)| \leq A(w),$$
$$\max_{|z| \leq w} |b(z)| \leq B(w);$$
2. *The coefficients of the Taylor series for $a(x)$, $b(x)$ are majorant-computable as constant functions.*

Proof. Let f satisfy the conditions of the theorem. We shell prove that the epigraph of f and the ordinate set of f are Σ-definable . Denote the Taylor series for $a(z)$ and $b(z)$ at the point 0 as follows:

$$a(z) = \sum_{k \in \omega} a_k z^k, \quad b(z) = \sum_{k \in \omega} b_k z^k.$$

Show that $a|_{\mathbf{R}}$ and $b|_{\mathbf{R}}$ are majorant-computable. Let us consider arbitrary disk $|z| \leq R$.

From the Cauchy inequality [2], for all $k \in \omega$ we have

$$|a_k| \leq \frac{M}{R^k}, \quad \text{where } M = \max_{|z| \leq R} a(z).$$

We also obtain the estimates

$$\left| \sum_{k=s+1}^{\infty} a_k z^k \right| \leq \sum_{k=s+1}^{\infty} |a_k||z^k| \leq \sum_{k=s+1}^{\infty} \frac{M|z^k|}{R^k}.$$

Denote

$$a_s(z) = \sum_{k \leq s} a_k z^k.$$

Let us consider

$$R = 2|z|.$$

We have

$$|a(z) - a_s(z)| \leq \sum_{k=s+1}^{\infty} \frac{A(2|z|)|z^k|}{(2|z|)^k} \leq A(2|z|)\frac{1}{2^s}.$$

Let us construct upper and lower approximations for $a|_{\mathbf{R}}$

$$a_s^*(x) = \max_{j \leq s}\left(a_j(x) + \tfrac{A(2|x|)}{2^j}\right),$$
$$a_s^{**}(x) = \min_{j \leq s}\left(a_j(x) - \tfrac{A(2|x|)}{2^j}\right).$$

By construction, $\{a_s^*\}_{s\in\omega}$ is a sequence of lower computable approximations for $a|_{\mathbf{R}}$ and $\{a^{**}\}_{s\in\omega}$ is a sequence of upper computable approximations for $a|_{\mathbf{R}}$. Corollary 2 implies that $a|_{\mathbf{R}}$ is computable. Similarly, $b|_{\mathbf{R}}$ is computable. It easy to proof that if $a|_{\mathbf{R}}$ and $b|_{\mathbf{R}}$ are majorant computable, then $|a||_{\mathbf{R}}$ and $|b||_{\mathbf{R}}$ are majorant-computable. Let $\{f_s^1\}_{s\in\omega}$ be a sequence of lower Σ-approximations uniformly converging to $|a||_{\mathbf{R}}$ and let $\{g_s^1\}_{s\in\omega}$ be a sequence of upper Σ-approximations uniformly converging to $|a||_{\mathbf{R}}$. Let $\{f_s^2\}_{s\in\omega}$ be a sequence of lower Σ-approximations uniformly converging to $|b||_{\mathbf{R}}$ and let $\{g_s^2\}_{s\in\omega}$ be a sequence of upper Σ-approximations uniformly converging to $|b||_{\mathbf{R}}$.

Then the ordinate set of $|f|$ can be defined as follows:

$$y \leq \frac{|a(x)|}{|b(x)|} \leftrightarrow$$

$$\mathbf{HF}(\mathbf{R}) \models (\exists s \in \mathbf{N})(\exists n \in \mathbf{N})(\exists k \in \mathbf{N})\left(\left(y \leq \frac{n}{k}\right) \wedge \left(n \leq g_s^1(x) \wedge k \geq g_s^2(x)\right)\right) \leftrightarrow$$

$$\mathbf{HF}(\mathbf{R}) \models \Phi(x, y).$$

By Corollary 2 , the function $\mathrm{sign}(f)$ is Σ-definable, and, by construction, the ordinate set of $|f|$ is Σ-definable.

Then the ordinate set of f is Σ-definable as follows:

$$y \leq \frac{a(x)}{b(x)} \leftrightarrow$$

$$\mathbf{HF}(\mathbf{R}) \models \left(\exists z\left(\mathrm{sign}(f(x)) = -1\right) \wedge \Phi(x, z) \wedge \left(y < -z\right)\right)$$

$$\vee\left(\left(\mathrm{sign}(f(x)) = 1\right) \wedge \Phi(x, y)\right).$$

So, the ordinate set of f is a Σ-definable set. Similarly, the epigraph of f is Σ-definable. □

Corollary 2 implies that f is a majorant-computable function.

Corollary 7. *If an analytical function f has Taylor-series expansion with majorant-computable coefficients and the function $|f|$ is bounded by an Σ-definable total real function then f is a majorant-computable function.*

Corollary 8. *The functions* sin, cos, *and* exp *are majorant-computable.*

Corollary 9. *For real-valued functions, we have the following inclusions:*

1. *The class of Σ-definable functions is contained in the class of majorant-computable functions.*
2. *The class of computable functions introduced by Moschovakis is contained in the class of majorant-computable functions.*
3. *The class of functions definable by dimensional machines of Blum et al. is contained the class of majorant-computable functions.*
4. *The class of computable functions introduced by Pour-El and Richards is contained in the class of majorant-computable functions.*

4. M.-c.e. sets

Definition 9. Let B be a set. A function χ is said to be the partial characteristic function of B if

$$\chi(x) = \begin{cases} 1 & \text{if } x \in B, \\ \text{undefined} & \text{otherwise.} \end{cases}$$

Definition 10. Let $B \subset \mathbf{R}^n$. The set B is said to be **m.-c.e.** if the partial characteristic function of B is majorant-computable.

Proposition 9. *If f is majorant-computable, then* $\mathrm{dom}(f)$ *is a* **m.-c.e.** *set.*

Proof. Let $f : \mathbf{R}^n \to \mathbf{R}$ be majorant-computable. Then there exist a prime extension of $\mathbf{R}$, $\bar{\mathbf{R}}$, and a sequence of lower Σ-approximations for f, $\{f_s\}_{s\in\omega}$, and a sequence of upper Σ-approximations for f, $\{g_s\}_{s\in\omega}$.

Put

$$\chi_{1s}(\mathbf{x}) = 1 + \frac{g_s(\mathbf{x}) - f_s(\mathbf{x})}{2},$$

$$\chi_{2s}(\mathbf{x}) = 1 - \frac{g_s(\mathbf{x}) - f_s(\mathbf{x})}{2}.$$

Then $\{\chi_{1s}\}_{s\in\omega}$, $\{\chi_{2s}\}_{s\in\omega}$ are sequences of approximations for the partial characteristic function of $\mathrm{dom}(f)$. So, $\mathrm{dom}(f)$ is **m.-c.e.**

Proposition 10. *If $B \subset \mathbf{R}^n$ is Π-definable, then B is* **m.-c.e.**

Proof. Let B be Π-definable. Proposition 3 implies that there exists an effective sequence of sets $\{B_s\}_{i\in\omega}$ such that

1. $B = \bigcap_{s\in\omega} B_s$,
2. $B_1 \supseteq B_2 \supseteq \cdots \supseteq B_n \supseteq \cdots$,
3. for all $s \in \omega$, B_s is definable by a quantifier free formula.

Put

$$\chi_{1s}(\mathbf{x}) = \begin{cases} 1 & \text{if } \mathbf{x} \in B_s, \\ -t & \text{otherwise,} \end{cases}$$

$$\chi_{2s}(\mathbf{x}) = \begin{cases} 1 & \text{if } \mathbf{x} \in B_s, \\ t & \text{otherwise,} \end{cases}$$

where $t > \mathbf{N}$.

Then $\{\chi_{1s}\}_{s\in\omega}$, $\{\chi_{2s}\}_{s\in\omega}$ are sequences of approximations for the partial characteristic function of B. So, B is **m.-c.e.**

Corollary 10. *Mandelbrot set M and the family of Julia sets $\{J_c\}_{c\in M}$ are* **m.-c.e.** □

Proof. The claim follows from the fact that M and $\{J_c\}_{c\in M} = \{(x, c) \mid x \in J_c\}$ are Π-definable. □

Proposition 11. 1. *If $B \subset \mathbf{R}^n$ is Σ-definable, then B is majorant-computable.*

2. *If $B \subset \mathbf{R}^n$ and $\mathbf{R}^n \setminus B$ are majorant-computable, then B is Σ-definable.*

Proposition 12. *If $B \subset \mathbf{R}^n$ and $A \subset \mathbf{R}^n$ are majorant-computable, then the sets $B \cup A$, $B \cap A$ are majorant-computable.*

5. Majorant-computable functionals

The aim of the following is to transfer our approach on the case of real-valued functionals.

Let $\mathcal{F} = \{f \mid f$ is a majorant-computable real-valued total function$\}$. We use the notation Σ for the set of Σ-formulas and Π for the set of Π-formulas.

Definition 11. Let g_1 be Gödel numbering of a set A_1, g_2 be Gödel numbering of a set A_2. A procedure $h : A_1 \to A_2$ is said to be effective if there exists a recursive function ξ such that the following diagram is commutative

$$\begin{array}{ccc} N & \xrightarrow{\xi} & N \\ g_1 \downarrow & & \downarrow g_2 \\ A_1 & \xrightarrow{h} & A_2 \end{array}$$

Definition 12. A functional $F : \mathbf{R}^n \times \mathcal{F} \to \mathbf{R}$ is *majorant-computable* if there exist an elementary proper extension $\bar{\mathbf{R}}$ and sequences of total functionals $\{F_s\}_{s\in\omega}$ and $\{G_s\}_{s\in\omega}$ such that:

1. For every $s \in \omega$, $F_s : \mathbf{R}^n \times \Sigma \to \bar{\mathbf{R}}$; $G_s : \mathbf{R}^n \times \Sigma \to \bar{\mathbf{R}}$.
2. For $\langle \bar{c}, f\rangle \in \mathrm{dom}(F)$ and for each Σ-formula $\Phi(m, x, y)$ with the following conditions:

   a. For all $m \in \omega$ $\Phi(m, x, y)$ defines the total function $f_m : \mathbf{R} \to \mathbf{R}$,
   b. The sequence $\{f_s\}_{s\in\omega}$ uniformly converge to f, $|f_s(x) - f(x)| \leq \frac{1}{s}$,

   the following properties hold:

   1. The sequence $\{F_s(\mathbf{c}, \Phi)\}_{s\in\omega}$ monotonically increases, the sequence $\{G_s(\mathbf{c}, \Phi)\}_{s\in\omega}$ monotonically decreases;
   2. $F_m(\mathbf{c}, \Phi) \leq F(\mathbf{c}, f) \leq G_m(\mathbf{c}, \Phi)$;
   3. $F(\mathbf{c}, f) = y \leftrightarrow \lim_{s\to\infty} \mathrm{sp}(F_s(\mathbf{c}, \Phi)) = y$ and $\lim_{s\to\infty} \mathrm{sp}(G_s(\mathbf{c}, \Phi)) = y$.

Definition 13. Let $\bar{\mathbf{R}}$ be an elementary proper extension of $\mathbf{R}$. A formula Φ is said to *determine* the functional F in the model $\mathbf{HF}(\bar{\mathbf{R}})$ if the following sentence holds

$$F(\mathbf{c}, f) = y \leftrightarrow \left(\mathbf{HF}(\bar{\mathbf{R}}) \models \Phi(\mathbf{c}, y) \text{ and } \{\mathrm{sp}(z) \mid \Phi(\mathbf{c}, z)\} = \{y\}\right).$$

Theorem 3. *A functional $F : \mathbf{R}^n \times \mathcal{F} \to \mathbf{R}$ is majorant-computable if and only if there exists an effective procedure $h : \Sigma \to \Pi$ such that for each Σ-formula $\Phi(m, x, y)$ with the following conditions:*

1. *For all $m \in \omega$, $\Phi(m, x, y)$ defines the total function $f_m : \mathbf{R} \to \mathbf{R}$.*
2. *The sequence $\{f_s\}_{s\in\omega}$ uniformly converge to f and $|f_s(x) - f(x)| \leq \frac{1}{s}$,*

the equality $h(\Phi) = \chi(\mathbf{c}, y)$ holds, where $\chi(\mathbf{c}, y)$ determines the functional $F(\mathbf{c}, f)$.

The proof is analogues to that of Theorem 1.

Let us consider the following example: the functional $I : \mathbf{R}^3 \times \mathcal{F} \to \mathbf{R}$ defined in the following way:

$$I(a, b, c, f) = \int_a^b f(x)\,dx, \quad \text{where } |f(x)| \leq c.$$

Theorem 4. *The functional $I(a, b, c, f)$ is majorant-computable.*

Proof. Corollary 5 imply that a function $f \in \mathcal{F}$ is piecewise continuous and, by definition, it is bounded. So, the functional $I(a, b, c, f)$ is defined for every $f \in \mathcal{F}$.

Let us suppose $f \in \mathcal{F}$. Take a Σ-formula $\Phi(m, x, y)$ with the following conditions:

1. For all $m \in \omega$, $\Phi(m, x, y)$ defines the total function $f_m : \mathbf{R} \to \mathbf{R}$.
2. The sequence $\{f_s\}_{s\in\omega}$ uniformly converge to f and $|f_s(x) - f(x)| \leq \frac{1}{s}$.

Let us construct approximations $w(s)$ such that

$$|I(a, b, c, f) - w(s)| \leq \frac{1 + |b - a|}{s}.$$

The properties of Borel sets and Proposition 5 imply that for the function f_s we can find the system of mutually disjoint intervals $\{[a_1, b_1], \ldots, [a_k, b_k]\}$ such that

1. f_s is continuous on each interval $[a_i, b_i]$, $1 \leq i \leq k$;
2. $a_i \in \mathbf{R_0}$ and $b_i \in \mathbf{R_0}$ for $1 \leq i \leq k$;
3. $|b - a| - \sum_{i=1}^{k} |b_i - a_i| \leq \frac{1}{2cs}$;
4. $\left(\sup_{x\in[a_i,b_i]} f_s(x) - \inf_{x\in[a_i,b_i]} f_s(x)\right) \leq \frac{1}{2|b-a|s}$.

Put

$$w(s) = \sum_{i=1}^{k} \left(|b_i - a_i| f_s\left(\frac{a_i + b_i}{2}\right)\right).$$

Then

$$\left|I(a, b, c, f) - w(s)\right| \leq \frac{1}{2s} + \frac{1}{2s} + \frac{|b - a|}{s} \leq \frac{|b - a| + 1}{s}.$$

Put

$$g(s) = w(s) - (1 + |a - b|)/s$$

and

$$v(s) = w(s) + (1 + |a - b|)/s.$$

Then $g(s)$ is a lower approximation of $I(a, b, c, f)$ and Σ-definable, $v(s)$ is an upper approximation of $I(a, b, c, f)$ and Σ-definable. So, the sets $\{x \mid x \leq I(a, b, c, f)\}$ and $\{x \mid x > I(a, b, c, f)\}$ are Σ-definable. Hence we can effectively write all the decimal points of $I(a, b, c, f)$. Then we construct F_s by this decimal points (to write only integer part of I and its decimal places from 1-st to s-th). We can see that $I(a, b, f) - 10^{-s} < F_s(a, b, c, \Phi) < I(a, b, f)$.

Put $G_s = F_s + 10^{-s}$. By construction, $\{F_s\}_{s\in\omega}$, $\{G_s\}_{s\in\omega}$ are required, so, the functionals $I(a, b, c, f)$ is majorant-computable. The theorem is proved.

References

[1] J. Barwise, Admissible sets and structures, Springer-Verlag, Berlin, 1975.

[2] A. V. Bitsadze, The bases of the analytic functions theory of complex variable, Nauka, Moscow, 1972.

[3] L. Blum and M. Shub and S. Smale, On a theory of computation and complexity over the reals: NP-completeness, recursive functions and universal machines, Bull. Amer. Math. Soc. (N.S.) 21 (1989), 1–46.

[4] L. van den Dries, Remarks on Tarski's problem concerning ($\mathbf{R}, +, *, \exp$), in: Proc. Logic Colloquium '82, Stud. Logic Found. Math. 112, North-Holland, 1984, 240–266.

[5] Yu. L. Ershov, Definability and computability, Nauka, Novosibirsk, 1996.

[6] H. Freedman and K. Ko, Computational complexity of real functions, Theoret. Comput. Sci. 20 (1982), 323–352.

[7] C. Gordon, A comparison of abstract computability theories, Compositio Math. 22 (1970), 333–346.

[8] N. Jacobson, Basic Algebra II, W. E. Freedman and Company, New York, 1995.

[9] R. Montague, Recursion theory as a branch of model theory, in: Proc. of the third international Congress on Logic, Methodology and the Philososophy of Science 1967, Amsterdam, 1968, 63–86.

[10] Y. N. Moschovakis, Abstract first order computability, Trans. Amer. Math. Soc. 138 (1969), 427–464.

[11] M. Korovina, Generalized computability of real functions, Siberian Adv. Math. 2 (4) (1992), 85–102.

[12] M. Korovina, About a universal recursive function and abstract machines over the reals with the list extension, Vychisl. Sistemy 156 (1996), 3–20.

[13] M. Korovina, O. Kudinov, New approach to computability over the reals, Technical report, Uppsala Univ., February, 1997, 1–18.

[14] M. B. Pour-El, J. I. Richards, Computability and noncomputability in classical analysis, Trans. Amer. Math. Soc. 275 (1983), 539–560.

[15] A. Robinson, Nonstandard Analysis, Stud. Logic and Found. Math., North-Holland, 1966.

Hyperarithmetical functions and algebraicity

*A. S. Morozov**

Abstract. First we prove that the classes of recursive and hyperarithmetical models have the same automorphism groups and describe countable automorphism groups of hyperarithmetical models.

Then characterizations of hyperarithmetical functions and permutations via the notion of algebraicity are given. We introduce natural notions of algebraicity over the group of recursive permutations and over the semigroup of recursive functions and prove that any hyperarithmetical permutation is generated by algebraic permutations as well as any hyperarithmetical function is a product of algebraic elements over recursive functions.

1. Automorphism groups of hyperarithmetical models

We consider *effective languages* only, i.e., first-order languages whose symbols are naturally numbered in such a way that given such a number one can effectively find the number of arguments of the corresponding symbol. If we consider a model whose elements are natural numbers, we can naturally extend this numbering of our language to a numbering of this language extended by individual constants for each element of the model, so that given a number of a formula we can effectively write down this formula and vice versa. If the diagram of this model (i.e., the set of atomic sentences and their negations true in $\mathcal{M}$) is recursive (hyperarithmetical) then we call this model *recursive (hyperarithmetical)*.

In what follows, $\mathrm{Sym}(\omega)$ stands for the group of all permutations of ω; Fin denotes the group of all finitary permutations, i.e., permutations which nontrivially move only a finite number of elements. We also use the notation $\mathrm{st}(f) = \{x \in \omega \mid f(x) = x\}$ and call this set the *stabilizer of* f; the set $\mathrm{sp}(f) = \{x \in \omega \mid f(x) \neq x\}$ is called the *support of* f.

We fix a triple of recursive functions $c : \omega^2 \xrightarrow{1-1} \omega$, $\ell : \omega \to \omega$, and $r : \omega \to \omega$, which one-to-one enumerate pairs of natural numbers and satisfy the following identities: $c(\ell(x), r(x)) = x$, $\ell(c(x, y)) = x$, $r(c(x, y)) = y$. The cardinality of a set A is denoted by $\mathrm{card}(A)$.

*Partially supported by Grant 96-15-96878 of the Russian Foundation for Basic Research and by the Alexander von Humboldt Foundation.

Theorem 1.1. *Let G be a group. Then the following statements are equivalent:*

1. *G is isomorphic to the group of all automorphisms of some recursive model;*

2. *G is isomorphic to the group of all automorphisms of some hyperarithmetical model.*

Proof. The nontrivial part is 2) $\Rightarrow$ 1). Suppose that $\mathcal{M}$ is hyperarithmetical and $G \cong \operatorname{Aut} \mathcal{M}$. Without loss of generality we can assume that the signature of $\mathcal{M}$ consists of one binary predicate symbol and its universe is ω (using the standard interpretation, for instance in [1], one can transform a hyperarithmetical model into another hyperarithmetical model whose signature contains only one binary predicate and automorphism group is isomorphic to the automorphism group of the former model). In [2] it has been proven that for each hyperarithmetical set A there exist ordinals α and β, $\alpha \neq \beta$ and a computable family of binary relations $\{P_n\}_{n<\omega}$ on ω such that for all $n \in \omega$

$$n \in A \Rightarrow \langle \omega;\ P_n \rangle \cong \langle \alpha;\ < \rangle$$

and

$$n \notin A \Rightarrow \langle \omega;\ P_n \rangle \cong \langle \beta;\ < \rangle$$

hold.

Define a recursive model $\mathcal{N}$. First, take some α, β, and a computable family $\{S_{x,y}\}_{x,y<\omega}$ of binary relations so that

$$\langle x, y \rangle \in R \Rightarrow \left\langle \omega;\ S_{x,y} \right\rangle \cong \langle \alpha;\ < \rangle,$$

$$\langle x, y \rangle \notin R \Rightarrow \left\langle \omega;\ S_{x,y} \right\rangle \cong \langle \beta;\ < \rangle.$$

The universe of our model is the set $\omega \cup \omega^3$, whose elements are naturally associated with natural numbers. Signature predicates Q^3 and P^2 are defined as follows:

$$Q(x, y, z) \Leftrightarrow x, y \in \omega \ \&\ \exists t\ (z = \langle x, y, t \rangle),$$

$$P(x, y) \Leftrightarrow \exists t_0 \exists t_1 \exists x_0 \exists y_0 (x = \langle t_0, t_1, x_0 \rangle \ \&\ y = \langle t_0, t_1, y_0 \rangle \ \&\ S_{t_0,t_1}(x_0, y_0)).$$

One can easily see that this model is isomorphic to a recursive model. Note also that the universe of our model $\mathcal{N}$ is a disjoint union of sets ω and $\{u \mid Q(x, y, u)\}$, $x, y \in \omega$.

Prove that $\operatorname{Aut} \mathcal{M} \cong \operatorname{Aut} \mathcal{N}$. Take a $\varphi \in \operatorname{Aut} \mathcal{M}$. Define an automorphism $\overline{\varphi}$ of $\mathcal{N}$ as follows. For $x \in \omega \subseteq |\mathcal{N}|$ we let $\overline{\varphi}(x) = \varphi(x)$. If $\langle x, y, z \rangle \in \omega^3 \subseteq |\mathcal{N}|$ then one can see by definition that if $\langle x, y \rangle \in R$ then P defines an ordering $S_{x,y}$ of type α on the set $\{u \mid Q(x, y, u)\} = \{\langle x, y, t \rangle \mid t \in \omega\}$ and an ordering of type $\beta \neq \alpha$ otherwise. Since φ is an automorphism, the orderings $S_{x,y}$ and $S_{\varphi(x),\varphi(y)}$ are isomorphic. Thus, there exists a unique isomorphism $\psi_{x,y} : \left\langle \{u \mid Q(x, y, u)\};\ S_{x,y} \right\rangle \rightarrow$

$\langle\{u \mid Q(\varphi(x), \varphi(y), u)\}; S_{\varphi(x),\varphi(y)}\rangle$. Define now $\overline{\varphi}(\langle x, y, z\rangle) = \psi_{x,y}(\langle x, y, z\rangle)$. Immediate check shows that $\overline{\varphi}$ is one-to-one and preserves the predicates P and Q of $\mathcal{N}$, i.e., $\overline{\varphi} \in \mathrm{Aut}\,\mathcal{N}$. Note also that $\overline{\varphi_0\varphi_1} = \overline{\varphi}_0\,\overline{\varphi}_1$ holds for all $\varphi_0, \varphi_1 \in \mathrm{Aut}\,\mathcal{M}$, since the automorphisms $\overline{\varphi_1}$, $i = 0, 1$ act on $\omega \subseteq |\mathcal{N}|$ in the same way as on $\omega = |\mathcal{M}|$ and their actions are uniquely defined on the rest part of $|\mathcal{N}|$.

To complete the proof, it remains to ascertain that each automorphism of $\mathcal{N}$ has the form $\overline{\varphi}$, for an appropriate $\varphi \in \mathrm{Aut}\,\mathcal{M}$. We can define ω in $\mathcal{N}$ as $\{x \mid \exists y \exists z Q(x, y, z)\}$. Henceforth, each automorphism $\psi \in \mathrm{Aut}\,\mathcal{N}$ takes ω onto itself. Prove that $\varphi = \psi|_\omega$ is an automorphism of $\mathcal{M}$ with the property $\overline{\varphi} = \psi$. It suffices to understand that $\varphi \in \mathrm{Aut}\,\mathcal{M}$. Indeed, let $\mathcal{M} \models R(x, y)$. Then $\langle\{u \mid Q(x, y, u)\}; S_{x,y}\rangle \cong \langle\alpha; <\rangle$. By this, $\langle\{u \mid Q(\psi(x), \psi(y), u)\}; S_{\psi(x),\psi(y)}\rangle \cong \langle\alpha; <\rangle$, which yields $R(\varphi(x), \varphi(y))$. Considering the automorphism ψ^{-1}, we obtain the other implication, i.e., $\mathcal{M} \models R(x, y) \Leftrightarrow \mathcal{M} \models R(\varphi(x), \varphi(y))$, whence $\varphi \in \mathrm{Aut}\,\mathcal{M}$. □

Theorem 1.2. *Let G be a countable group. Then the following statements are equivalent:*

1. *G is isomorphic to the group of all automorphisms of some recursive model;*
2. *G is isomorphic to the group of all automorphisms of some hyperarithmetical model;*
3. *G is isomorphic to a hyperarithmetical group.*

Proof. 1) ⇔ 2) by Theorem 1.1.

3) ⇒ 2). We can assume that G is the set ω of natural numbers with a hyperarithmetical group operation. Then by [3], the model $\langle\omega; \ell_a\rangle_{a\in\omega}$, where $\ell_a(x) = a \cdot x$, is a hyperarithmetical model whose automorphism group is isomorphic to G.

2) ⇒ 3). In [4] it was proven that in this case $G \in \mathbf{HYP}_\omega$. We can assume G to be infinite. Since for each nonempty set $a \in \mathbf{HYP}_\omega$ there exists a function $f \in \mathbf{HYP}_\omega$, $f : \omega \to \omega$ such that $f(\omega) = a$, there exists an operation $\circ : \omega^2 \to \omega$ in $\mathbf{HYP}_\omega$ such that $\langle\omega; \circ\rangle$ is a group isomorphic to G. By [5], the operation $\circ$ is Δ^1_1 on ω, and thus G is hyperarithmetical. This completes the proof. □

2. The group of all hyperarithmetical permutations is generated by algebraic elements

Let $\langle \mathrm{Sym}(\omega), \ldots\rangle$ be an arbitrary many-sorted structure whose first universe is $\mathrm{Sym}(\omega)$. The *restricted language* of this structure is defined as the usual first order many-sorted language with the only exception, that for $\mathrm{Sym}(\omega)$ the quantification over finitary permutations is allowed only, while there are no restrictions for the other sorts. We denote such quantifiers as $\exists f \in \mathrm{Fin}$ and $\forall f \in \mathrm{Fin}$. Formulas of restricted languages are called *restricted formulas.*

We define

$$\mathcal{M}_0 = \langle \mathrm{Sym}(\omega), \circ, {}^{-1}, 1 \rangle,$$

$$\mathcal{M}_1 = \langle \mathrm{Sym}(\omega), \omega; \circ, {}^{-1}, 1, \mathrm{ap} \rangle,$$

where $\circ$ is the usual composition, 1 stands for id_ω, $\mathrm{ap} : \mathrm{Sym}(\omega) \times \omega \to \omega$, and $\mathrm{ap}(f, x) = f(x)$.

A permutation $\tau \in \mathrm{Sym}(\omega)$ is called a *transposition* provided that it moves exactly two elements. If π and τ are two transpositions with $[\pi, \tau] \neq 1$ then there exists a unique element x which is moved by both of them. We denote this element by $\alpha(\pi, \tau)$. It has been known that transpositions in $\mathrm{Sym}(\omega)$ are distinguished by the formula $\neg(x = 1) \mathbin{\&} (x^2 = 1) \mathbin{\&} \forall y([x, y]^6 = 1)$ (see [6]). Carefully examining the proof of this result in [6], one can easily ascertain that this formula can be also replaced by the restricted formula $\mathrm{tr}(x) \rightleftharpoons \neg(x = 1) \mathbin{\&} (x^2 = 1) \mathbin{\&} \forall y \in \mathrm{Fin}([x, y]^6 = 1)$.

Assume that $A \subseteq \mathrm{Sym}(\omega)$. An element $f \in \mathrm{Sym}(\omega)$ is called *algebraic over* A provided that there exists a restricted formula $\varphi(x, \overline{y})$ and a tuple $\overline{a}$ of elements of A such that $\mathrm{Sym}(\omega) \models \varphi(f, \overline{a})$ and $\mathrm{Sym}(\omega) \models \exists! x \varphi(x, \overline{a})$. As it will be easily seen below, in our case the latter is equivalent to the existence of a formula $\varphi(x, \overline{y})$ together with a tuple $\overline{a}$ of elements of A such that the set $\{x \mid \mathrm{Sym}(\omega) \models \varphi(x, \overline{a})\}$ is finite.

Theorem 2.1. *The group of all hyperarithmetical permutations is generated by permutations which are algebraic over recursive permutations in* $\mathrm{Sym}(\omega)$.

Proof. We need several lemmas.

Lemma 2.2. *Assume* $\varphi(x_1, \dots, x_m, y_1, \dots, y_n)$ *is a formula in the restricted language of* $\mathcal{M}_1$ *in which* $x_1, \dots, x_m$ *have type* $\mathrm{Sym}(\omega)$ *and* $y_1, \dots, y_n$ *have type* ω. *Then there exists a formula* $\overline{\varphi}(x_1, \dots, x_m, y_1', y_1'', \dots, y_n', y_n'')$ *in the restricted language of* $\mathcal{M}_0$ *such that for all* $f_1, \dots, f_m \in \mathrm{Sym}(\omega)$ *and all pairs of transpositions* $\langle \tau_1', \tau_1'' \rangle, \dots, \langle \tau_n', \tau_n'' \rangle$ *such that* $[\tau_i', \tau_i''] \neq 1$ *for* $i = 1, \dots, n$ *the equivalence*

$$\mathcal{M}_0 \models \overline{\varphi}(x_1, \dots, x_m, \tau_1', \tau_1'', \dots, \tau_n', \tau_n'') \Leftrightarrow$$
$$\mathcal{M}_1 \models \varphi(x_1, \dots, x_m, \alpha(\tau_1', \tau_1''), \dots, \alpha(\tau_n', \tau_n''))$$

holds.

Proof. We use induction on the complexity of a formula. First we reduce our consideration to some class of atomic formulas. An atomic formula in $\mathcal{M}_1$ can be of one of the two kinds: $t_1 = t_2$, where t_1 and t_2 have group type, or $t_3 = t_4$, where t_3 and t_4 are of type ω. We will simplify the second case. In this case $t_3 = t_4 \Leftrightarrow \exists n(t_3 = n \mathbin{\&} t_4 = n)$. If t_3 or t_4 have the form $\mathrm{ap}(u, v)$ and u is a nontrivial term of kind $u_0 u_1$, then we replace the corresponding subformulas $\mathrm{ap}(u_0 u_1, v) = n$

with $\exists m(\mathrm{ap}(u_1, v) = m \ \&\ \mathrm{ap}(u_0, m) = n)$ until all atomic subformulas will have the form $\mathrm{ap}(w, v) = n$ or $\mathrm{ap}(w^{-1}, v) = n$, where w is a term of length 1. The latter can be replaced by $\exists k(\mathrm{ap}(w, n) = k \ \&\ v = k)$ and so on, until all atomic subformulas will be of kind $m = n$ or $\mathrm{ap}(q, v) = n$, where q and n are terms of length 1. Next, we replace all atomic subformulas of kind $\mathrm{ap}(u, \mathrm{ap}(v, w)) = n$ by $\exists l(\mathrm{ap}(v, w) = l \ \&\ \mathrm{ap}(u, l) = n)$, etc., until we arrive at a formula whose all atomic subformulas are of kind $t_1 = t_2$, where t_1 and t_2 are group terms, or of kind $m = n$, where m and n are variables of type ω, or $\mathrm{ap}(u, v) = w$, where u, v, w are terms of length 1. Note that after all these transformations restricted formulas remain restricted.

By the above, we may assume our formula to contain only such simple atomic subformulas. Prove the lemma for these atomic formulas.

If φ asserts the equality of two group terms then it can itself serve as $\overline{\varphi}$.

If φ is the equality of two variables for natural numbers of kind $a = b$ then $\overline{\varphi}(a', a'', b', b'')$ is $E(a', a'', b', b'')$, where $E(p_0, p_1, q_0, q_1)$ is the formula

$$\textstyle\bigwedge_{i,j=0,1}(p_i q_j)^3 = 1 \ \&$$
$$\textstyle\bigwedge_{i,j=0,1}\big((p_i = q_j \ \&\ p_{1-i} \neq q_{1-j}) \Rightarrow p_{1-i} \neq p_i^{-1} q_{1-j} p_i\big),$$

which expresses the fact that the common moved elements of the pairs of transpositions $\langle p_0, p_1\rangle$ and $\langle q_0, q_1\rangle$ coincide, for all pairs of transpositions $\langle p_0, p_1\rangle$, $\langle q_0, q_1\rangle$ such that $[p_0, p_1] \neq 1 \neq [q_0, q_1]$; this can be verified by careful examining of all possible cases.

If φ is $\mathrm{ap}(v, a) = b$ then the formula $\overline{\varphi}(v, a', a'', b', b'')$ can be taken in the form $E(va'v^{-1}, va''v^{-1}, b', b'')$.

The cases of propositional connectives and restricted quantifiers are trivial.

Consider the case when φ is $\exists z \psi(x_1, \ldots, x_m, z, y_1, \ldots, y_n)$, i.e., the existential quantifier over natural numbers. Here we can take $\overline{\varphi}$ in the form

$$\exists \tau_0 \in \mathrm{Fin}\, \exists \tau_1 \in \mathrm{Fin}(\mathrm{tr}(\tau_0) \ \&\ \mathrm{tr}(\tau_1) \ \&\ [\tau_0, \tau_1] \neq 1 \ \&$$
$$\overline{\psi}(x_1, \ldots, x_m, \tau_0, \tau_1, y_1', y_1'', \ldots, y_n', y_n'')).$$

The case of universal quantifier is handled similarly.

The proof of the lemma is complete. □

Define a further extension of the initial model $\mathcal{M}_0$:

$$\mathcal{M}_2 = \langle \mathrm{Sym}(\omega), \omega, \mathcal{F}; \circ, {}^{-1}, 1, \mathrm{ap}, +, \cdot, \in \rangle,$$

where $\mathcal{F}$ is the set of all finite subsets of ω, $+$ and $\cdot$ are the usual addition and multiplication on natural numbers, and $\in$ is the membership relation on $\omega \times \mathcal{F}$.

In what follows, we will use some natural abbreviations like $x \in \mathrm{sp}(f)$ for $\mathrm{ap}(f, x) \neq x$, $x \in \mathrm{st}(f)$ for $\mathrm{ap}(f, x) = x$, etc. All these notations can be easily rewritten as usual (restricted) formulas.

Lemma 2.3. *Assume that $\varphi(x_1, \dots, x_m, y_1, \dots, y_n, z_1, \dots, z_k)$ is a restricted formula in the language of $\mathcal{M}_2$ in which $x_1, \dots, x_m$ have type* Sym(ω)*, $y_1, \dots y_n$ have type ω, and $z_1, \dots, z_k$ have type $\mathcal{F}$. Then there exists a restricted formula $\overline{\varphi}(x_1, \dots, x_m, y_1, \dots, y_n, z'_1, z''_1, \dots, z'_k, z''_k, u_0, u_1)$ in the language of $\mathcal{M}_1$ such that for all $f_1, \dots, f_m \in \mathrm{Sym}(\omega)$, $l_1, \dots, l_n \in \omega$, $h'_1, \dots, h'_k, h''_1, \dots, h''_k \in \mathrm{Fin}$*

$$\mathcal{M}_1 \models \overline{\varphi}(f_1, \dots, f_m, l_1, \dots l_n, h'_1, h''_1, \dots h'_k, h''_k, g_0, g_1) \Leftrightarrow$$

$$\mathcal{M}_2 \models \varphi(f_1, \dots, f_m, l_1, \dots l_n, \mathrm{sp}(h'_1) \cap \mathrm{sp}(h''_1), \dots, \mathrm{sp}(h'_k) \cap \mathrm{sp}(h''_k))$$

holds, where $g_0 = \prod_{i<\omega}(2i, 2i+1)$ and $g_1 = \prod_{i<\omega}(2i+1, 2i+2)$.

Proof. We outline the basic moments of the proof only. As before, we can restrict ourselves to the atomic subformulas of the following kinds:

1. $n \in F$, where n has type ω and F has type $\mathcal{F}$;

2. $x + y = z$, where x, y, z have type ω;

3. $x \cdot y = z$, where x, y, z have type ω;

4. $F_1 = F_2$, where F_1 and F_2 have type $\mathcal{F}$;

5. $\mathrm{ap}(x, y) = z$, where y and z have type ω and x has type Sym(ω);

6. $t_1 = t_2$, where t_1 and t_2 are group terms.

We consider here the nontrivial cases for items 1–4 only.

If φ is $n \in F$ then the formula $n \notin \mathrm{sp}(h')$ & $n \notin \mathrm{sp}(h'')$ can be taken as $\overline{\varphi}$.

Assume that φ is $x + y = z$. We define some auxiliary formulas:

$$\begin{aligned} x < y \rightleftharpoons\ & \exists h \in \mathrm{Fin}[\exists n_0(u_1(n_0) = n_0 \ \&\ n_0 \in \mathrm{sp}(h)) \ \& \\ & \exists! z(z \in \mathrm{sp}(h) \ \&\ (u_0(z) \notin \mathrm{sp}(h) \vee u_1(z) \notin \mathrm{sp}(h))) \ \& \\ & x \in \mathrm{sp}(h) \ \&\ y \notin \mathrm{sp}(h)], \end{aligned}$$

$$s(x, y) \rightleftharpoons x < y \ \&\ \forall z \neg(x < z < y).$$

Then

$$\begin{aligned} x + y = z \rightleftharpoons\ & \exists h \in \mathrm{Fin}\, \exists n_0(u_1(n_0) = n_0 \ \&\ h(n_0) = x \ \& \\ & \forall t < y \forall t_1(s(t, t_1) \Rightarrow s(h(t), h(t_1))) \ \&\ h(y) = z), \end{aligned}$$

and the right-hand formula can be taken as $\overline{\varphi}$.

If φ is $x \cdot y = z$ then the following formula can be taken as $\overline{\varphi}$ (we assume that the subformula with "+" is replaced with appropriate formula from above):

$$\begin{aligned} x \cdot y = z \rightleftharpoons \exists h \in \mathrm{Fin}\, \exists n_0 (u_1(n_0) = n_0 \;\&\; h(n_0) = n_0 \;\&\; \\ \forall t < y \forall t_1 (s(t, t_1) \Rightarrow h(t) + x = h(t_1)) \;\&\; \\ h(y) = z). \end{aligned}$$

If φ is the equality of finite sets $F_1 = F_2$ then $\overline{\varphi}$ is $\forall n((n \in \mathrm{sp}(h_1') \;\&\; n \in \mathrm{sp}(h_1'')) \Leftrightarrow (n \in \mathrm{sp}(h_2') \;\&\; n \in \mathrm{sp}(h_2'')))$.

If φ is $\exists F \psi(F, \dots)$ then $\overline{\varphi}$ is $\exists z_1 \in \mathrm{Fin}\, \exists z_2 \in \mathrm{Fin}\, \overline{\psi}(z_1, z_2, \dots)$.

The rest of the proof is clear. □

These two lemmas define two translations of the above mentioned languages. We will use the composition of these translations, i.e., in order to prove that some property is expressible in the language of groups with parameters, we will express it in the language of $\mathcal{M}_2$ by a formula with group parameters.

Lemma 2.4. *Assume that* $g_0, g_1 \in A \subseteq \mathrm{Sym}(\omega)$ *and* $f \in \mathrm{Sym}(\omega)$. *Then* f *is algebraic over* A *if and only if* $\{f\}$ *is an arithmetical class over* A.

Proof. Assume that f is algebraic over A and is the unique solution for a restricted formula $\psi(f, \overline{a})$ with parameters $\overline{a}$ in A. Fix some recursive function $d(e, x)$ so that $\mathrm{Fin} = \{\lambda x.d(e, x) \mid e < \omega\}$. It suffices to prove that each restricted formula $\varphi(x_1, \dots, x_m, f_1, \dots, f_n)$ in the language of $\mathcal{M}_1$ can be translated into a formula $\overline{\varphi}(x_1, \dots, x_m)$ of the language of arithmetics with additional unary functional symbols $f_1, \dots, f_n$ so that

$$\mathcal{M}_1 \models \varphi(k_1, \dots, k_m, f_1, \dots, f_n) \Leftrightarrow \langle \omega; +, \cdot, f_1, \dots, f_n \rangle \models \overline{\varphi}(k_1, \dots, k_m)$$

holds for all $k_1, \dots, k_m \in \omega$ and $f_1, \dots, f_n \in \mathrm{Sym}(\omega)$. To make it true for all functions $f : \omega \to \omega$, it remains only to add a new conjunctive member $\forall u \exists! v(\mathrm{ap}(f, v) = u)$ to the formula $\psi(f, \overline{a})$; this remark will be needed in the next section.

We can assume without loss of generality that all atomic subformulas of φ are of one of the two kinds: $a = b$ and $\mathrm{ap}(h, a) = b$, where a and b are variables of type ω and h has length 1 (equalities of group terms $t_1 = t_2$ can be replaced by formulas $\forall n \exists m(\mathrm{ap}(t_1, n) = m \;\&\; \mathrm{ap}(t_2, n) = m)$ and then, as in the previous lemma, we can reduce formulas $\mathrm{ap}(t_i, n)$, $i = 1, 2$ to one of the above mentioned forms).

If φ is atomic and has the form $a = b$, then it itself can be used as $\overline{\varphi}$. If φ is $\mathrm{ap}(h, a)$ then let $\overline{\varphi}$ be $h(a) = b$. The inductive steps related to propositional connectives and quantifiers over ω are trivial. It remains to consider quantifiers over Fin. Suppose that φ is $\exists h \in \mathrm{Fin}\, \psi(h, \overline{x}, \overline{f})$. Then by induction

$$\mathcal{M}_2 \models \exists h \in \mathrm{Fin}\, \psi(h, \overline{x}, \overline{f}) \Leftrightarrow \big(\langle \omega; +, \cdot, h, \overline{f} \rangle \models \overline{\psi}(\overline{x}) \text{ for some } h \in \mathrm{Fin}\big).$$

We assume as above that all atomic subformulas of $\overline{\psi}$ containing h are of kind $h(a) = b$, where a and b are terms of length 1. The latter is equivalent to $\langle \omega; +, \cdot, \overline{f} \rangle \models$

$\exists e\psi^*(\overline{x})$, where ψ^* is obtained from $\overline{\psi}$ by replacing all subformulas of kind $h(a) = b$ by arithmetical formulas which assert that $d(e, a) = b$.

Now suppose that $\{f\}$ is an arithmetical class over A. Then there exists a formula $\varphi(h)$ in the arithmetical language extended by a unary functional symbol such that

$$\{f\} = \{h \mid \langle \omega; +, \cdot, h \rangle \models \varphi(h)\}.$$

We have

$$\langle \omega; +, \cdot, h \rangle \models \varphi(h) \Leftrightarrow \mathcal{M}_2 \models \varphi(h).$$

Now it remains to use Lemmas 2.2 and 2.3. □

This lemma implies that all algebraic permutations are hyperarithmetical as well as their products and inverses. On the other hand, it explains why not all hyperarithmetical permutations are algebraic. This is so, because there exists a hyperarithmetical function which fails to form an arithmetical class, see [7, p. 452, exercise 16–72]; this was proven by P. Cohen and S. Feferman.

Suppose that $A, S \subseteq \omega$ and $\mathrm{card}(A) = \mathrm{card}(S) = \omega$. We define a permutation $p(A, S)$ as follows. Assume that $S = \{s_0 < s_1 < \cdots\}$. The permutation $p(A, S)$ will be the union of the infinite chain of permutations p_i, $i < \omega$:

$$p_0 \subseteq p_1 \subseteq p_2 \subseteq \cdots \bigcup_{i \in \omega} p_i = p(A, S).$$

We let $p_0 = \mathrm{id}_{\omega \setminus A}$. Suppose that p_i is already defined. Take the first $s_i + 3$ elements in the set $A \setminus \delta p_i$ and denote them as $a_0 < \cdots < a_{s_i} < a_{s_i+1} < a_{s_i+2}$. Now we define the permutation p_{i+1}:

$$p_{i+1}(x) = \begin{cases} p_i(x), & \text{if } x \in \delta p_i \\ a_{k+1}, & \text{if } x = a_k \text{ and } k < s_{i+2} \\ a_0, & \text{if } x = a_{s_i+2}. \end{cases}$$

Finally, let $p(A, S) = \bigcup_{i \in \omega} p_i$.

Recall that a set S is called an *arithmetical singleton* provided that the class $\{S\}$ is arithmetical. It is known that hyperarithmetical functions are exactly the functions that are m-reducible to some arithmetical singleton [7].

If π is a permutation, we denote its *graph* $\{c(x, y) \mid \pi(x) = y\}$ by Γ_π.

Lemma 2.5. *Let $S \subseteq \omega$ be an infinite arithmetical singleton, $\pi \in \mathrm{Sym}(\omega)$, $\pi^2 = 1$, $\Gamma_\pi \leq_m S$, and let* $\mathrm{card}(\mathrm{st}(\pi)) = \omega$. *Then the permutation $j(\pi, S) = \pi \circ p(\mathrm{st}(\pi), S)$ is an algebraic element over $\{g_0, g_1\}$.*

Proof. Take a recursive function g so that $x \in \Gamma_\pi \Leftrightarrow g(x) \in S$. The permutation $f = j(\pi, S)$ is uniquely defined by the conjunction of the following family of conditions:

1. $\forall x (f(x) \neq x)$;

2. each $n \in \omega$ is a member of some nontrivial finite cycle of f;

3. f contains a cycle of cardinality $n + 3$ if and only if $n \in S$;

4. if we define a $\tau \in \mathrm{Sym}(\omega)$ as

$$\tau(x) = y \Leftrightarrow g(c(x, y)) \in \{n \mid f \textit{ has a cycle of cardinality } n + 3\}$$

   then $f^2(m) = m$ implies $\tau(m) = f(m)$ and $f^2(m) \neq m$ implies $\tau(m) = m$, for all $m \in \omega$;

5. if $f^2(n) \neq n$ then all m that belong to an f-cycle of cardinality greater than 2 and less than that of the cycle for n, satisfy $m < n$;

6. all cycles of f of cardinality greater than 2 have distinct numbers of elements;

7. if F is a finite cycle of f with $n \in F$, then all its elements m except for the greatest one have the property $f(m) > m$.

We prove now that all these properties can be expressed by first order formulas over $\mathcal{M}_0$ with parameters g_0, g_1, which yields the lemma. Since there exists a translation from the language of $\mathcal{M}_2$ into the language of $\mathcal{M}_0$, it suffices to write down these statements in the extended language. The following formula expresses the property that $F \in \mathcal{F}$ is a finite cycle of f:

$$f(F) = F \;\&\; \exists m(m \in F) \;\&\; \forall F' \subseteq F\ (f(F') = F' \Rightarrow F' = \emptyset \vee F' = F).$$

In what follows, this formula will be referred to as $\mathrm{FC}(F, f)$.

In the definition of this translation we follow the enumeration above.

1. In fact, we have already written it.

2. For instance, $\forall n(f(n) \neq n \;\&\; \exists F\ (\mathrm{FC}(F, f) \;\&\; n \in F))$.

3. Suppose that $\theta(X)$ is a formula of arithmetics with added unary predicate X that is true only on S. Without loss of generality we can assume that all occurrences of X have the form $X(x)$, where x is a variable. To obtain the required formula, replace all occurrences of $X(x)$ by the following formula, which states that the set $\{1, \ldots, x, x + 1, x + 2, x + 3\}$ and some cycle of f have the same cardinalities:

$$\exists F \exists h \in \mathrm{Fin}(\mathrm{FC}(F, f) \;\&\; h(\{z \in \omega \mid 1 \leq z \leq x + 3\}) = F). \qquad (1)$$

   Denote the formula (1) by $C(x, f)$.

4. This can be written down as

$$\forall m[(f^2(m) = m \Rightarrow \exists n(C(g(c(m, n))) \;\&\; n = f(m)))$$
$$\& \;(f^2(m) \neq m \Rightarrow C(g(c(m, m))))].$$

Recall that recursive functions can be expressed by formulas of arithmetics. By this, the last formula is equivalent to a formula of the extended language.

5. This can be expressed by a formula as follows:

$$\forall m \forall n (f^2(m) \neq m \ \& \ f^2(n) \neq n \ \&$$
$$\exists F_0 \exists F_1 (\mathrm{FC}(F_0, f) \ \& \ \mathrm{FC}(F_1, f) \ \& \ F_0 \neq F_1 \ \&$$
$$m \in F_0 \ \& \ n \in F_1 \ \& \ \exists h \in \mathrm{Fin}(h(F_0) \subseteq F_1)) \Rightarrow m < n).$$

6. Note that this property follows from the previous formula, because if there were two distinct cycles of the same cardinality greater than 2, then taking their elements m and n we obtain by the above $m < n$ and $n < m$ at once, which is impossible.

7. We merely rewrite what was in this item:

$$\forall n \forall F [f^2(n) \neq n \ \& \ n \in F \ \& \ \mathrm{FC}(F, f) \ \& \ \exists m \in F(m > n) \Rightarrow f(n) > n].$$

The lemma is proven. □

Lemma 2.6. *Assume that $A \subseteq \omega$ is an infinite hyperarithmetical set with infinite complement and S is an arithmetical singleton with $A \leq_m S$. Then the permutation $p(\omega \setminus A, S)$ is an algebraic element over $\{g_0, g_1\}$.*

Proof. Take a recursive function g so that $x \in A \Leftrightarrow g(x) \in S$. The permutation $f = p(\omega \setminus A, S)$ is uniquely defined by the conjunction of the following conditions:

1. each $n \in \omega$ is a member of a finite (maybe trivial) cycle of f;

2. f does not contain cycles of cardinality 2;

3. $\{n \mid f$ contains a cycle of cardinality $n + 3\} = S$;

4. $\mathrm{st}(f) = g^{-1}\{n \mid f$ contains a cycle of cardinality $n + 3\}$;

5. if $f(n) \neq n$ then all m in a nontrivial f-cycle of cardinality less than that of the cycle containing n, satisfy $m < n$;

6. if numbers m and n are in distinct nontrivial cycles of f then the cardinalities of these cycles are distinct;

7. if $n \in \omega$ is contained in a nontrivial cycle for f then all elements m in this cycle except for the maximal one, satisfy $f(m) > m$.

To prove the lemma it suffices to express these properties in the language of $\mathcal{M}_2$. In the proof we use the formulas $\mathrm{FC}(F, f)$ and $C(x, f)$ from the proof of Lemma 2.5.

Thus:

1. $\forall n \exists F(\mathrm{FC}(F, f) \;\&\; n \in F)$;

2. $\forall n(f^2(n) = n \Rightarrow f(n) = n)$;

3. see the proof of Lemma 2.5;

4. $\forall n(f(n) = n \Leftrightarrow C(g(n), f))$.

The remaining items are handled just as in the proof of Lemma 2.5. □

Lemma 2.7. *Suppose that h is a hyperarithmetical permutation which has the properties* $\mathrm{card}(\mathrm{st}(h)) = \mathrm{card}(\mathrm{sp}(h)) = \omega$ *and $h^2 = 1$. Then h is a product of two algebraic elements over $\{g_0, g_1\}$.*

Proof. Inasmuch as h is hyperarithmetical, there exists an arithmetical singleton S such that $\Gamma_h \leq_m S$ and $\mathrm{st}(h) \leq_m S$ [8]. Then by Lemmas 2.5 and 2.6, the permutations $j(h, S)$ and $p(\mathrm{st}(h), S)$ are algebraic over $\{g_0, g_1\}$ and immediate check shows that $h = j(h, S) \circ p(\mathrm{st}(h), S)^{-1}$.

It remains to note that for all $A \subseteq \mathrm{Sym}(\omega)$ and all $f \in \mathrm{Sym}(\omega)$, f is algebraic over A iff f^{-1} is algebraic over A. □

Now we prove the theorem. Take an arbitrary hyperarithmetical permutation g and consider an arbitrary recursive permutation e for which the sets $\mathrm{sp}(e)$ and $\mathrm{st}(e)$ are infinite and $e^2 = 1$.

C. F. Kent proved in [9] that any *effectively closed group of permutations* (i.e., a group that together with any finite number of permutations contains all permutations recursive in them) has no nontrivial normal subgroups which contain a non-finitary permutation.

The group of all hyperarithmetical permutations is evidently effectively closed. By this, e has a finite number of hyperarithmetical conjugates whose product is g. By Lemma 2.7, each of them is a product of algebraic elements over $\{g_0, g_1\}$. The theorem is proven. □

Remarks. 1. We see that the theorem remains true if we replace the group of all recursive permutations by the set $\{g_0, g_1\}$, where g_0 and g_1 are defined as in Lemma 2.3.

2. The minimal number of parameters needed is two: if we consider only one parameter a, all algebraic permutations f defined with this parameter must be invariant under all automorphisms of kind $f \mapsto h^{-1} f h$, where h satisfies $a = h^{-1} a h$. It is clear that this is always a nontrivial family of automorphisms, while even the set $\{h \mid h^{-1} g_0 h = g_0 \;\&\; h^{-1} g_1 h = g_1\}$ consists of the identity permutation only.

3. Hyperarithmetical functions are generated by algebraic elements

In this section we prove a similar result for hyperarithmetical functions. Here we consider the semigroup $\mathcal{W}_0 = \langle \omega^\omega, \circ, 1\rangle$ whose universe is the set ω^ω of all functions from ω to ω, $\circ$ is the usual composition, and 1 is the identity mapping.

We define the sets $\mathrm{sp}(f)$ and $\mathrm{st}(f)$ for a function $f : \omega \to \omega$ formally just as above. A function f is called *finitary* provided that the set $\mathrm{sp}(f)$ is finite. The family of all finitary functions is denoted by Ff.

Restricted quantifiers and restricted languages are understood here in a similar way, i.e., the *restricted language* of a many-sorted structure of kind $\langle \omega^\omega; \circ, 1, \ldots\rangle$ is defined as the usual first order many-sorted language with the only exception, that for ω^ω the quantification is allowed over finitary functions only, while there are no any restrictions for the other sorts.

We could use quantifiers over finitary permutations as well, since they can be easily expressed via quantifiers over finitary functions. On the other hand, note, that in the forthcoming proof we need the quantifiers over finitary permutations only; we need this to use some previous results.

The text of the definition of *algebraic elements* formally remains the same, we have replaced the base model and the notion of restricted formula only.

Our main result here is the following theorem.

Theorem 3.1. *The semigroup of all hyperarithmetical functions is generated by algebraic elements over recursive functions in* ω^ω.

Proof. We give here a sketch proof. The first thing we have to do is the interpretation of the action of functions on the elements. Here we naturally associate elements of ω with constant functions. Unfortunately, these functions are not in Ff. So we use a recursive function $\mathbf{0}(x) \equiv 0$ as a new parameter, and all the other natural numbers are associated with finitary permutations as follows: a number n is associated with all finitary permutations g such that $g(0) = n$.

Define $\mathcal{W}_1 \rightleftharpoons \langle \omega^\omega, \omega; \circ, 1, \mathrm{ap}\rangle$, where $\mathrm{ap} : \omega^\omega \times \omega \to \omega$ and $\mathrm{ap}(f, n) = f(n)$, like above.

Lemma 3.2. *Assume that* $\varphi(x_1, \ldots, x_m, y_1, \ldots, y_n)$ *is a formula in the restricted language of* $\mathcal{W}_1$ *in which* $x_1, \ldots, x_m$ *have type* ω^ω *and* $y_1, \ldots, y_n$ *have type* ω. *Then there exists a formula* $\overline{\varphi}(x_1, \ldots, x_m, y_1, \ldots, y_n, z)$ *in the restricted language of* $\mathcal{W}_0$ *such that for all* $f_1, \ldots, f_m, h_1, \ldots, h_n \in \omega^\omega$

$$\mathcal{W}_0 \models \overline{\varphi}(f_1, \ldots, f_m, h_1, \ldots, h_n, \mathbf{0}) \Leftrightarrow \mathcal{W}_1 \models \varphi(f_1, \ldots, f_m, h_1(0), \ldots, h_n(0))$$

holds.

Sketch of the proof. Without loss of generality, we assume that all the atomic subformulas of our formulas have one of the following forms:

1. $t_1 = t_2$, where t_1 and t_2 are semigroup terms;

2. $x = y$, where x and y are variables of type ω;

3. $\mathrm{ap}(f, x) = y$, where f is a semigroup term, and x, y are variables of type ω.

The transformation $\varphi \mapsto \overline{\varphi}$ is defined by induction as follows:
$\overline{x = y} \rightleftharpoons (x \circ z = y \circ z)$, if x and y are of type ω;
$\overline{\mathrm{ap}(f, x) = y} \rightleftharpoons (f \circ x \circ z = y \circ z)$;
$\overline{t_1 = t_2} \rightleftharpoons (t_1 = t_2)$, if t_1 and t_2 are semigroup terms;
The cases of restricted quantifiers and propositional connectives are trivial;
Consider now the case when φ has the form $\exists x \psi(x_1, \dots x_m, x, y_1, \dots y_n)$. In this case we define $\overline{\varphi}$ as $\exists x \in \mathrm{Fin}\, \overline{\psi}(x_1, \dots x_m, x, y_1, \dots y_n, z)$.
The proof of the lemma is complete. □

Define now a new model $\mathcal{W}_2 \rightleftharpoons \langle \omega^\omega, \omega, \mathcal{F}; \circ, ^{-1}, 1, \mathrm{ap}, +, \cdot, \in \rangle$, where $\mathcal{F}$ is the family of all finite subsets of ω as above, $+$ and $\cdot$ are usual arithmetical operations, and $\in$ is the usual membership relation on $\omega \times \mathcal{F}$.

With the use of the same tricks as in the previous section, we can prove the following lemma.

Lemma 3.3. *Assume that $\varphi(x_1, \dots x_n, y_1, \dots, y_n, z_1, \dots, z_k)$ is a formula in the language of $\mathcal{W}_2$ whose variables $x_1, \dots x_n$ have type ω^ω, $y_1, \dots, y_n$ have type ω, and $z_1, \dots, z_k$ have type $\mathcal{F}$. Then there exists a restricted formula*

$$\overline{\varphi}(x_1, \dots x_n, y_1, \dots, y_n, z_1, \dots, z_k, u_0, u_1, u_2)$$

in the language of $\mathcal{W}_1$ such that for all $f_1, \dots, f_m \in \mathrm{Sym}(\omega)$, $l_1, \dots, l_n \in \omega$, $h'_1, \dots, h'_k, h''_1, \dots, h''_k \in \mathrm{Fin}$

$$\mathcal{M}_1 \models \overline{\varphi}(f_1, \dots, f_m, l_1, \dots l_n, h'_1, h''_1, \dots h'_k, h''_k, g_0, g_1, \mathbf{0}) \Leftrightarrow$$

$$\mathcal{M}_2 \models \varphi(f_1, \dots, f_m, l_1, \dots l_n, \mathrm{sp}(h'_1) \cap \mathrm{sp}(h''_1), \dots, \mathrm{sp}(h'_k) \cap \mathrm{sp}(h''_k))$$

holds, where $g_0 = \prod_{i<\omega}(2i, 2i+1)$ and $g_1 = \prod_{i<\omega}(2i+1, 2i+2)$.

The proof is just the same as for Lemma 2.3.

Lemma 3.4. *Assume that $f : \omega \to \omega$ is a function such that there exists a maximal finite family of pairwise disjoint sets $\{x, y\}$ with the properties $x \neq y$ and $f(x) = y$. Then there exist a permutation $\sigma \leq_T f$ and a recursive function $g : \omega \to \omega$ such that $f = \sigma^{-1} g \sigma$.*

Proof. Take such a maximal finite family $M = \{\{x_0, y_0\}, \dots, \{x_{n-1}, y_{n-1}\}\}$ so that $x_i \neq y_i$ and $f(x_i) = y_i$ for all $i < n$ and $\{x_i, y_i\} \cap \{x_j, y_j\} = \emptyset$ for all $i < j < n$. Denote $M_0 = \{x_0, y_0, \dots, x_{n-1}, y_{n-1}\}$. Consider the set $M_1 = \{f(y_0), \dots, f(y_{n-1})\} \setminus M_0$. Evidently, $f(M_0 \cup M_1) \subseteq M_0 \cup M_1$; otherwise we would have some new pair $\{x^*, y^*\}$ not in M with the properties $x^* \neq y^*$ and $f(x^*) = y^*$, which contradicts maximality of M. Further, by the same reason, $f^{-1}(M_1) \subseteq M_0 \cup M_1$. Consider now the sets

$$B_i = \{z \notin M_0 \cup M_1 \mid f(z) = x_i\} \quad \text{and} \quad C_i = \{z \notin M_0 \cup M_1 \mid f(z) = y_i\}$$

for $i < n$. By maximality of M_0 we have $f^{-1}(B_i) \setminus B_i = f^{-1}(C_i) \setminus C_i = \emptyset$. Further, we see that $f(x) = x$ holds for all $x \notin \bigcup_{i<n} B_i \cup \bigcup_{i<n} C_i \cup M_0 \cup M_1$. Now it is clear that, except for the finite set $M_0 \cup M_1$, f acts as follows: it takes all elements of sets B_i to x_i, all elements of sets C_i to y_i, and all the other elements are not moved by f. By this, f is isomorphic to some recursive function g and one can effectively with respect to f construct a permutation σ so that $f = \sigma^{-1} g \sigma$. □

Note now, that if a permutation is algebraic over recursive permutations in the sense of the previous section then it is also algebraic in the sense of this section. Indeed, assume that $\varphi(x, \overline{a})$ is a formula of the language of $\mathcal{M}_0$ with recursive permutations $\overline{a}$ as parameters, whose only solution is f. Consider the formula $\lambda(x, \overline{z}) = \varphi(x, \overline{a})$ & $\forall u \exists! v$ ap$(x, v) = u$ in the language of $\mathcal{M}_1$. One can easily verify by induction, that for all formulas $\psi(\overline{x}, \overline{y})$ of language of $\mathcal{M}_1$ which do not contain $^{-1}$, where $\overline{x}$ is a tuple of variables of type ω and $\overline{y}$ is a tuple of variables of type Sym(ω), and for all tuples $\overline{n} \in \omega^{<\omega}$, $\overline{h} \in \mathrm{Sym}(\omega)^{<\omega}$ of appropriate lengths

$$\mathcal{M}_1 \models \psi(\overline{n}, \overline{h}) \iff \mathcal{W}_1 \models \psi(\overline{n}, \overline{h})$$

holds, if we consider quantifiers over Fin for $\mathcal{W}_1$ as appropriate expressions via quantifiers over Ff. To eliminate $^{-1}$, we consider first all atomic subformulas of our formula of the form $t_1 = t_2$, where t_1 and t_2 are group terms. We can replace them by the equivalent formulas $\forall m \exists n(\mathrm{ap}(t_1, m) = n \ \& \ \mathrm{ap}(t_2, m) = n)$. Thus, we reduce our consideration to formulas of kind ap$(t, m) = n$. Further, we assume that all $^{-1}$ in t are situated at variables only. If $t = q_1 q_2$ then we replace this formula by the equivalent formula $\exists k(\mathrm{ap}(q_2, m) = k \ \& \ \mathrm{ap}(q_1, k) = n)$ and so on. At last, we arrive at a formula of kind ap$(b^{-1}, x) = y$ or ap$(b, x) = y$, where b, x, and y are terms of length 1; the first such formula is equivalent to ap$(b, y) = x$. Thus, we can assume that $\lambda(x, \overline{z})$ does not contain $^{-1}$. Now it remains to use Lemmas 3.2 and 3.3 to translate it into the language of $\mathcal{W}_0$; probably, some new recursive parameters will appear in the result.

Thus, if a function satisfies the condition of Lemma 3.4, it can be represented as a product of algebraic elements, since any recursive function is trivially algebraic and, by the remark above, all hyperarithmetical functions are also representable as products of algebraic elements.

Take an arbitrary $n \in \omega$. Any set of kind

$$\{x \mid \exists m \exists y_0 \dots y_{m-1}(y_0 = n \ \& \ y_{m-1} = x \ \& \forall i < m(f(x_i) = x_{i+1} \vee f(x_{i+1}) = x_i))\}$$

is called a *block*.

Now we assume that a hyperarithmetical function f does not satisfy the condition of Lemma 3.4. In this case, we can effectively in f enumerate an infinite family of pairwise disjoint pairs $\{\langle x_i, y_i\rangle\}_{i<\omega}$ so that $f(x_i) = y_i$ for all $i < \omega$. Clearly, the set $\{x_i, y_i \mid i < \omega\}$ is $f^{(1)}$-recursive. In this case, the permutation $\zeta = \prod_{i<\omega}(x_i, y_i)$ (the infinite product of 2-cycles) is $f^{(1)}$-recursive and the function $f_1 = \zeta f$ has the property that the set $\{x \in \omega \mid f_1(x) = x\}$ contains the set $\{x_i \in \omega \mid i < \omega\}$ and is, therefore, infinite. So f can be thought of as the union of an infinite family of blocks; therefore f_1 can be presented as a product of two hyperarithmetical functions h_0 and h_1 (arithmetical in f), for which the sets $\mathrm{st}^*(h_i) \rightleftharpoons \{x \mid h_i(x) = x \;\&\; h_i^{-1}(x) = \{x\}\}$, $i = 0, 1$ are infinite. We have $f = \zeta^{-1}h_0h_1 = \zeta h_0 h_1$ (note that $\zeta = \zeta^{-1}$), and it remains to prove the theorem for a function f with infinite $\mathrm{st}^*(f)$.

Without loss of generality we assume that $\mathrm{st}^*(f)$ has infinite complement. First, we separate all blocks on which f is a permutation and naturally decompose f into a product of some permutation and a function h whose each element is either in $\mathrm{st}^*(h)$ or is in a block on which h is not a permutation. These functions are arithmetical in f and therefore they are hyperarithmetical. Since all hyperarithmetical permutations are representable as products of algebraic elements, it remains to prove the theorem for h.

The further proof uses the same ideas as the proofs above, so we outline the basic steps only.

First we take an arithmetical singleton S so that $h \leq_m S$.

We define a function q_0 so that

1. q_0 is a permutation on $\mathrm{st}^*(h)$ that contains at most one cycle of each length; it contains a cycle of length $n+2$ if and only if $n \in S$, each element of $\mathrm{st}^*(h)$ belong to some nontrivial cycle; these cycles have the property, that each element of a cycle with less length is less than each element of a cycle of a greater length; elements in these cycles are mapped by q_0 in the increasing order and then the greater element is taken back to the smallest element in the cycle;

2. q_0 coincides with h on $\mathrm{st}^*(h)$.

Prove that q_0 is a singleton. Note that q_0 is uniquely defined by the following conditions:

1. Each element is either in a block on which q_0 is not a permutation or belongs to a block which is a finite cycle.

2. The function q_0 takes the maximal element of each such cycle to a minimal one; q_0 is monotonous on the rest elements of the cycle.

3. All elements of such a cycle with greater number of elements are greater than all elements of such a cycle with smaller number of elements.

4. The set $\{n \mid n+2$ is a cardinality of a finite block which is a cycle$\}$ coincides with S.

5. Fix a recursive function p that reduces h to S. The condition is: on the blocks, where q_0 is not a permutation, q_0 is reducible to S by means of p.

It remains to express this in $\mathcal{W}_2$. To do so, we need to express the relation "x and y are in the same block" in the language of $\mathcal{W}_2$. This can be done as follows:

$$\exists s(\mathrm{seq}(s) \ \& \ (s)_0 = x \ \& \ (s)_{\mathrm{lh}(s)-1} = y \ \&$$
$$\forall i < \mathrm{lh}(s)((f((s))_i = (s)_{i+1}) \vee (f((s)_{i+1}) = (s)_i)),$$

where $\mathrm{seq}(s)$ means that s is a code of a sequence, $\mathrm{lh}(s)$ denotes the length of the sequence coded by s, and $(s)_i$ stands for the i-th element of this sequence. All these predicates and functions are recursive, i.e., they can be expressed by formulas of arithmetics. The rest can be done as in the proof of Lemma 2.5.

On the next step, we define a function q_1 which coincides with q_0^{-1} on elements of blocks on which q_0 is a permutation. This can be done as in Lemma 2.6.

Now it remains to note that $f = q_0 q_1$. The proof of the theorem is complete. □

Remarks. 1. In fact, we have proved that all hyperarithmetical functions are generated by algebraic functions that are defined with parameters $\mathbf{0}$, g_0, g_1; i.e., we have added the new parameter $\mathbf{0}$ into the consideration. The author believes that it can be eliminated and thus the number of parameters will be the minimal possible.

2. We arrive at the same results if we consider quantifiers over recursive functions (or recursive permutations in the preceding section).

References

[1] S. S. Goncharov, On the problem of the number non-self-equivalent constructivizations, Algebra and Logic 19 (1980), 401–414.

[2] C. J. Ash, J. F. Knight, Pairs of recursive structures, Ann Pure Appl. Logic 46 (1990), 211–234.

[3] G. Birkhoff, Sobre los grupos de automorfismos, Rev. Un. Mat. Argentina 11 (1946), 155–157.

[4] A. S. Morozov, Functional trees and automorphisms of models, Algebra and Logic 32 (1993), 28–38.

[5] J. Barwise, Admissible Sets and Structures, Springer-Verlag, Berlin, Heidelberg, 1975.

[6] R. McKenzie, Automorphism groups of denumerable Boolean algebras, Canad. J. Math. 29 (1977), 466–471.

[7] H. Rogers, Theory of Recursive Functions and Effective Computability, McGraw-Hill Book Company, New York, St. Louis, San Francisco, Toronto, London, Sydney, 1967.

[8] G. Sacks, Higher Recursion Theory, Springer-Verlag, Heidelberg, 1990.

[9] C. F. Kent Constructive analogues of the group of permutations of the natural numbers, Trans. Amer. Math. Soc. 104 (1962), 347–362.

Weak presentations of fields not extendible to recursive presentations

*Alexandra Shlapentokh**

Abstract. A weak presentation of a field C is an isomorphism from C to a field whose universe is a subset of the natural numbers such that all the field operations are interpreted by restrictions of total recursive functions. We show that if C is any infinite recursive field, it has a weak presentation which is not extendible to a recursive presentation of any field containing C.

1. Introduction

Presentations were originally introduced to formalize the notion of an algorithm on countable mathematical structures. The object under consideration was mapped into the natural numbers and a function on the object was considered recursive if itstranslation was recursive on the image. The original definition of presentations, as can be found in [2] and [3], required that all the operations of the structure be translated by total recursive functions and also required the image of the presentation to be recursive. The structures that have such presentations are usually called *recursive*.

On the other hand, there are naturally arising algebraic structures which do not have recursive presentations but are embedded in structures which are recursive. The operations of these non-recursive structures can be represented by restrictions of total recursive functions. Existence of such structures can motivate one to define another class of presentations: *weak presentations*, which do not require the image of the presentation map to be recursive but do require that all the operations associated with the object are interpreted by functions extendible to total recursive ones. We state the definition explicitly for fields.

Definition 1.1. Let K be a countable field. Let $j : K \longrightarrow \mathbb{N}$ be an injective map such that there exist total recursive functions $P_+, P_-, P_\times, P_/ : \mathbb{N}^2 \longrightarrow \mathbb{N}$ with the property thatfor all $x, y \in K$, $P_+(j(x), j(y)) = j(x + y)$, $P_-(j(x), j(y)) = j(x - y)$, $P_\times(j(x), j(y)) = j(x \times y)$, and if $y \neq 0$, $P_/(j(x), j(y)) = j(x/y)$. Then j is called a weak presentation of K.

*The author would like to thank Carl Jockusch for giving her the idea for this paper as well as help in realizing it. The research for this paper has been partially supported by NSA grant MDA904-96-1-0019

Given such a definition, one can consider the following class of problems: given a recursive object, what kind of weak presentations does such an object have ? Solving a problem of this type should illuminate the relationship between the algebraicand logical structures of the object under consideration. The study of weak presentations of computable objects can also shed some light on the problems of Diophantine definability. For more detailed discussion of these issues see [5].

In previous work, Carl Jockusch and the author have investigated various weak presentations of computable fields (see [1], [6], [7], [8], [9]). The investigation in these papers was mainly directed at the studyof the great variety of the Turing degrees of the image of weak presentations of a given computable field. In this paper we consider another aspect of weak presentations. As we have mentioned above, it is clear that weak presentations arisenaturally as presentations of structures embedded in bigger recursive ones. In this paper we will show that weak presentations can also arise under different circumstances. That is, for any infinite recursive field C, it is possible to constructa weak presentation of C which does not extend to a recursive presentation of any field containing C.

The problem discussed in this paper is related to work done by Remmel in [4], where he proved the following result. There exists a field whose universe consists of natural numbers and whose operations are extendible to partial recursivefunctions whose domain includes the field, such that this field is not contained in any r.e. field whose operations satisfy the same condition. (Such a field, of course, has to be infinitely generated. We, on the other hand, consider arbitraryrecursive fields, including finitely generated ones. Therefore, we can only claim that there is no recursive extension.)

We construct our presentation using a diagonalization argument. We will make sure that any total recursive function restricting to the addition on the image of our field violates the associative law infinitely many times outside the image of thefield. Let P_+ be a total recursive function of two variables which restricts to the image of addition of the field under consideration. For each natural number k we will define

$$P_+(8k+1, 8k+2) = 8k+3,$$

$$P_+(8k+3, 8k+4) = 8k+5,$$

$$P_+(8k+2, 8k+4) = 8k+6,$$

$$P_+(8k+1, 8k+6) = 8k+7.$$

It is clear that not all of the seven integers $8k+1, \ldots, 8k+7$ can be in the image of our field. If they were, the associative law would be violated. We will make sure that integers occurring in at most *one* of the equations will be in the image. On the other hand, let ϕ be any partial recursive function of two variables, and consider how ϕ acts on the pairs $(8k+1, 8k+2)$, $(8k+3, 8k+4)$, $(8k+2, 8k+4)$, $(8k+1, 8k+6)$. If ϕ diverges on one of the pairs, it is not a total function and thus we don't have to consider it. If ϕ is a total function then we will eventually know its actions on the pairs. If it agrees with P_+ on all the pairs then ϕ violates the associative law on this septuplet of integers. If ϕ does not agree with P_+ on a pair, we will make sure that the pair on which ϕ disagrees with P_+ is included in the image of the field and thus ϕ is not an extension of P_+'s restriction to the image of the field. For each ϕ we will consider infinitely many septuplets to make sure that any ϕ extending P_+'s restriction to the image of the field violates the associative law infinitely many times. This will be enough to make sure that the image of our field is not contained in an image of a bigger field under a recursive presentation. If that were the case, the complement of the image of the bigger field would have to be infinite and the recursive presentation of the bigger field could then be extended to a recursive presentation of a field whose image is all of $\mathbb{N}$. The addition function of that field would have to be a total recursive extension ofP_+'s restriction to the image of our original field without violations of associative law. Since we will make sure that such a function does not exist, we will be done.

The main difficulty that one encounters during the construction of the required weak presentation stems from the necessity of making the images of field operations restrictions of total recursive functions. This requirement leads us to define the values of P_{op}, where op stands for a field operation, for pairs of the form $(8k+m, 8r+s)$, where $0 \leq m, r \leq 7$, before we know what field element, if any, will be mapped to these integers. (Actually, we will have to define the values of the functions representing field operations on the pairs whose elements are the results of *several* previous applications of these functions to the integers not equivalent to 0 modulo 8, before we know what elements of the field C, if any, are mapped to these elements.) To overcome these difficulties we treat integers not equivalent to 0 modulo 8 as variables algebraically independent over C, and construct a map from a rational function field over C into the natural numbers.

Once we know that a certain integer not equivalent to 0 modulo 8 has to be added to the image of the field, we will seek an appropriate element of the field to map to this integer. To make sure this assignment is consistent with all the previous definitions and assignments we have made, we perform a substitution: we replace the variable corresponding to the integer to be added to the image of the field by the chosen element of the field.

As we have pointed above, out of any septuplet $(8k+1, \ldots, 8k+7)$ at most three elements, all occurring in one equality, will be added to the image of the field. Thus, for example $8k+5$ and $8k+7$ can never be both added to the image, since each of these integers is the result of addition of $8k+1$, $8k+2$, $8k+3$ with different positioning of parentheses. On the other hand, consider $8k+3$. This integer can be added to the image of the field either because the triple $(8k+1, 8k+2, 8k+3)$ was added or because the triple $(8k+3, 8k+4, 8k+5)$ was added. In the first case, if $a_1, a_2, a_3 \in C$ were mapped to $8k+1$, $8k+2$, $8k+3$ respectively, then clearly $a_3 = a_1 + a_2$. Similarly, in the second case, if a_3, a_4, a_5 were mapped to $8k+3$, $8k+4$, $8k+5$respectively then $a_3 + a_4 = a_5$. Thus, even though $8k+1$, $8k+2$ will never be added to the image together with $8k+4, 8k+5$, we have to keep in mind that, for example, $P_-(8k+5, 8k+4) = P_+(8k+1, 8k+2) = 8k+3$. If we translate these requirements into the language of rational functions and adopt the convention of assigning x_t to the integer $t \not\equiv 0$ modulo 8, then we can say that we have to coordinate the values of the rational functions contained in a certain subring of a rational function field over C if these functions are equivalent modulo the ideal generated by all the polynomials corresponding to the initial assignments of P_+. This coordination is done using "normal forms" described in the section 3 of the paper. A normal form of a rational function is a rational function equivalent to the original one modulo the ideal mentioned above and containing the smallest number of variables possible. In our particular case, both polynomials: $x_{8k+1}+x_{8k+2}$ and$x_{8k+5}-x_{8k+4}$ are "replaced" or "linked" to their normal form: x_{8k+3}.

Our proof is contained in the following three sections. The first two deal with the necessary algebraic preliminaries. The third one describes the construction of the desired weak presentation.

2. Some ideals in a rational function field of several variables

In this section we discuss properties of certain ideals over a rational function field. The field C, whose weak presentation we are aiming to construct, is the field of constants of this rational function field. Before we can proceed we need to introduce a rather lengthy list of notations.

Notations 2.1.

- C denotes an infinite field.
- $X = \{x_{i,j}\}, i \in \mathbb{N}, j = 1, \ldots, 7$ denotes a set of variables algebraically independent over C.
- $K = C(X)$.
- $R = C[X]$.
- For every $i \in \mathbb{N}$, let

$$\begin{aligned} L_{i,1}(x_{i,1}, x_{i,2}, x_{i,3}) &= x_{i,1} + x_{i,2} - x_{i,3}, \\ L_{i,2}(x_{i,3}, x_{i,4}, x_{i,5}) &= x_{i,3} + x_{i,4} - x_{i,5}, \\ L_{i,3}(x_{i,2}, x_{i,4}, x_{i,6}) &= x_{i,2} + x_{i,4} - x_{i,6}, \\ L_{i,4}(x_{i,1}, x_{i,6}, x_{i,7}) &= x_{i,1} + x_{i,6} - x_{i,7}. \end{aligned}$$

- P denotes the ideal of R generated by all $L_{i,j}, i \in \mathbb{N}, j = 1, 2, 3, 4$. For each $i \in \mathbb{N}$, P_i denotes the ideal generated by $(L_{i,1}, \ldots, L_{i,4})$.

- R_P denotes the localization of R at P. (This notations pre-supposes that P is a prime ideal which will be shown below.)

- Let $F(x_{1,1}, \ldots, x_{n,7}) \in K$, let $\{f_{i,j}\}_{i \in I, j \in J} \subset K$, where I and J are some sets of natural numbers. Then $F(x_{i,j}/f_{i,j})$ denotes the result of substitutions of $f_{i,j}$ for $x_{i,j}$ in F.

- Let $F \in K$. Then $V(F)$ denotes the subset of X such that $F \in C(V(F))$ but for any $V_0 \subset V(F)$, $F \notin C(V_0)$.

- $V_i = \bigcup_{j=1}^{4} V(L_{i,j})$.

- $V_{i,j} = V(L_{i,j})$.

- Let $\{F_r\} \subset K$. Then $V_i(\{F_r\}) = \bigcup_r V(F_r) \cap V_i$, $V_{i,j}(\{F_r\}) = \bigcup_r V(F_r) \cap V_{i,j}$.

The following lemma can be verified by a direct computation which we carry out in the appendix.

Lemma 2.1. *Let $x_{i,j_i}, x_{i,k_i}, x_{i,r_i} \in V_i$, assume it is not the case that $\{x_{i,5}, x_{i,7}\} \subset \{x_{i,j_i}, x_{i,k_i}, x_{i,r_i}\}$, $\{x_{i,j_i}, x_{i,k_i}, x_{i,r_i}\} \not\subset V_{i,j}$ for any $j = 1, 2, 3, 4$, and $\{x_{i,j_i}, x_{i,k_i}, x_{i,r_i}\} \neq \{x_{i,3}, x_{i,4}, x_{i,7}\}$ or $\{x_{i,1}, x_{i,6}, x_{i,5}\}$. Finally assume, for some $j \in \{1, 2, 3, 4\}$, $x_{i,j_i}, x_{i,k_i} \in V_{i,j}$. Then $x_{i,j_i}, x_{i,k_i}, x_{i,r_i}$ constitute a basis of the solution space for the system $\{L_{i,j} = 0\}$, $j = 1, 2, 3, 4$. Alternatively, for each i, there exist a collection of 7 polynomials*

$$S_{i,u,j_i,k_i,r_i} = S_{i,u} = b_{i,u,j_i} x_{i,j_i} + b_{i,u,k_i} x_{i,k_i} + b_{i,u,r_i} x_{i,r_i}, \quad u = 1, \ldots, 7,$$

with the property that for any $a_{i,1}, a_{i,2}, a_{i,3} \in C$, $\{x_{i,u} = S_{i,u}(a_{i,1}, a_{i,2}, a_{i,3}), u = 1, \ldots, 7\}$ is a set of solutions to $\{L_{i,j} = 0\}$, $j = 1, \ldots, 4$, or, in other words, for all i, j, $L_{i,j}(x_{i,u}/S_{i,u}(x_{i,j_i}, x_{i,k_i}, x_{i,r_i})) = 0$.

Definition 2.1. A triple $(x_{i,j_i}, x_{i,k_i}, x_{i,r_i})$ satisfying the requirements described above will be called a *solution triple*. For each i, the substitution $x_{i,u}/S_{i,u}(x_{i,j_i}, x_{i,k_i}, x_{i,r_i})$ will be called an *i-th standard substitution*.

Note also that in the above notations,

$$S_{i,j_i} = x_{i,j_i}, \quad S_{i,k_i} = x_{i,k_i}, \quad S_{i,r_i} = x_{i,r_i}$$

Finally, we note that for any pair $(x_{i,j_i}, x_{i,k_i}) \neq (x_{i,5}, x_{i,7})$, there exists x_{i,r_i} such that $(x_{i,j_i}, x_{i,k_i}, x_{i,r_i})$ is a solution triple. (See Appendix for a proof.)

Notations 2.2.

- For each $i \in \mathbb{N}$, $u = 1, \ldots, 7$, let
$$M_{i,u} = x_{i,u} - S_{i,u}.$$
It is easy to see (see Appendix) that P is generated by $M_{i,j}$, $i \in \mathbb{N}$, $j = 1, \ldots, 4$ and for each $i \in \mathbb{N}$, P_i is generated by $M_{i,1}, \ldots, M_{i,4}$.

- For each triple of sequences $\mathcal{J} = \{j_i\}_{i\in\mathbb{N}}$, $\mathcal{K} = \{k_i\}_{i\in\mathbb{N}}$, $\mathcal{R} = \{r_i\}_{i\in\mathbb{N}}$, with $j_i, k_i, r_i \in \{1, \ldots, 7\}$, such that for each $i \in \mathbb{N}$, $(x_{i,j_i}, x_{i,k_i}, x_{i,r_i})$ is a solution triple, let
$$R_{\mathcal{J},\mathcal{K},\mathcal{R}} = C[x_{1,j_1}, x_{1,k_1}, x_{1,r_1}, \ldots, x_{m,j_m}, x_{m,k_m}, x_{m,r_m}, \ldots].$$

Definition 2.2. A set $A \subseteq C^3$ is *fat* if there is a finite, possibly empty set $V \subset C$, a function f from $C \setminus V$ into finite, possibly empty subsets of C, and a function g such that for each $a_1 \in C \setminus V$ and $a_2 \in C \setminus f(a_1)$, $g(a_1, a_2)$ is a finite, possibly empty subset of C for which
$$A = \{(a_1, a_2, a_3) \in C^3 \mid a_1 \notin V, a_2 \notin f(a_1), a_3 \notin g(a_1, a_2)\}.$$

Lemma 2.2. *If A is fat and W_1, W_2, W_3 are finite subsets of C, then $A \setminus W_1 \times W_2 \times W_3$ is infinite.*

Note 2.1. Let $B = A_1 \cap \cdots \cap A_k$, where for all $i = 1, \ldots, k$, A_i is fat. Then B is also fat.

Lemma 2.3. *For $k \geq 3$, let $\{F_t(z_1, \ldots, z_k) \mid t = 1, \ldots, r\}$ be a finite collection of non-zero polynomials over C. Consider the following subsets of C^3:*

1. $S_1 = \{(a_1, a_2, a_3) \mid a_1, a_2, a_3 \in C, F_t(z_1/a_1, z_2/a_2, z_3/a_3, z_4, \ldots, z_k) \neq 0, t = 1, \ldots, r\}$.

2. $$S_2 = \{(a_1, a_2, a_3) \mid a_1, a_2, a_3 \in C,$$
$$V(F_t(z_1/a_1, z_2/a_2, z_3/a_3, z_4, \ldots, z_k)) = V(F_t) \setminus \{z_1, z_2, z_3\}, t = 1, \ldots, r.$$

Then S_1 and S_2 both contain fat sets.

Proof. 1. For each t,
$$F_t = \sum A_{t,i_4,\ldots,i_k}(z_1, z_2, z_3) z_4^{i_4} \ldots z_k^{i_k},$$
where for each t, for some $(i_4, \ldots, i_k)$,
$$A_{t,i_4,\ldots,i_k}(z_1, z_2, z_3) \neq 0.$$

On the other hand,

$$A_{t,i_4,\dots,i_k}(z_1, z_2, z_3) = \sum B_{t,i_4,\dots,i_k,j_2,j_3}(z_1) z_2^{j_2} z_3^{j_3},$$

where for each t, for some $(i_4, \dots, i_k, j_2, j_3)$,

$$B_{t,i_4,\dots,i_k,j_2,j_3}(z_1) \neq 0.$$

Let V be the set of C-roots of all the non zero polynomials $B_{t,i_4\dots i_k,j_1,j_2}(z_1)$ for all the indices. This set is clearly finite. Next let $z_1 = a_1 \in C \setminus V$. Then for all t, for some $(i_4, \dots, i_k)$,

$$0 \neq \sum_{j_1,j_2} B_{t,i_4,\dots i_k,j_1,j_2}(a_1) z_2^{j_2} z_3^{j_3} = A_{t,i_4,\dots,i_k}(a_1, z_2, z_3) = \sum_{r_3} C_{t,i_4,\dots i_k,r_3}(a_1, z_2) z_3^{r_3}.$$

Thus, for each t, for some $(i_4, \dots, i_k, r_3)$,

$$C_{t,i_4,\dots,i_k,r_3}(a_1, z_2) \neq 0.$$

Define $f(a_1)$ to be the set of all C-roots of all the non-zero polynomials $C_{t,i_4\dots i_k,r_3}(a_1, z_2)$. Furthermore, let $z_2 = a_2 \notin f(a_1)$, and let $g(a_1, a_2)$ be the set of all C-roots of all the non-zero polynomials $A_{t,i_4,\dots,i_k}(a_1, a_2, z_3)$. (As above, we know that for each t, for some $(i_1, \dots, i_4)$, $A_{t,i_4\dots,i_k}(a_1, a_2, z_3) \neq 0$.) It is clear that $g(a_1, a_2)$ is also finite. Finally, note that for all $t = 1, \dots, r$, for all $a_1 \in C \setminus V$, for all $a_2 \in C \setminus f(a_1)$, for all $a_3 \in C \setminus g(a_1, a_2)$, $A_{t,i_1,\dots,i_k}(a_1, a_2, a_3) \neq 0$, assuming $A_{t,i_1,\dots,i_k}(z_1, z_2, z_3) \neq 0$ to begin with. Thus, for all t,

$$F_t(a_1, a_2, a_3, z_4, \dots, z_4) \neq 0.$$

2. For each $s = 4, \dots, k$, write

$$F_t(z_1, \dots, z_k) = \sum A_{t,s,r}(z_1, \dots, z_{s-1}, z_{s+1}, \dots, z_k) z_s^r,$$

where for each t, for each s, for some $r > 0$, $A_{t,s,r} \neq 0$. (Otherwise, $z_s \notin V(F_t)$.) In order to make sure that z_s still occurs in

$$F_t(z_1/a_1, z_2/a_2, z_3/a_3, z_4, \dots, z_k)$$

it is enough to ensure that for all $r > 0$,

$$A_{t,s,r}(a_1, a_2, a_3, z_4, \dots, z_{s-1}, z_{s+1}, \dots, z_k) \neq 0$$

for all t, s such that $A_{t,s,r}(z_1, \dots, z_k) \neq 0$. Thus, we can apply part 1 of the lemma and Note 2.1 to reach the desired conclusion. □

Corollary 2.4. *Let $k \geq 3$ and let $\{F_t(z_1, \dots, z_k), t = 1, \dots, r\}$ be a finite set of non-zero rational functions over C, such that for all t, $z_4, \dots, z_k$ do occur as variables in F_t. Consider the following two sets.*

1. $Z_1 = \{(a_1, a_2, a_3) \mid a_1, a_2, a_3 \in C, F_t(z_1/a_1, z_2/a_2, z_3/a_3, \dots, z_k) \neq 0$ *and is well defined* $\}$.

2. $Z_2 = \{(a_1, a_2, a_3) \mid a_1, a_2, a_3 \in C, V(F_t(z_1/a_1, z_2/a_2, z_3/a_3, z_4, \ldots, z_k)) = V(F_t) \setminus \{z_1, z_2, z_3\}, t = 1, \ldots, r$.

Then both Z_1 and Z_2 contain fat sets.

Proof. 1. For each $t = 1, \ldots, r$, let $F_t = \frac{F_{1,t}}{F_{2,t}}$, where $F_{1,t}$, $F_{2,t}$ are relatively prime polynomials over C. To make sure that $F_t(z_1/a_1, z_2/a_2, z_3/a_3, \ldots, z_k) \neq 0$ and is well defined, it is enough to make sure that neither $F_{1,t}(z_1/a_1, z_2/a_2, z_3/a_3, \ldots, z_k)$ nor $F_{2,t}(z_1/a_1, z_2/a_2, z_3/a_3, \ldots, z_k)$ is equal to zero. Thus, this part of the corollary is true by Lemma 2.3.

2. Let $s \in \{4, \ldots, k\}$ and let $F_t = \frac{F_{1,t}}{F_{2,t}}$, where $F_{1,t}$, $F_{2,t}$ are relatively prime polynomials over C. Since the same variables occur in a rational function and its reciprocal, without loss of generality, we can assume that $\text{degree}_{z_s}(F_{1,t}) \geq \text{degree}_{z_s}(F_{2,t})$. First assume the inequality is strict. Then $F_t = G_t + H_t$, where G_t is a polynomial over $C(z_1, \ldots, z_{s-1}, z_{s+1}, \ldots, z_k)$ whose degree in z_s is greater than zero, and $H_t = 0$ or is a rational function over C such that the z_s-degree of its reduced denominator is higher than the z_s-degree of its reduced numerator. Thus, it is enough to make sure that after the substitution $F_{2,t}$ does not become zero and the z_s-degree of G_t is preserved. The last requirement can be satisfied by making sure that the rational functions which are non-zero coefficients of G_t corresponding to positive powers of z_s remain non zero after the substitution. So our goal can be reached by applying the first part of this corollary. □

Suppose now that $\text{degree}_{z_s}(F_{1,t}) = \text{degree}_{z_s}(F_{2,t})$. Then $F_t = G_t + H_t$, where G_t is a rational function with no occurrences of z_s and $H_t = \frac{H_{1,t}}{H_{2,t}}$ with $\text{degree}_{z_s}(H_{1,t}) < \text{degree}_{z_s}(H_{2,t})$. First of all, we note that it is enough to make sure that z_s still occurs in H_t after the substitution. Secondly, we note that by considering $1/H = H_{2,t}/H_{1,t}$ we can reduce this case to the previous one.

Lemma 2.5. *Suppose $F \in R$, and that for some collection of solution triples $\{x_{i,j_i}, x_{i,k_i}, x_{i,r_i}\}$, $i = 1, \ldots, m$,*

$$F(x_{i,u}/S_{i,u}(x_{i,j_i}, x_{i,k_i}, x_{i,r_i})) = 0.$$

Then $F \in P$.

Proof. Using the remark in Notations 2.2, it is not difficult to see that $F - F(x_{i,u}/S_{i,u}(x_{i,j_i}, x_{i,k_i}, x_{i,r_i})) \in P$. Thus, our assumptions imply $F \in P$. □

Lemma 2.6. *Every equivalence class of R modulo P contains a polynomial F such that for each i,*

$$V_i(F) \subseteq \{x_{i,j_i}, x_{i,k_i}, x_{i,r_i}\}, \tag{2.1}$$

where $\{x_{i,j_i}, x_{i,k_i}, x_{i,r_i}\}$ is a fixed solution triple.

Proof. Let $G \in R$. Let F be the polynomial which is obtained from G by substituting $S_{i,u} = S_{i,u,j_i,k_i,r_i}$ for $x_{i,u}$. Then F satisfies Condition (2.1) and is equivalent to G modulo P.

Lemma 2.7. *The following statements are true.*

1. *Let $F \in R_{\mathcal{J},\mathcal{K},\mathcal{R}}$, then $F \in P$ if and only if F is identically zero.*

2. *For any $\mathcal{J}$, $\mathcal{K}$, $\mathcal{R}$ as described in Notations 2.2, $R_{\mathcal{J},\mathcal{K},\mathcal{R}} \cong R/P$*

3. *P is a prime ideal.*

Proof. 1. Suppose $F \in P \cap R_{\mathcal{J},\mathcal{K},\mathcal{R}}$. Then $F = \sum_{i,j} A_{i,j} L_{i,j}$. Furthermore,

$$F = F(x_{i,u}/S_{i,u}(x_{i,j_i}, x_{k_i}, x_{i,r_i})).$$

On the other hand, $F(x_{i,u}/S_{i,u}(x_{i,j_i}, x_{k_i}, x_{i,r_i}), i = 1, \ldots, m) = 0$, by Lemma 2.1. Thus, we have the desired conclusion.

2. Consider a map $\phi : R/P \longrightarrow R_{\mathcal{J},\mathcal{K},\mathcal{R}}$ defined as follows. Let $F \in R$ and let $[F]$ be the equivalence class of F modulo P. By Lemma 2.6, this equivalence class contains an element $G \in R_{\mathcal{J},\mathcal{K},\mathcal{R}}$, and by the first part of this lemma this element is unique. Thus, we can let $\phi([F]) = G$ and conclude that ϕ is well-defined and injective. Since for every $G \in R_{\mathcal{J},\mathcal{K},\mathcal{R}}$, $\phi([G]) = G$, ϕ is also surjective. Finally, ϕ is clearly a homomorphism.

3. Since $R_{\mathcal{J},\mathcal{K},\mathcal{R}}$ is an integral domain, P must be a prime ideal. □

Corollary 2.8. *Suppose $F \in PR_P$. Let m be a positive integer such that for any $s > m$, $V_s(F) = \emptyset$. Then for any collection of solution triples $\{x_{i,j_i}, x_{i,k_i}, x_{i,r_i}\}$, $i = 1, \ldots, m$,*

$$F(x_{i,u}/S_{i,u}(x_{i,j_i}, x_{i,k_i}, x_{i,r_i}), i = 1, \ldots, m) = 0.$$

Proof. If $F \in PR_P$, then $F = \frac{F_1}{F_2}$, where $F_1 \in P$ and $F_2 \notin P$. The rest of the proof follows from Lemma 2.5 and Lemma 2.1. □

Lemma 2.9. *Let $\{G_t\}_{t \in \mathbb{N}} \subset P$. Fix an $i \in \mathbb{N}$ and let $a_{i,1}, a_{i,2}, a_{i,3} \in C$. Then for each t, $G_t(x_{i,u}/S_{i,u}(a_{i,1}, a_{i,2}, a_{i,3})) \in P$.*

Proof. For each t, $G_t = \sum_{s \neq i} A_{s,j,t} L_{s,j} + \sum A_{i,j,t} L_{i,j}$, where for all r, $A_{r,j,t} \in R$. Thus

$$G_t(x_{i,u}/S_{i,u}(a_{i,1}, a_{i,2}, a_{i,3})) = A_{s,j,t}(x_{i,u}/S_{i,u}(a_{i,1}, a_{i,2}, a_{i,3}))L_{s,t}$$

$$+ \sum A_{i,j,t}(x_{i,u}/S_{i,u}(a_{i,1}, a_{i,2}, a_{i,3})L_{i,j}(x_{i,u}/S_{i,u}(a_{i,1}, a_{i,2}, a_{i,3})) = A_{s,j,t} L_{s,t} \in P,$$

by Lemma 2.1. □

Lemma 2.10. *Let* $\{F_t\} \subset R \setminus P$. *Then for each* $i \in \mathbb{N}$, *there exists a fat set* $A_i \subset C^3$, *such that if* $(a_{i,1}, a_{i,2}, a_{i,3}) \in A_i$ *then for all* t,

$$F_t(x_{i,u}/S_{i,u}(a_{i,1}, a_{i,2}, a_{i,3})) \in R \setminus P.$$

Proof. Fix sequences $\mathcal{J}$, $\mathcal{K}$, $\mathcal{R}$ defined in Notations 2.2. For each t, let $F_t \cong \bar{F}_t$ modulo P where $\bar{F} \in R_{\mathcal{J},\mathcal{K},\mathcal{R}}$. We need to find $a_{i,1}, a_{i,2}, a_{i,3} \in C$ such that for all t, $\bar{F}_t(x_{i,j_i}/a_{i,1}, x_{i,k_i}/a_{i,2}, x_{i,r_i}/a_{i,3}) \neq 0$. By Corollary 2.4, the set of triples from C satisfying this condition contains a fat set A_i. On the other hand, by Lemma 2.9, for all$t$, and any triple $(a_{i,1}, a_{i,2}, a_{i,3}) \in C^3$

$$F_t(x_{i,u}/S_{i,u}(a_{i,1}, a_{i,2}, a_{i,3})) \cong \bar{F}(x_{i,j_i}/a_{i,1}, x_{i,k_i}/a_{i,2}, x_{i,r_i}/a_{i,3}),$$

and thus our lemma is true, since any element of $R_{\mathcal{J},\mathcal{K},\mathcal{R}}$ remains an element of $R_{\mathcal{J},\mathcal{K},\mathcal{R}}$ after substituting any constant for any of its variables. □

Corollary 2.11. *Let* $\{F_t\} \subset PR_P$, *let* $\{G_s\} \subset R_P \setminus PR_P$. *Then for each* $i \in \mathbb{N}$ *there exists a fat set* $A_i \subset C^3$ *such that for all* $(a_{i,1}, a_{i,2}, a_{i,3}) \in A_i$, *for all* t,

$$F_t(x_{i,u}/S_{i,u}(a_{i,1}, a_{i,2}, a_{i,3})) \in PR_P,$$

and for all s,

$$G_s(x_{i,u}/S_{i,u}(a_{i,1}, a_{i,2}, a_{i,3})) \in R_P \setminus PR_P.$$

Lemma 2.12. *Assume* C *is a recursive field. Let* $F \in R$. *Then there is a recursive procedure to determine whether* $F \in P$.

Proof. Let m be a positive integer such that for any $s > m$, $V_s(F) = \emptyset$. List the first m triples of some fixed sequences $\mathcal{J}$, $\mathcal{K}$, $\mathcal{R}$. Since C is recursive, using the listing above, we can effectively determine $\bar{F} \in R_{\mathcal{J},\mathcal{K},\mathcal{R}}$ such that $F \cong \bar{F}$ and then check whether $\bar{F}$ is zero.

3. Normal forms

In this section we will define various normal forms of rational functions and investigate their properties.

Definition 3.1. Let $\{F_r\} \subset K$. Then we will call $\{F_r\}$ *proper* if for every i there exists a j such that $V_i(\{F_r\}) = V_{i,j}(\{F_r\})$

If the requirement described above is violated for $i = t$, we will say that $\{F_r\}$ is improper with respect to t.

Definition 3.2. Fix $i \in \mathbb{N}$, $j \in \{1, 2, 3, 4\}$ and let $L_{i,j}(x_{i,t_1}, x_{i,t_2}, x_{i,t_3}) = x_{i,t_1} + x_{i,t_2} - x_{i,t_3}$. Next, consider the following three substitutions: $x_{i,t_1}/(x_{i,t_3} - x_{i,t_2})$; $x_{i,t_2}/(x_{i,t_1} - x_{i,t_3})$; $x_{i,t_3}/(x_{i,t_2} + x_{i,t_1})$. These three substitutions will be called (i, j)-th *partial standard substitutions.*

Lemma 3.1. *Let $\{F_r\} \subset R_P$ be proper so that for all i, for some j, $V_i(\{F_r\}) \subset V_{i,j}$. Then the action of an (i, j)-th partial standard substitution on $\{F_r\}$ is equivalent to the action of an i-th standard substitution on this set of rational functions.*

Proof. Let $\{x_{i,j_i}, x_{i,k_i}\} \subset \{x_{i,t_1}, x_{i,t_2}, x_{i,t_3}\}$. Then by Lemma 2.1, there exists $r_i \in \{1, \ldots, 7\}$ such that $(x_{i,j_i}, x_{i,k_i}, x_{i,r_i})$ is a solution triple. Furthermore, since $x_{i,r_i} \notin V_{i,j}$, it does not occur in F_r for any r. Taking this into consideration and using the definition of an i-th standard substitution, it is clear that for all r,

$$F_r(x_{i,u}/S_{i,u}(x_{i,j_i}, x_{i,k_i}, x_{i,r_i}))$$

is equal to one of the following functions:

$$F_{r1} = F_r(x_{i,t_1}/x_{i,t_3} - x_{i,t_2});$$

$$F_{r,2} = F_r(x_{i,t_2}/x_{i,t_3} - x_{i,t_1});$$

$$F_{r,3} = F_r(x_{i,t_3}/x_{i,t_1} + x_{i,t_2}).$$

□

Definition 3.3. Let n be a positive integer, let

$$F(x_{1,1}, \ldots, x_{7,n}), G(x_{1,1}, \ldots, x_{7,n}) \in R_P,$$

and assume that the pair is proper with respect to i. Then $G(x_{1,1}, \ldots, x_{7,n})$ is called the i-th normal form of $F(x_{1,1}, \ldots, x_{7,n})$ if the following conditions are satisfied.

1. $G - F \in P_i$.

2. For every $H \in R_P$ such that $F \cong H$ modulo P_i and the pair (F, H) is proper, for $j = 1, \ldots, 4$,

$$|V_{i,j}(G)| \leq |V_{i,j}(H)|$$

3. Either $|V_i(G)| \leq 1$ or for some j,

$$|V_{i,j}(G) = 2$$

and

$$V_{i,j}(G) = \{x_{i,t_1}, x_{i,t_2}\},$$

where $L_{i,j} = x_{i,t_1} + x_{i,t_2} - x_{i,t_3}$

We denote the i-th normal form of F by $N_i(F)$.

The definition assumes that thus defined i-th normal form is unique. This will be shown below.

Proposition 3.2. *Let F be a proper element of R_P. Then for every $i \in \mathbb{N}$, F has the i-th normal form.*

Proof. Assume $V_i(F) \subset V_{i,j}$. Using notations from Part 3 of Definition 3.3, consider

$$\begin{aligned} F_1 &= F(x_{i,t_1}/(x_{i,t_3} - x_{i,t_2})), \\ F_2 &= F(x_{i,t_2}/(x_{i,t_3} - x_{i,t_1})), \\ F_3 &= F(x_{i,t_3}/(x_{i,t_1} + x_{i,t_2})). \end{aligned}$$

(Note here that by Lemma 3.1 and Lemma 2.5, $F_1, F_2, F_3 \in R_P$.) We claim that for some $u \in \{1, 2, 3\}$, $F_u = N_i(F)$. First of all, we note that for all u, $F_u \cong F$ modulo P_i. Second, we have to consider three cases: for some $u \in \{1, 2, 3\}$, $V_i(F_u) = \emptyset$; for some $u \in \{1, 2, 3\}$, $|V_i(F_u)| = 1$; for all u, $|V_i(F_u)| \geq 2$. In the first case, any F_u satisfying the described condition will clearly be the i-th normal form. Next, assume we are in the second case, $s \in \{1, 2, 3\}$, and $|V_i(F_s)| = 1$. Suppose F_s is not the i-th normal form of F. Then there exists $G \cong F$ modulo P_i such that $V_i(G) = \emptyset$. Let $V_i(F_s) = \{x_{i,j_i}\}$. Given our assumptions on G, $F_s \cong G$ modulo P_i. Furthermore, $(F_s - G)(x_{i,u}/S_{i,u}(x_{i,j_i}, x_{i,k_i}, x_{i,r_i})) = 0$, where $(x_{i,j_i}, x_{i,k_i}, x_{i,r_i})$ is a solution triple. Therefore, $F_s = G$, since an i-th standard substitution$x_{i,u}/S_{i,u}(x_{i,j_i}, x_{i,k_i}, x_{i,r_i})$ does not change F_s, by a remark at the end of Lemma 2.1, and does not change G, since G does not have any occurrences of the variables involved in the substitution. But this equality contradicts our assumptions on F_s and G.

Finally, suppose we are in the third case. In this case we can assume that $V_i(F) = V_{i,j}(F)$ for some j contains at least two variables. On the other hand, $V_{i,j}(F_1) = \{x_{i,t_2}, x_{i,t_3}\}$, $V_{i,j}(F_2) = \{x_{i,t_3}, x_{i,t_1}\}$, $V_{i,j}(F_1) = \{x_{i,t_1}, x_{i,t_2}\}$. Suppose, there exists $G \in R_P$ such that the pair (F, G) is proper, $F \cong G$ and $|V_i(G)| \leq 1$. Since (F, G) is proper, we must assume that either $V_i(G) \subset \{x_{i,t_1}\}$ or $V_i(G) \subset \{x_{i,t_2}\}$, or $V_i(G) \subset \{x_{i,t_3}\}$. Thus, by selecting an appropriate s, without loss of generality, we can assume that $V_i(F_s) \cup V_i(G) \subset \{x_{i,j_i}, x_{i,k_i}, x_{i,r_i}\}$, where $(x_{i,j_i}, x_{i,k_i}, x_{i,r_i})$ is a solution triple. Next consider $(F_s - G)(x_{i,u}/S_{i,u}(x_{i,j_i}, x_{i,k_i}, x_{i,r_i})) = 0$ as before. Since this substitution does not change F_s or G, as above we must conclude $F_s = G$. This equality however contradicts our assumptions on $V(F_s)$. □

Corollary 3.3.

- *If $F \in R_P$ and $|V_i(F)| = 1$ or $|V_i(F)| = 0$ then $F = N_i(F)$.*
- *If F is proper with respect to i and $F \in R_P$ then $N_i(F) = F_1$ or F_2 or F_3, where we use notations F_1, F_2, F_3 in the same way they were used in the preceding lemma.*

- *If F is proper with respect to i, $F \in R_P$ and $|V_i(F_1)| = |V_i(F_2)| = |V_i(F_3)| = 2$, then $F_3 = N_i(F)$.*

- *Let $F \in R_P$ be proper with respect to i and let $G = N_i(F)$. Then for any $H \in R_P$ such that (F, H) is proper with respect to i, (G, H) is proper with respect to i.*

- *Let $s \neq i$. Let $F \in R_P$ be proper with respect to i and s and let $G = N_i(F)$. Then for any $H \in R_P$ such that (F, H) is proper with respect to s, (G, H) is proper with respect to s.*

- *If C is recursive then there is an effective procedure for computing i-th normal forms of elements of R_P.*

Proof. The only point which needs to be clarified here is the fact that for any $F \in R_P$, $V_{i,j}(F)$ can be determined effectively. In general, if F is a rational function over C presented as a ratio of two polynomials and x is a variable occurring in the denominator or numerator, then $x \in V(F)$ if and only if for any (or all) $a \in C$ such that the substitution of a for x does not make denominator zero, $F \neq F(x/a)$ as rational functions. Since for any $a \in C$ it can be determined effectively whether or not substituting a for x makes the denominator zero, this procedure for determining whether $x \in V(F)$ is effective. □

Lemma 3.4. *Let $F \in R_P$ be proper. Then its i-th normal form is unique.*

Proof. Let G_1, G_2 be two i-th normal forms of F. Again we need to consider three cases:

$$V_i(G_1) = \emptyset; \ |V_i(G_1)| = 1; \ |V_i(G_1)| = 2.$$

In the first case, we have

$$G_1 - G_2 = (G_1 - G_2)(x_{i,u}/S_{i,u}(x_{i,j_i}, x_{i,k_i}, x_{i,r_i})) = 0$$

for any solution triple $(x_{i,j_i}, x_{i,k_i}, x_{i,r_i})$, and we are done. In the second case, G_1 cannot have an occurrence of $x_{i,5}$ while G_2 has an occurrence of $x_{i,7}$ because both G_1 and G_2 are proper with F. (This is so because $V_{i,2} \cap V_{i,4} = \emptyset$.) Thus, without loss of generality we can still assume that $V_i(G_1) \cup V_i(G_2) \subseteq \{x_{i,j_i}, x_{i,k_i}, x_{i,r_i}\}$ for some solution triple $(x_{i,j_i}, x_{i,k_i}, x_{i,r_i})$. Therefore neither G_1 nor G_2 will be altered by the standard substitution $x_{i,u}/S_{i,u}(x_{i,j_i}, x_{i,k_i}, x_{i,r_i})$, and thus $G_1 - G_2 = 0$.

Finally, suppose $V_i(G_1) = \{x_{i,j_i}, x_{i,k_i}\}$. Then since (F, G_1) is proper and (F, G_2) is proper, and given the definition of an i-th normal form, we must conclude that $V_i(G_2) = \{x_{i,j_i}, x_{i,k_i}\}$. The rest of the proof proceeds as in the preceding cases. □

We will next define the global normal form with respect to the whole ideal P and relate it to i-th normal forms. As was the case with the i-th normal form, a priori it is not clear that the normal form is unique. However, this will be shown below.

Definition 3.4. Let $(F, G) \in R_P$ be a proper pair satisfying the following conditions.

1. $G \cong F$ modulo P.

2. For every $H \in R_P$ such that $F \cong H$ modulo P, and the pair (F, H) is proper, for every $i \in \mathbb{N}$, $j = 1, \ldots, 4$, $|V_{i,j}(G)| \leq |V_{i,j}(H)|$

3. If for some i, j, $|V_i(G)| = |V_{i,j}(G)| = 2$, then $V_i(G) = \{x_{i,t_1}, x_{i,t_2}\}$, where $L_{i,j} = x_{i,t_1} + x_{i,t_2} - x_{i,t_3}$.

Then G will be called the normal form of F. The normal form of F will be denoted by $N(F)$.

If $G \in R_P$ satisfies Requirements 1 and 2 with respect to some $F \in R_P$ but not the Requirement 3, then G will be called a *pseudo-normal* form of F.

The following lemma is a direct consequence of the definition of the normal form.

Lemma 3.5. *Suppose $F \in R_P$ is proper and assume $F = N(F)$. Then for every i, $F = N_i(F)$. Further, for any $H, F \in R_P$ such that (H, F) is proper, $(H, N(F))$ is also proper.*

Proof. It is enough to show that the first assertion of the lemma holds, since the second assertion follows from the first one and properties of i-th normal forms listed in Corollary 3.3. So suppose $F = N(F)$ but for some i, $F \neq N_i(F)$. Then $F \cong N_i(F)$ modulo P_i and thus $F \cong N_i(F)$ modulo P. At the same time, one of the following statements is true.

1. $|V_i(N_i(F))| < |V_i(F)|$;

2. $|V_i(N_i(F))| = |V_i(F_i)| = 2$, for some $j = 1, \ldots, 4$, $V_i(F) = V_{i,j}(F)$ and $V_i(F) \neq \{x_{i,j_i}, x_{i,k_i}\}$, where $L_{i,j} = x_{i,j_i} + x_{i,k_i} - x_{i,r_i}$;

3. $|V_i(N_i(F))| = |V_i(F_i)| = 0$;

4. $|V_i(N_i(F))| = |V_i(F_i)| = 1$;

The first case will violate part 2 of the definition of the normal form. The second case will violate part 3 of the definition of the normal form. Finally, by Corollary 3.3 each of the last two cases implies $F = N_i(F)$.

Proposition 3.6. *Let $F \in R_P$ be proper. Then F has the normal form.*

Proof. We will describe a construction, and prove that the result of this construction is the normal form of F. Let m be a positive integer such that for any $s > m$, $V_s(F) = \emptyset$. Let $G_0 = F$, let $G_n = N_i(G_{n-1})$, where $n \cong i$ modulo m. Then $N(F) = G_r$ such that for all $i = 1, \ldots, m$, $N_i(G_r) = G_r$. First of all, we should note here that it is easy to see that such a G_r exists. Indeed, at every step of the construction we either reduce the number of variables or replace a pair of variables by another pair. Furthermore, if a step corresponding to a construction of the i-th normal form for some i was a replacement step, then in the next round the step corresponding to this normal form must either be a variable reduction or result in no action. Since variable reduction cannot continue indefinitely, eventually steps corresponding to$i$-th normal forms will produce no action for all $i = 1, \ldots, m$.

Next suppose G_r is not the normal form of F. Then there exists $H \in R_P$ such that $F \cong H \cong G_r$ modulo P, the pair (H, F) is proper, and there exists i such that $|V_i(G_r)| > |V_i(H)|$. By repeated application of Corollary 3.3, we know that (H, G_r) is also proper. Let $s \neq i$. Since (H, G_r) is a proper pair, $V_s(H) \cup V_s(G_r) \subseteq V_{s,j}$for some $j \in \{1, 2, 3, 4\}$. Suppose we actually have the equality. Since G_r is in s-th normal form, $|V_s(G_r)| \leq 2$. Thus, $|V_s(H)| \geq 1$, and at least one variable occurring in $V_s(H)$ does not occur in $V(G_r)$. This variable can be replaced by sum or difference (whichever one is appropriate) of the other two variables occurring in the same $L_{s,j}$. Let H_s be the result of this substitution. Note that $H_s \cong H \cong G_r$ modulo P. Then for some solution triple $(x_{s,j_s}, x_{s,k_s}, x_{s,r_s})$, $V_s(H_s) \cup V_s(G_r) \subset \{x_{s,j_s}, x_{s,k_s}, x_{s,r_s}\}$. On the other hand, if the inclusion above was strict to begin with, then $V_s(H) \cup V_s(G_r) \subset \{x_{s,j_s}, x_{s,k_s}, x_{s,r_s}\}$for some solution triple $(x_{s,j_s}, x_{s,k_s}, x_{s,r_s})$ right away.

Thus, performing such a substitution for all $s \neq i$ if necessary, we can let $\bar{H}$ be the function obtained from H by this sequence of substitutions intended to assure that for every $l \neq i$, $V_l(\bar{H}) \cup V_l(G_r) \subset \{x_{l,j_l}, x_{l,k_l}, x_{l,r_l}\}$ for some solution triple. Observe that $\bar{H} \cong H \cong G_r$ modulo P or, in other words,

$$G_r - \bar{H} = \Big(\sum_{t \neq i} A_{t,j} L_{t,j}\Big) + \sum_j A_{i,j} L_{i,j},\ A_{s,j} \in R_P.$$

Further,

$$|V_i(\bar{H})| \leq |V_i(H)| < |V_i(G_r)|. \tag{3.1}$$

On the other hand, note that $G_r - \bar{H} = (G_r - \bar{H})(x_{l,u}/S_{l,u}(x_{l,j_l}, x_{l,k_l}, x_{l,r_l}), l \neq i)$. That is, $G_r - \bar{H}$ is not changed by the above described sequence of standard substitutions. Therefore,

$$\begin{aligned}
&G_r - \bar{H} \\
&= \Big(\sum_{t \neq i} A_{t,j}(x_{l,u}/S_{l,u}(x_{l,j_l}, x_{l,k_l}, x_{l,r_l}), l \neq i) L_{t,j}(x_{l,u}/S_{l,u}(x_{l,j_l}, x_{l,k_l}, x_{l,r_l}), l \neq i)\Big) \\
&\qquad + \sum_j A_{i,j}(x_{l,u}/S_{l,u}(x_{l,j_l}, x_{l,k_l}, x_{l,r_l}), l \neq i) L_{i,j} \\
&= \sum_j A_{i,j}(x_{l,u}/S_{l,u}(x_{l,j_l}, x_{l,k_l}, x_{l,r_l}), l \neq i) L_{i,j} \in P_i.
\end{aligned}$$

The last equality together with the Equation (3.1) contradicts our assumption that $N_i(G_r) = G_r$. □

In the course of proving the existence of the global normal form we have also proved the following corollary which is the converse of Lemma 3.5.

Corollary 3.7. *Let $F \in R_P$ be proper and such that for all $i \in \mathbb{N}$, $N_i(F) = F$, then $N(F) = F$.*

Corollary 3.8. *If C is a recursive field, then there is a recursive procedure to construct the normal form of a given rational function.*

Proposition 3.9. *The following statements are true.*

1. *Let $F \in R_P$ be proper. Then $N(F)$ is unique.*

2. *Let $F, G \in R_P$. Assume F is proper. Assume G is proper. Further assume $F \neq G$, $F = N(F)$, $G = N(G)$, and $F \cong G$ modulo P. Then the pair (F, G) is improper.*

Proof. 1. Let $G_1, G_2 \in R_P$ be normal forms of F. Then for every $i \in \{1, \dots, m\}$, $|V_i(G_1)| = |V_i(G_2)|$. In the case, $|V_i(G_1)| = |V_i(G_2)| = 0$ or $|V_i(G_1)| = |V_i(G_2)| = 2$, by definition of normal form, we actually have $V_i(G_1) = V(G_2)$. Suppose now $|V_i(G_1)| = |V_i(G_2)| = 1$. Here we note that since both pairs (G_1, F) and (G_2, F) are proper, it is not the case that $x_{i,5} = V_i(G_1), x_{i,7} = V_i(G_2)$ or vice versa. Thus, for all i, in any case, there exist a solution triple $(x_{i,j_i}, x_{i,k_i}, x_{i,r_i})$ such that $V_i(G_1) \cup V_i(G_2) \subset \{x_{i,j_i}, x_{i,k_i}, x_{i,r_i}\}$. Thus, using the standard substitution $x_{i,u}/S_{i,u}(x_{i,j_i}, x_{i,k_i}, x_{i,r_i})$ for all $i = 1, \dots, m$, we will be able to conclude that $G_1 = G_2$.

2. We will assume that the pair (F, G) is proper and conclude that $F = G$. If we assume that (F, G) is proper, then by definition of normal form we must conclude, as above, that for all $i = 1, \dots, m$, there exist a solution triple $(x_{i,j_i}, x_{i,k_i}, x_{i,r_i})$ such that $V_i(F) \cup V_i(G) \subset \{x_{i,j_i}, x_{i,k_i}, x_{i,r_i}\}$. The conclusion now follows, again as above, by the standard substitution. □

Corollary 3.10. *Let $F \in R_P$. Assume $F = N(F)$ and $F \cong c$ modulo P, for some $c \in C$. Then $F = c$.*

Corollary 3.11. *Let $G \in R_P$ and let F be either the normal or a pseudo-normal form of G. Let i be fixed and let $(x_{i,j_i}, x_{i,k_i}, x_{i,r_i})$ be any solution triple. Then the following statements are true.*

1. *For any $s \neq i$, $V_s(F(x_{i,u}/S_{i,u}(x_{i,j_i}, x_{i,k_i}, x_{i,r_i}))) = V_s(F)$.*

2. *There exists a fat set B_i, such that for any $a_1, a_2, a_3 \in B_i$ and any $s \neq i$,*

$$V_s(F(x_{i,u}/S_{i,u}(a_1, a_2, a_3))) = V_s(F).$$

Proof. 1. Suppose for some $s \neq i$, $V_s(F(x_{i,u}/S_{i,u}(x_{i,j_i}, x_{i,k_i}, x_{i,r_i}))) \neq V_s(F)$.

Since no new variables of V_s can be added by an i-th standard substitution, we must conclude that $|V_s(F(x_{i,u}/S_{i,u}(x_{i,j_i}, x_{i,k_i}, x_{i,r_i})))| < |V_s(F)|$, while one has $F(x_{i,u}/S_{i,u}(x_{i,j_i}, x_{i,k_i}, x_{i,r_i}) \cong F$ modulo P. This however would contradict our assumption that F is either a normal or a pseudo-normal form.

2. The second assertion of the corollary now follows from Corollary 2.4. □

Proposition 3.12. *Let $\{G_t = N(G_t)\}, t = 1, \ldots, m$ be a finite set of elements of R_P. Let i be fixed. Let $(x_{i,j_i}, x_{i,k_i}, x_{i,r_i})$ be a solution triple. Then there exists a fat set $B_i \subset C^3$ such that for all triples$(a_{i,1}, a_{i,2}, a_{i,3}) \in B_i$ for all $t = 1, \ldots, m$,*

$$G_t(x_{i,u}/S_{i,uj_ik_ir_i}(a_{i,1}, a_{i,2}, a_{i,3})) = N(G_t(x_{i,u}/S_{i,u}(a_{i,1}, a_{i,2}, a_{i,3})),$$

and no variable from V_s, $s \neq i$, vanishes from the functions under consideration.

Proof. From Corollary 3.7, it follows that it is enough to insure that for all triples $(a_{i,1}, a_{i,2}, a_{i,3}) \in B_i$, for all $s \neq i$,

$$G_t(x_{i,u}/S_{i,u}(a_{i,1}, a_{i,2}, a_{i,3})) = N_s(G_t(x_{i,u}/S_{i,u}(a_{i,1}, a_{i,2}, a_{i,3}))).$$

If $V(G_t) \cap V_s = \emptyset$, then substitution will not affect the status of G_t as the s-th normal form. If $|V_s(G_t)| = 1$, then by Corollary 3.3, $N_s(G_t(x_{i,u}/S_{i,u}(a_{i,1}, a_{i,2}, a_{i,3})) = G_t(x_{i,u}/S_{i,u}(a_{i,1}, a_{i,2}, a_{i,3})$, since the substitution cannot add new s-variables, and it is enough to make sure that after the substitution

$$V_s(G_t) = V_s(G_t(x_{i,u}/S_{i,u}(a_{i,1}, a_{i,2}, a_{i,3}))). \tag{3.2}$$

Finally, if $|V(G_t) \cap V_s| = 2$, then consider $G_{t,1}, G_{t,2}, G_{t,3}$ defined in a manner analogous to F_s in Proposition 3.2. (Note that since $G_t = N(G_t)$, by Corollary 3.3, $G_t = G_{t,3}$.) Note further that $G_{t,1}, G_{t,2}, G_{t,3}$ are either normal or pseudo-normal forms of G_t. It is enough to make sure that

$$|V_s(G_{t,1}(x_{i,u}/S_{i,u}(a_{i,1}, a_{i,2}, a_{i,3}))| = 2, \tag{3.3}$$

$$|V_s(G_{t,2}(x_{i,u}/S_{i,u}(a_{i,1}, a_{i,2}, a_{i,3}))| = 2, \tag{3.4}$$

$$|V_s(G_{t,3}(x_{i,u}/S_{i,u}(a_{i,1}, a_{i,2}, a_{i,3}))| = 2 \tag{3.5}$$

after the substitution. (Note that making a substitution leading from G_t to $G_{t,j}$ for some $j \in \{1, 2, 3\}$, and then substituting $S_{i,u}(x_{i,j_i}, x_{i,k_i}, x_{i,r_i})$ for $x_{i,u}$ is equivalent to substituting $S_{i,u}(x_{i,j_i}, x_{i,k_i}, x_{i,r_i})$ for $x_{i,u}$ first and then making a substitution leading from $G_t(x_{i,u}/S_{i,u}(x_{i,j_i}, x_{i,k_i}, x_{i,r_i}))$ to $G_{t,j}(x_{i,u}/S_{i,u}(x_{i,j_i}, x_{i,k_i}, x_{i,r_i}))$ for some $j \in \{1, 2, 3\}$ due to the fact that the sets of variables involved in these two substitutions are disjoint. Thus, all the s-th pseudo-normal forms of $G_t(x_{i,u}/S_{i,u}(x_{i,j_i}, x_{i,k_i}, x_{i,r_i}))$ are

$[G_{t,j}](x_{i,u}/S_{i,u}(x_{i,j_i}, x_{i,k_i}, x_{i,r_i}))$, where "[ ]" signifies that the substitution involving variables from V_s was done first.)

By Corollary 3.11, for each equation (3.3)–(3.5), the set of triples of values which can be used as values of $(a_{i,1}, a_{i,2}, a_{i,3})$ to assure that this equation is satisfied, contains a fat set. Thus, by Note 2.1, the set of triples of values which can be used as values of $(a_{i,1}, a_{i,2}, a_{i,3})$ to assure that all the equations (3.2)–(3.5) are satisfied simultaneously, also contains a fat set.

4. Construction

In this section we will construct a map from R_P into natural numbers. This map restricted to C will be the presentation we are seeking.

Notations 4.1. Let $q : \mathbb{N} \longrightarrow C$ be a recursive enumeration of C so that $q^{-1} : C \longrightarrow \mathbb{N}$ is a recursive presentation of C. Let $(\pi_1, \pi_2) : \mathbb{N}^2 \longrightarrow \mathbb{N}$ be a recursive bijection. The functions P_+, P_-, $P_\times$, $P_/$ will eventually be recursive translations of the field operations for the field C. The function "inv" will be a function from $\mathbb{N}$ into the set of rational functions of R_P in normal form. (For future reference, we note here that all the elements of C are automatically in normal form and, therefore, C will be included in the range of "inv".) Our goal is to construct a function $f : C \longrightarrow \mathbb{N}$ which will be a weak presentation of C not extendible to a recursive presentation of any field containing C. The function "inv" restricted to the range of f will be the inverse of f.

Let $\{\phi_i\}$ be an effective listing of all partial recursive functions of two variables with every partial recursive function appearing infinitely many times on this list.

Step 0. Start a parallel computation of all partial recursive functions of two variables. Define initial values of inv: $\mathbb{N} \longrightarrow R_P$. For each $k \in \mathbb{N}$, $j = 1, \ldots, 7$ let $\text{inv}(8k+j) = x_{k,j}$. Next define the initial values of P_+.

$$P_+(8k+1, 8k+2) = 8k+3,$$

$$P_+(8k+3, 8k+4) = 8k+5,$$

$$P_+(8k+2, 8k+4) = 8k+6,$$

$$P_+(8k+1, 8k+6) = 8k+7.$$

For $n \in \mathbb{N}$ do the following:

Step $7n+1$. Let m be the smallest non-negative integer such that $f(q(m))$ has not been defined yet. Let $f(q(m)) = t \cong 0$ modulo 8, where t is a fresh natural number, i.e. a natural number which has not been used in the procedure by this step. Set $\text{inv}(t) = q(m)$.

Steps $7n+2, 7n+3, 7n+4, 7n+5$. Let op stand for $+, -, \times, /$ in steps $7n+2, 7n+3, 7n+4$, and $7n+5$ respectively. Let m be the smallest non-negative integer such that $P_{\text{op}}(\pi_1(m), \pi_2(m))$ has not been defined yet, while $\text{inv}(\pi_1(m))$ and $\text{inv}(\pi_2(m))$ have been defined. Let $T_1 = \text{inv}(\pi_1(m))$ and let $T_2 = \text{inv}(\pi_2(m))$. First we consider an element of K, $T = T_1 \text{ op } T_2$. If op = "/" and $\text{inv}(\pi_2(m)) \in P$ or if either of $\text{inv}(\pi_1(m))$, $\text{inv}(\pi_2(m))$ contains discarded variables (this term will be explained below), or if the pair ($\text{inv}(\pi_1(m))$,$\text{inv}(\pi_2(m))$) is improper then set $P_{\text{op}}(\pi_1(m), \pi_2(m)) = 0$. Otherwise we have the following two cases.

Case 1. There exists a natural $r \in \mathbb{N}$ such that $\text{inv}(r) = N(T)$. In this case set $P_{\text{op}}(\pi_1(m), \pi_2(m)) = r$.

Case 2. There is no $r \in \mathbb{N}$ such that $\text{inv}(r) = N(T)$. Then find a natural number $r \cong 0$ modulo 8 which has not been used in the construction so far. Set $\text{inv}(r) = N(T)$, $P_{\text{op}}(\pi_1(m), \pi_2(m)) = r$. If $\text{inv}(r) \in C$ then define $f(\text{inv}(r)) = r$.

We should note here, that assuming inductively that $T_1, T_2 \in R_P$, (T_1, T_2) is a proper pair and op $\neq$ /" if $T_2 \in P$, we are guaranteed that $T \in R_P$ and proper. Thus, $N(T)$ is defined.

Step $7n+6$. Let k be the smallest integer such that ϕ_k, is defined at this point of the construction on the pairs $(8k+1, 8k+2)$, $(8k+3, 8k+4)$,$(8k+2, 8k+4)$, $(8k+1, 8k+6)$, and for at least one of the above listed pairs the image under ϕ_k is different from the image under P_+. Without loss of generality we can assume that the "disagreement" triple is $(8k+1, 8k+2, 8k+3)$. Then *discard* variables $x_{k,4}, \dots, x_{k,7}$ and find two previously unused values $a, b \in C$ satisfying the following requirements for all functions in the range of inv containing no discarded variables.

1. For each function T, $T(x_{k,1}/a, x_{k,2}/b, x_{k,3}/(a+b))$ is well defined after the substitution, i.e. its denominator (if any) does not become an element of P.

2. For every pair of functions T_1, T_2 such that such that $T_1 - T_2 \notin PR_P$,

   $$T_1(x_{k,1}/a, x_{k,2}/b, x_{k,3}/(a+b)) - T_2(x_{k,1}/a, x_{k,2}/b, x_{k,3}/(a+b)) \in R_P \setminus PR_P.$$

3. The number of occurrences of all the other variable remains the same in all the functions after the substitution.

4. All the functions remain in normal form after the substitution.

5. If T_1, T_2, T_3 are all in the range of inv, contain no discarded variables, and for some op $\in \{+, -, \times, /\}$, $T_1 \text{ op } T_2 \cong T_3$ modulo P, then

   $$T_1(x_{k,1}/a, x_{k,2}/b, x_{k,3}/(a+b)) \text{ op } T_2(x_{k,1}/a, x_{k,2}/b, x_{k,3}/(a+b))$$

   $$\cong T_3(x_{k,1}/a, x_{k,2}/b, x_{k,3}/(a+b)) \text{ modulo } P.$$

Carry out the substitutions in the functions described above. If, after the substitutions, a function in the range of inv becomes an element of C, make the corresponding f-definition.

Given a pair of values $a, b \in C$, checking whether this pair satisfies these requirements will take only finitely many steps, since we only have to check finitely many functions. On the other hand, to establish that the required values always exist, consider the following. Let $x_{k,j}$, $j \neq 3$ be an element of V_k such that $x_{k,1}$, $x_{k,2}$ and $x_{k,j}$ form a solution triple. Then in the k-th standard substitution corresponding to this solution triple, $x_{k,3}$ is replaced by $x_{k,1} + x_{k,2}$. What we are actually doing in this step is carrying out a partial standard substitution. For the functions in the range of inv which only have occurrences of $x_{k,1}, x_{k,2}, x_{k,3}$, this partial standard substitution is equivalent to the full standard substitution, as has been shown in Lemma 3.1. On the other hand, if for some $m \in \mathbb{N}$, $\mathrm{inv}(m)$ is a rational function which has occurrences of any of $x_{k,4}, \ldots, x_{k,7}$ then m will never be added to the range of f.

By Corollaries 3.11 and 2.11, we know that the set of values for $x_{k,1}, x_{k,2}, x_{k,j}$ satisfying Requirements 1, 2 and 3 contains a fat set. Furthermore, by Proposition 3.12, the same statement applies to Requirement 4. Finally, Note 2.1 guarantees that the set of values simultaneously satisfying Requirements 1–4 also contains a fat set. Finally, Requirement 5 will be automatically satisfied by Lemma 2.9. Thus, the set of possible values for a and b satisfying all the requirements of this step is infinite.

Lemma 4.1. *The following statements are true.*

1. *For all* $m \in \mathbb{N}$, $\mathrm{inv}(m)$, *if it is defined, is an element of* R_P *in normal form.*

2. *For any* $\mathrm{op} \in \{+, -, \times, /\}$, $m_1, m_2 \in \mathbb{N}$, $\mathrm{inv}(m_1)\ \mathrm{op}\ \mathrm{inv}(m_2) \cong \mathrm{inv}(P_{\mathrm{op}}(m_1, m_2))$ *modulo* P, *unless* $\mathrm{inv}(m_2) \in P$ *and* $\mathrm{op} =$ "/", $(\mathrm{inv}(m_1), \mathrm{inv}(m_2))$ *is an improper pair or* $\mathrm{inv}(m_1)$ *or* $\mathrm{inv}(m_2)$ *contains discarded variables.*

3. *If at some point of the construction the pair* $(\mathrm{inv}(m_1)\ \mathrm{inv}(m_2))$ *is improper, then after any later step, either the pair is improper or one of the* inv*'s contains discarded variables.*

4. *If at some point of the construction the pair* $(\mathrm{inv}(m_1), \mathrm{inv}(m_2))$ *is improper, then after any later step, it is not the case that* $\mathrm{inv}(m_1) \in C$, *and* $\mathrm{inv}(m_2) \in C$.

5. *If* $m_1 \neq m_2$ *then either* $\mathrm{inv}(m_1) \not\cong \mathrm{inv}(m_2)$ *modulo* P *or the pair* $(\mathrm{inv}(m_1), \mathrm{inv}(m_2))$ *is improper. Furthermore,* inv *restricted to* R_P *functions in normal form without discarded variables is injective and for all* $c \in C$ *and all* $m \in \mathbb{N}$, $f(c) = m$ *if and only if* $\mathrm{inv}(m) = c$. *(At this point we do not assert that* f *is single valued. That will follow however from the injectivity of* inv.*)*

6. *For any* m, *the value of* $\mathrm{inv}(m)$ *can change only finitely many times during the construction and* $\mathrm{inv} : \mathbb{N} \longrightarrow \{R_P\text{-elements in normal form}\}$ *is single-valued.*

Proof. 1. We prove this statement by induction. It is certainly true after initialization. Suppose it is true after r steps. Let $r+1 \cong 1$ modulo 7. A new definition of inv in this step will produce a value in C and thus one in normal form. If $r+1 \cong 2,3,4,5$ modulo 7 then the new definition of inv made in this step will produce a value in normal form by construction. Finally, if $r+1 \cong 6$ modulo 7, then all modifications to values of inv result in values in normal form by Requirement 4 of this step.

2. We again proceed by induction. This assertion is true after initialization. Suppose it is true after r steps. Let $r+1 \cong 1$ modulo 7. No new values of P_{op} are defined in this step. Furthermore, if for some $m \in \mathbb{N}$, $\text{inv}(m)$ has been defined in this step, then m had not been used in the construction so far and thus could not have been a part of the definition of any P_{op} so far. If $r+1 \cong 2,3,4,5$ modulo 7, then we have to consider two cases. If anew definition of P_{op} is made using an old value of inv, then the statement under consideration will be true after this step by construction, since for any $T \in R_P$, $N(T) \cong T$ modulo P. If, on the other hand, a P_{op} definition has been made using a new value of inv, then the statement under consideration is true because, as above, $N(T) \cong T$ modulo P and due to the fact that the number added to the domain of inv had not been used in the construction before, and thus isn't a part of any other P_{op} definition. Finally, if $r+1 \cong 6$ modulo 7, then Requirement 5 guarantees that the statement under consideration is true.

3. If $\text{inv}(m_1)$, $\text{inv}(m_2)$ is an improper pair after some step of the procedure, then for some i, $\text{inv}(m_1)$, $\text{inv}(m_2)$ are improper with respect to i. Previously defined inv's are modified in steps $7n+6$ only. Thus, it is enough to show that the assertion under consideration is true after any step of the form $7n+6$. Given a step of the form $7n+6$, we have to consider whether the substitution taking place in the step is replacing variables from V_i or not. In the first case, after this step either $\text{inv}(m_1)$ or $\text{inv}(m_2)$ will have discarded variables and thus will not be modified in any subsequent steps. In the second case, Requirement 3 assures, that $V_i(\text{inv}(m_1))$ and $V_i(\text{inv}(m_2))$ remain unchanged after this step, and thus the pair will remain improper with respect to i.

4. If $(\text{inv}(m_1), \text{inv}(m_2))$ is an improper pair after some step of the procedure, then by the preceding argument, the pair will remain improper or one of the members of the pair will have discarded variables. Thus not both $\text{inv}(m_1)$,$\text{inv}(m_2)$ can become elements of C, since this would make the pair proper and would mean that neither function has occurrences of discarded variables.

5. We proceed by induction again. The assertion is true after initialization. Assume it is true after r steps. Let $r+1 \cong 1$ modulo 7. A new inv and f definition is made at this step. It will add an element c of C to the range of inv and to the domain of f, and a new integer m to the domain of inv and the range of f. Since $f(c)$ has not been defined before this step, by the inductive hypothesis c is not in the range of inv before this step. Thus, inv will remain injective and the statement "$\text{inv}(m) \in C$ if and only if $f(\text{inv}(m)) = m$" will remain true. Injectivity of inv together with the fact that all functions in the range of inv are in normal form, will continue to make inv injective modulo P when restricted to proper pairs without discarded variables, by Proposition 3.9.

If $r+1 \cong 2,3,4,5$ modulo 7, and for some m, inv(m) has been defined in this step, then, m is a new element of the domain of inv and inv(m) is a new element of the range of inv. Thus injectivity and injectivity mod P on proper pairs without discarded variables are preserved, since inv(m) and all the other elements of the range of inv are in normal form. If inv$(m) \in C$, then a new f-definition is made in this step.

Finally, suppose $r+1 \cong 6$. Requirements 2 and 3, assure that proper pairs of values of inv are inequivalent or one element of the pair has discarded variables after the substitution. Thus the assertion concerning the injectivity, and injectivity modulo P, of inv restricted to proper pairs without discarded variables, will remain true after this step. On the other hand, if for some m, inv(m) becomes an element $c \in C$ in this step, we set $f(c) = m$.

6. For any natural number m, the number of times inv(m) can change is less or equal to the number of variables in inv(m) at the time of its first definition. Thus, inv(m) is defined uniquely for any m by the construction. □

Proposition 4.2. *The constructed function $f : C \to \mathbb{N}$ is a weak presentation of C.*

Proof. Steps $7n+1$ assure that f is defined for every $c \in C$. Since inv is one-to-one and single valued, f is one-to-one and single valued. Next let op $\in \{+, -, \times, /\}$, let $c_1, c_2 \in C$ and consider $P_{\mathrm{op}}(f(c_1), f(c_2))$ assuming op $\neq$ "/" unless $c_2 \neq 0$. By Lemma 4.1, $\mathrm{inv}(P_{\mathrm{op}}(f(c_1), f(c_2))) \cong \mathrm{inv}(f(c_1))\ \mathrm{op}\ \mathrm{inv}(f(c_2)) = c_1\ \mathrm{op}\ c_2$ modulo P. On the other hand, again by Lemma 4.1, $\mathrm{inv}(P_{\mathrm{op}}(f(c_1), f(c_2))$ is an element of R_P in normal form. Thus, by Corollary 3.10, $\mathrm{inv}(P_{\mathrm{op}}(f(c_1), f(c_2)) = c_1\ \mathrm{op}\ c_2$ and consequently, by Lemma 4.1 again, $P_{\mathrm{op}}(f(c_1), f(c_2)) = f(c_1\ \mathrm{op}\ c_2)$. □

Proposition 4.3. *There is no recursive field M containing C such that f can be extended to a recursive presentation of M.*

Proof. From the construction it is clear that every total recursive function to which P_+ is extendible will violate associative law infinitely many times. Thus, if M and an extension of f as described above existed, $\mathbb{N} \setminus f(M)$ would have to be infinite. In this case however, it is not hard to take a finitely generated extension of M and extend f recursively to this extension so that the image is all of $\mathbb{N}$. This however will produce a total recursive extension of P_+respecting associative law. □

5. Appendix

Lemma 5.1. *Consider the following system.*

$$\begin{cases} x_1 + x_2 - x_3 = 0 \\ x_3 + x_4 - x_5 = 0 \\ x_2 + x_4 - x_6 = 0 \\ x_1 + x_6 - x_7 = 0 \end{cases} \tag{5.1}$$

Then two variables from any of the equations and a variable from another equation will constitute a basis for the solution space except for the following cases:

- *The triple cannot contain x_5 and x_7 at the same time.*
- *Triples (x_3, x_4, x_7) and (x_1, x_6, x_5) are excluded.*

Proof. 1. (x_1, x_2, x_4), or (x_1, x_3, x_4) or (x_2, x_3, x_4):

$$x_3 = x_1 + x_2;\ x_5 = x_3 + x_4;\ x_6 = x_2 + x_4;\ x_7 = x_1 + x_6.$$

2. (x_1, x_2, x_5), or (x_1, x_3, x_5) or (x_2, x_3, x_5):

$$x_3 = x_1 + x_2;\ x_4 = x_5 - x_3;\ x_6 = x_2 + x_4;\ x_7 = x_1 + x_6.$$

3. (x_1, x_2, x_6), or (x_1, x_3, x_6) or (x_2, x_3, x_6):

$$x_3 = x_1 + x_2;\ x_4 = x_6 - x_2;\ x_5 = x_3 + x_4;\ x_7 = x_1 + x_6.$$

4. (x_1, x_2, x_7), or (x_1, x_3, x_7) or (x_2, x_3, x_7):

$$x_3 = x_1 + x_2;\ x_6 = x_7 - x_1;\ x_4 = x_6 - x_2;\ x_5 = x_3 + x_4.$$

5. (x_3, x_4, x_6), or (x_3, x_5, x_6) or (x_4, x_5, x_6):

$$x_3 + x_4 = x_5;\ x_2 = x_6 - x_4;\ x_1 = x_3 - x_2;\ x_7 = x_1 + x_6.$$

6. (x_2, x_4, x_7), or (x_2, x_6, x_7) or (x_4, x_6, x_7):

$$x_2 + x_4 = x_6;\ x_1 = x_7 - x_6;\ x_3 = x_1 + x_2;\ x_5 = x_3 + x_4.$$

7. (x_1, x_6, x_4), or (x_1, x_7, x_4) or (x_6, x_7, x_4):

$$x_1 + x_6 = x_7;\ x_2 = x_6 - x_4;\ x_3 = x_1 + x_2;\ x_5 = x_3 + x_4. \qquad \square$$

Corollary 5.2. *Let $x_{i,j_i}, x_{i,k_i} \in V_i$ be such that $\{x_{i,j_i}, x_{i,k_i}\} \neq \{x_{i,5}, x_{i,7}\}$. Then there exists $x_{i,r_i} \in V_i$ such that $(x_{i,j_i}, x_{i,k_i}, x_{i,r_i})$ is a solution triple.*

Proof. If $x_{i,j_i}, x_{i,k_i} \in V_{i,j}$ for some j then let $x_{i,k_i} \in V_i \setminus V_{i,j}$ and assume $x_{i,k_i} \notin \{x_{i,5}, x_{i,7}\}$. (Such a x_{i,k_i} always exists because V_i contains seven variables out of which at most five cannot be used.) Then $(x_{i,j_i}, x_{i,k_i}, x_{i,r_i})$ is a solution triple by Lemma 5.1. If for all $j = 1, 2, 3, 4$, $x_{i,j_i}, x_{i,k_i} \notin V_{i,j}$, then we can assume that $x_{i,j_i} \notin \{x_{i,5}, x_{i,7}\}$ and for some $j = 1, 2, 3, 4$, $x_{i,k_i} \in V_{i,j}$. Let $x_{i,r_i} \in V_{i,j} \setminus \{x_{i,k_i}\}$. Then, by Lemma 5.1 again, $(x_{i,j_i}, x_{i,k_i}, x_{i,r_i})$ is a solution triple. $\square$

Lemma 5.3. *Let $i \in \mathbb{N}$ and let*

$$M_{i,u} = x_{i,u} - S_{i,u}(x_{i,j_i}, x_{i,k_i}, x_{i,r_i}),$$

where $(x_{i,j_i}, x_{i,k_i}, x_{i,r_i})$ is a solution triple. Then $(M_{i,1}, \ldots, M_{i,7}) = P_i$.

Proof. First of all note that by Lemma 2.1, $L_{i,j}(x_{i,u}/S_{i,u}) = 0$ for all $i = 1, \ldots, m$ and all $j = 1, 2, 3, 4$. Secondly, consider the following equations.

$$L_{i,j} - L_{i,j}(x_{i,u}/S_{i,u}) = L_{i,j}, \, j = 1, 2, 3, 4.$$

Taking into account that $L_{i,j}$ is a linear equation we obtain

$$L_{i,j}(x_{i,u}/(x_{i,u} - S_{i,u})) = L_{i,j}, \, j = 1, 2, 3, 4.$$

$$L_{i,j}(x_{i,u}/M_{i,u}) = L_{i,j}, \, j = 1, 2, 3, 4.$$

In fact we can consider the last four equations as a system of linear equations over the function field:

$$\begin{cases} M_{i,1} + M_{i,2} - M_{i,3} = L_{i,1} \\ M_{i,3} + M_{i,4} - M_{i,5} = L_{i,2} \\ M_{i,2} + M_{i,4} - M_{i,6} = L_{i,3} \\ M_{i,1} + M_{i,6} - M_{i,7} = L_{i,4} \end{cases}$$

Next note that $M_{i,j_i} = M_{i,k_i} = M_{i,r_i} = 0$. Assuming, x_{i,s_i} belonged to the same triple as $x_{i,j_i}, x_{i,k_i}, M_{i,s_i} \in P_i$. Next note that either the solution triple contained $x_{i,5}$ or $x_{i,7}$ or among the four equations of the system there is one, different from the equation containing $M_{i,j_i}, M_{i,k_i}, M_{i,s_i}$, where one of $M_{i,j_i}, M_{i,k_i}, M_{i,s_i}$ occurs together with M_{i,r_i}. In the first case we note that $M_{i,5} - M_{i,7}\, in P_i$, and thus both must be in P_i. In the latter case, the third variable in the above mentioned equation must be in P_i. In either case the remaining two equations will contain at least two variables from P_i, thus forcing all the variables into P_i. Therefore, we can conclude that $(M_{i,1}, \ldots, M_{i,7}) \subseteq P_i$. On the other hand, the system above shows that $P_i \subseteq (M_{i,1}, \ldots, M_{i,7})$. □

References

[1] C. Jockusch, A. Shlapentokh, Weak Presentations of Computable Fields, J. Symbolic Logic 60, (1995), 199–208.

[2] M. Fried, M. Jarden, Field Arithmetic, Springer-Verlag, New York, 1986.

[3] M. Rabin, Computable Algebra, Trans. Amer. Math. Soc. 95 (1960), 341–360.

[4] J. B. Remmel, Effective Structures not Contained in Recursively Enumerable Structures, Aspects of Effective Algebra (ed. J. N. Crossley), 1985, 206–225.

[5] A. Shlapentokh, Diophantine Equivalence and Countable Rings, J. Symbolic Logic 59 (1994), 1068–1095.

[6] A. Shlapentokh, Non-Standard Extensions of Weak Presentations, J. Algebra. 176 (1995), 735–749.

[7] A. Shlapentokh, Algebraic and Turing Separability of Rings, J. Algebra 185 (1996), 229–257.

[8] A. Shlapentokh, Rational Separability over Global Fields, Ann. Pure Appl. Logic 79 (1996), 93–108.

[9] A. Shlapentokh, Weak Presentations of Non-finitely Generated Fields, Ann. Pure Appl. Logic. 94 (1998), 223–252.

Jumps of Σ_2^0-high e-degrees and properly Σ_2^0 e-degrees

Richard Shore and Andrea Sorbi†*

Abstract. We show that the Σ_2^0 high e-degrees coincide with the high e-degrees. We also show that not every properly Σ_2^0 e-degree is high.

1. Introduction

Enumeration reducibility is the notion of relative enumerability of sets: a set A is *enumeration reducible* (or simply *e-reducible*) to a set B, in symbols, $A \leq_e B$, if there is an effective procedure for enumerating A given *any* enumeration of B. Formally, we define $A \leq_e B$ if there is some computably enumerable set Φ (called in this context an *enumeration operator* or simply an *e-operator*) such that

$$A = \{x : (\exists \text{ finite } D)[\langle x, D\rangle \in \Phi \;\&\; D \subseteq B]\}$$

(throughout the paper we identify finite sets with their canonical indices). We denote by $\equiv_e$ the equivalence relation generated by the preordering relation $\leq_e$ and $\deg_e(A)$ denotes the equivalence class (or the *e-degree*) of A. The partially ordered structure of the e-degrees is denoted by $\mathfrak{D}_e$; its partial ordering is denoted by $\leq$. $\mathfrak{D}_e$ is, in fact, an upper semilattice with least element $\mathbf{0}_e$.

One of the most interesting features of the e-degrees is that they extend the structure $\mathfrak{D}_T$ of the Turing degrees ([Med55] and [Rog67]). Indeed if we define $\iota : \mathfrak{D}_T \to \mathfrak{D}_e$ by $\iota(\deg_T(A)) = \deg_e(\chi_A)$ (where $\deg_T(A)$ is the Turing degree of the set A and χ_A denotes the characteristic function of A) then ι is the desired embedding. (In fact, it preserves joins and least element.) The verification that ι is well defined relies on the following lemma:

*Department of Mathematics, Cornell University, Ithaca NY 14853, USA. The first author was partially supported by NSF Grant DMS-9503503 and a CNR visiting professorship at the University of Turin.

†Department of Mathematics, University of Siena, Italy. The second author was partially supported by the HC&M research network *Complexity, Logic and Recursion Theory* (*COLORET*), contract no. ERBCHRXCT930415; MURST 60%, and CNR-GNSAGA. During the preparation of part of this paper the second author was visiting at the University of Wisconsin, Madison.

Lemma 1.1. *For every total function f and g, we have*

$$f \leq_T g \Leftrightarrow f \leq_e g.$$

Proof. See e.g. [Rog67, p. 153]. □

One can define (see below) a jump operation $'$ on the e-degrees, and therefore introduce the notions of a *low* e-degree (i.e. an element of the class $\mathbf{L}_1 = \{\mathbf{a} \leq \mathbf{0}'_e : \mathbf{a}' = \mathbf{0}'_e\}$); and that of a *high* e-degree (i.e. an element of $\mathbf{H}_1 = \{\mathbf{a} \leq \mathbf{0}'_e : \mathbf{a}' = \mathbf{0}''_e\}$). Moreover, since ι preserves jump, we have that low Turing degrees are mapped to low e-degrees and high Turing degrees are mapped to high e-degrees.

A nice, useful characterization of the class $\mathbf{L}_1$ of the low e-degrees is given in [MC85]: $\mathbf{a} \in \mathbf{L}_1$ if and only if $\mathbf{a}$ contains a set A such that, for every $B \leq_e A$, $B \in \Delta^0_2$. Thus $\mathbf{a}$ and all the e-degrees below $\mathbf{a}$ consist entirely of Δ^0_2 sets. In this paper, we characterize the class $\mathbf{H}_1$ of high e-degrees by a result analogous to the one characterizing the high Turing degrees as those containing a set with an approximation whose associated computation function dominates every total recursive function (Theorem 2.1). The relevant definitions of a Σ^0_2 approximation and a computation function are given below. The e-degrees of sets with such approximations are known as the Σ^0_2-high e-degrees (Definition 1.7). This characterization answers Question 7.3 of [Coo90].

Since the e-degrees below $\mathbf{0}'_e$ are exactly the e-degrees consisting of Σ^0_2 sets, in view of the above cited characterization of the low e-degrees, a natural question to ask is where an e-degree $\mathbf{a} \leq \mathbf{0}'_e$ which contains no Δ^0_2 set (such an e-degree is called a *properly* Σ^0_2 e-degree) lies in the low/high hierarchy. A natural conjecture might be that the properly Σ^0_2 e-degrees are all in $\mathbf{H}_1$. Cooper and Copestake ([CC88]) show that there exist properly Σ^0_2 e-degrees that are Σ^0_2-high, and thus lie in $\mathbf{H}_1$ by Theorem 2.1. However, in Theorem 3.1 we show that the properly Σ^0_2 e-degrees are not contained in $\mathbf{H}_1$.

Our notations and terminology are mostly based on [Soa87]. The reader is referred to [Coo90] for an introduction and extensive bibliography on enumeration reducibility. We will be mostly working with Σ^0_2 sets. We recall that a Σ^0_2 *approximation* to a Σ^0_2 set A is computable sequence of computable sets $\{A^s : s \in A^s\}$ such that $A = \{x : (\exists t)(\forall s \geq t)[x \in A^s]\}$. See [LS92] for an introduction to Σ^0_2 approximations, and for a proof that every Σ^0_2 set has a *good* Σ^0_2 approximation $\{A^s : s \in A^s\}$, i.e. a computable sequence of computable (in fact, finite) sets such that $\{s : A^s \subseteq A\}$ is infinite.

Let X be any set of natural numbers; if x is a number, then $X^{[x]} = \{z \in X : (\exists y)[z = \langle x, y\rangle]\}$, and $X \restriction x = \{y \in X : y < x\}$. If σ is a string and $x < |\sigma|$ (where $|\sigma|$ denotes the length of σ), then $\sigma \restriction x$ denotes the initial segment of σ having length x; likewise, if f is a function, then $f \restriction x$ denotes the initial segment of f having length x.

Let $\{\varphi_i\}_{i\in\omega}$ be the standard enumeration of all partial computable functions with corresponding enumerations $\{W_i\}_{i\in\omega}$ and $\{\Phi_i\}_{i\in\omega}$ of the computably enumerable sets and the enumeration operators, respectively. Let us fix, as in [Soa87, p. 16], computable approximations $\{\varphi_{i,s}\}_{i,s\in\omega}$ to the partial computable functions. Without loss of generality, we may assume that if $\varphi_{i,s}(x)\downarrow$ then $\varphi_{i,s}(x) < s$. Correspondingly, we get computable finite approximations $\{W_{i,s}\}_{i,s\in\omega}$ and $\{\Phi_{i,s}\}_{i,s\in\omega}$ to the computably enumerable sets and the enumeration operators, respectively.

Let

$$K^A = \{x : x \in \Phi_x^A\}.$$

Lemma 1.2 ([McE85]). *Let A, B be sets; then*

$$A \leq_e B \Leftrightarrow A \leq_1 K^B \Leftrightarrow K^A \leq_1 K^B.$$

Define the *jump* of a set A ([McE85]) to be the set $J_e(A) = \chi_{K^A}$. (Note that we identify functions with their graphs.) Clearly $J_e(A) \equiv_e K^A \oplus \overline{K^A}$. If $\mathbf{a}$ is an e-degree, then we can define $\mathbf{a}'$ as $\deg_e(J_e(A))$ for any $A \in \mathbf{a}$ since, by the previous lemma,

$$A \equiv_e B \Rightarrow K^A \oplus \overline{K^A} \equiv_e K^B \oplus \overline{K^B}$$

this gives a well defined unary operation on the e-degrees. Moreover, $\mathbf{a} < \mathbf{a}'$ for every e-degree $\mathbf{a}$.

The following lemma records two important properties of the jump operation.

Lemma 1.3 ([McE85]). *For every set A,*

1. $\iota((\deg_T(A))') = (\deg_e(A))'$;

2. *if A is* total *(i.e. the graph of some total function), then $J_e(A) \equiv_e \overline{K^A}$.*

Definition 1.4. A set A is called *e-high* if $A \in \Sigma_2^0$ and $J_e^{(2)}(\emptyset) \leq_e J_e(A)$. An e-degree $\mathbf{a}$ is called *high*, if $\mathbf{a}$ contains an e-high set (hence $\mathbf{a}' = \mathbf{0}_e''$).

By Lemma 1.3, the embedding ι preserves highness, i.e. it maps high T-degrees to high e-degrees.

The following is a useful characterization of the e-high sets.

Lemma 1.5. *For every set A,*

$$A \textit{ is e-high} \Leftrightarrow \text{Tot} \leq_e J_e(A) \Leftrightarrow \text{Tot} \leq_T K^A.$$

Proof. First notice that, by Lemma 1.3, A is e-high if and only if $\overline{K^{\overline{K}}} \leq_e J_e(A)$. Moreover, A is e-high if and only if, by Lemma 1.3, $\chi_{K^{\overline{K}}} \leq_e \chi_{K^A}$, if and only if, by Lemma 1.1, $\overline{K^{\overline{K}}} \leq_T K^A$. On the other hand, $\overline{K^{\overline{K}}} \equiv_1 \text{Tot}$. Indeed, $\overline{K^{\overline{K}}} \leq_1 \text{Tot}$

follows from the fact that Tot is Π^0_2-complete. To show that Tot $\leq_1 \overline{K^{\overline{K}}}$ simply observe that $\overline{\text{Tot}} \in \Sigma^0_2$, hence $\overline{\text{Tot}} \leq_e \overline{K}$, and thus, by Lemma 1.2, $\overline{\text{Tot}} \leq_1 K^{\overline{K}}$.

It follows that A is e-high if and only if Tot $\leq_e J_e(A)$ if and only if Tot $\leq_T K^A$. □

Finally notice the following:

Lemma 1.6. *For every total function f, $J_e(f) \equiv_T f'$.*

Proof. Since K^f is computably enumerable in f, we have $K^f \leq_1 f'$, and so $\overline{K^f} \leq_T f'$. On the other hand, f' is computably enumerable in f, hence, by the totality of f, $f' \leq_e f$, and thus $f' \leq_1 K^f$, by Lemma 1.2, which shows that $f' \leq_1 J_e(f)$.

Hence we conclude that $J_e(f) \equiv_T f'$. □

McEvoy ([McE85]) defines the notion of a Σ^0_2-high e-degree:

Definition 1.7. A Σ^0_2-*high* approximation $\{A^s\}_{s\in\omega}$ to a set A is a Σ^0_2 approximation such that the function

$$c(x) = \mu s(s > x \;\&\; A^s \upharpoonright x \subseteq A)$$

(called the *computation function for* A relative to the given approximation) is total and dominates every computable function. A set A is called Σ^0_2-*high*, if it has a Σ^0_2-high approximation. Finally an e-degree is said to be Σ^0_2-*high*, if it contains a Σ^0_2-high set.

Using Lemma 1.3 (1), it is shown in [McE85] that if $\mathbf{a} \leq \mathbf{0}'_e$ is total (hence a degree of the form $\iota(\mathbf{b})$, for some Turing degree $\mathbf{b} \leq_T \mathbf{0}'_T$), and $\mathbf{a}$ is high, then $\mathbf{a}$ is Σ^0_2-high. On the other hand, the class of Σ^0_2-high e-degrees properly extends the class of all high total e-degrees (in fact, there exist quasi-minimal Σ^0_2-high e-degrees, [McE85]).

In the next section we show that the high *e*-degrees and the Σ^0_2-high e-degrees coincide. This answers Question 7.3 of [Coo90].

2. The Σ^0_2-high e-degrees coincide with the high e-degrees

Theorem 2.1. *An e-degree* **a** *is Σ^0_2-high if and only if* **a** *is high.*

Proof. (⇒) Let A be Σ^0_2-high; let $\{A^s\}_{s\in\omega}$ be a Σ^0_2-high approximation to A, and let c be the computation function for A relative to this approximation.

By Lemma 1.5 it is enough to show that Tot $\leq_e J_e(A)$. We claim that, for every i,

$$
\begin{aligned}
i \in \text{Tot} &\Leftrightarrow (\forall y)[\varphi_i(y)\downarrow] \\
&\Leftrightarrow (\exists x)(\exists s)[(\forall y < x)\varphi_{i,s}(y)\downarrow \ \& \\
&\qquad (\forall y \geq x)(\forall t > y)[A^t \upharpoonright y \subseteq A \Rightarrow \varphi_{i,t}(y)\downarrow]].
\end{aligned}
$$

Indeed, if φ_i is total, then the function

$$\hat{\varphi}_i(x) = \mu s(\varphi_{i,s}(x)\downarrow)$$

is a total computable function, and thus dominated by c. Let x be such that $c(y) > \hat{\varphi}_i(y)$ for every $y \geq x$; then

$$(\forall y \geq x)(\forall t > y)[A^t \upharpoonright y \subseteq A \Rightarrow \varphi_{i,t}(y)\downarrow].$$

This establishes the left-to-right implication in the claimed equivalence. The right-to-left implication is trivial.

Now let

$$B = \{\langle i, x\rangle : (\exists y \geq x)(\exists t > y)[A^t \upharpoonright y \subseteq A \ \& \ \varphi_{i,t}(y)\uparrow]\}.$$

Clearly $B \leq_e A$. Hence, by Lemma 1.2, $B \leq_1 K^A$, via, say, the computable function h.

Then

$$
\begin{aligned}
i \in \text{Tot} &\Leftrightarrow (\exists x)(\exists s)[(\forall y < x)\varphi_{i,s}(y)\downarrow \ \& \ \langle i, x\rangle \notin B] \\
&\Leftrightarrow (\exists x)(\exists s)[(\forall y < x)\varphi_{i,s}(y)\downarrow \ \& \ h(\langle i, x\rangle) \in \overline{K^A}].
\end{aligned}
$$

It follows that Tot $\leq_e \overline{K^A}$, hence Tot $\leq_e J_e(A)$, as desired.

($\Leftarrow$) Assume that A is e-high. Then, by Lemma 1.5, we have that Tot $\leq_T K^A$. Let $Z = K^A$ and let $\{Z^s\}_{s\in\omega}$ be a good Σ_2^0-approximation to K^A. Let ψ be some Turing functional such that Tot $= \psi^Z$.

We now define an enumeration operator Θ by stages, and show that Θ^Z is Σ_2^0-high. Since the Σ_2^0-high e-degrees are closed upwards in Σ_2^0-e-degrees (see [BCS97]), it follows that $\deg_e(A)$ is Σ_2^0-high as well.

Construction:

Stage 0: Let $\Theta_0 = \emptyset$.

Stage $s+1$: For every $i \leq s$, we distinguish the following two cases:

(a) if $\psi_s^{Z^s}(i) = 0$, then for every $x \leq s$, we enumerate the axiom

$$\langle\langle i, x\rangle, Z^s\rangle \in \Theta_{s+1};$$

(b) otherwise, do nothing.

Let Θ^{s+1} consist of Θ^s plus all the axioms enumerated at stage $s+1$ and let $\Theta = \bigcup_{s\in\omega} \Theta_s$. We now prove that Θ is the desired enumeration operator by verifying a series of claims.

Verifications:

Claim 1. For every number i,

$i \in \text{Tot} \Rightarrow (\Theta^Z)^{[i]}$ finite;

$i \notin \text{Tot} \Rightarrow (\Theta^Z)^{[i]} = \omega^{[i]}$.

Proof. If $i \in \text{Tot}$ then at all sufficiently large good stages of the approximation $\{Z^s\}_{s\in\omega}$ we do nothing on behalf of i (i.e. case (b) of the definition of Θ applies to i). To see this, assume that $i \in \text{Tot}$, and let $\sigma \subset \chi_Z$ be such that $\psi^\sigma(i) = 1$. Let t_0 be such that

$$(\forall x < |\sigma|)[\sigma = 1 \Rightarrow (\forall s \geq t_0)[x \in Z^s]].$$

Then

$$(\forall s \geq t_0)[s \text{ is good } \Rightarrow \psi_s^{Z^s}(i) = 1].$$

It follows that at stages $s \geq t_0$, if (a) holds then s is not good, hence $\Theta_s^{Z^s} \not\subseteq \Theta^Z$. Therefore $(\Theta^Z)^{[i]}$ is finite.

If $i \notin \text{Tot}$, then at all sufficiently large good stages, case (a) of the definition of Θ applies to i. Indeed, if $i \notin \text{Tot}$, then $\psi^Z(i) = 0$. One thus argues as in the preceding case, but starting with a string $\sigma \subset \chi_Z$ such that $\psi^\sigma(i) = 0$.

Since $\Theta_s^{Z_s} \subseteq \Theta^Z$, for all good stages s, it follows in this case that $(\Theta^Z)^{[i]} = \omega^{[i]}$. □

Now let $Y = \Theta^Z$. We want to show that Y has a Σ_2^0-high approximation $\{\hat{Y}^s\}_{s\in\omega}$. Let $\{Y^s\}_{s\in\omega}$ be any good Σ_2^0-approximation to Y. Given a partial function φ and a number u, define $\varphi \restriction u \downarrow$, if $\varphi(v)\downarrow$ for all $v < u$.

Define

$$\langle i, x\rangle \in \hat{Y}^s \Leftrightarrow [\langle i, x\rangle \in Y^s \vee \varphi_{i,s} \restriction \langle i, x+1\rangle \uparrow].$$

Claim 2. $\{\hat{Y}^s\}_{s\in\omega}$ is a Σ_2^0-approximation to Y.

Proof. Let $\hat{Y} = \{y : (\exists t)(\forall s \geq t)[y \in \hat{Y}^s]\}$. If $i \notin \text{Tot}$ then $Y^{[i]} = \omega^{[i]} = \hat{Y}^{[i]}$. On the other hand, assume that $i \in \text{Tot}$. If $\langle i, x\rangle \in Y$, then clearly $\langle i, x\rangle \in \hat{Y}$. If $\langle i, x\rangle \notin Y$, then at all sufficiently large stages $\varphi_{i,s} \restriction \langle i, x+1\rangle \downarrow$ and so when $\langle i, x\rangle \notin Y^s$, we have that $\langle i, x\rangle \notin \hat{Y}^s$. □

Next, let c be the computation function for Y relative to the Σ_2^0 approximation $\{\hat{Y}^s\}_{s\in\omega}$. The following claim completes the proof of the Theorem.

Claim 3. The function c is total and dominates all total computable functions.

Proof. Let us first show that c is total. To this end, let $z \in \omega$ be given. Let $t > z$ be a stage such that

$$(\forall \langle i, x\rangle < z)[\varphi_i \upharpoonright \langle i, x+1\rangle \downarrow \Leftrightarrow \varphi_{i,t} \upharpoonright \langle i, x+1\rangle \downarrow].$$

Then if $s \geq t$ is a good stage of the enumeration $\{Y^s\}_{s\in\omega}$, we have that $\hat{Y}^s \upharpoonright z \subseteq Y$. Therefore $c(z)$ is defined.

Now consider any total φ_i. Let x be such that $\langle i, y\rangle \notin Y$, for every $y \geq x$. Let $z \geq \langle i, x\rangle$, and let y be the least number such that $\langle i, y\rangle \leq z < \langle i, y+1\rangle$. Let s be the least stage such that $\varphi_{i,s}(z)\downarrow$, hence $\varphi_{i,t} \upharpoonright \langle i, y+1\rangle \uparrow$ for every $t < s$. Then $\langle i, y\rangle \in \hat{Y}^t$, for every $t < s$. Therefore $\varphi_i(z) < s \leq c(z)$. □

3. Jumps of properly Σ_2^0 e-degrees

A Σ_2^0 e-degree **a** is called *properly* Σ_2^0 ([CC88]) if **a** contains no Δ_2^0 set. Copestake and Cooper, [CC88, Theorem 1], show that there exist e-degrees that are properly Σ_2^0 and Σ_2^0-high. Since every high computably enumerable Turing degree corresponds, under the embedding ι, to a high e-degree, it follows that not every Σ_2^0-high e-degree is properly Σ_2^0-high. (A trivial counterexample is $\mathbf{0}_e' = \deg_e(\overline{K})$). It is shown in [MC85] that $\deg_e(A)$ is low if and only if $B \in \Delta_2^0$, for every $B \leq_e A$. This characterization of the low e-degrees seems to suggest the possibility that the properly Σ_2^0 e-degrees are all high. We show in this section that this is not the case.

Theorem 3.1. *Let C be such that C is computably enumerable in $\emptyset'$, $\emptyset' \leq_T C <_T \emptyset''$ and $C' \equiv_T \emptyset'''$. Then there exists a set A of properly Σ_2^0 e-degree, such that $J_e(A) \leq_e \chi_C$.*

Corollary 3.2. *There exist properly Σ_2^0 e-degrees that are not high.*

Proof. Let C and A be as in the previous theorem. If A were e-high, then $J_e^{(2)}(\emptyset) \leq_e \chi_C$, from which, by totality, $J_e^{(2)}(\emptyset) \leq_T \chi_C$; but $J_e^{(2)}(\emptyset) \equiv_T \emptyset''$, by Lemma 1.6. Hence $\emptyset'' \leq_T C$, contradiction. □

3.1. Proof of Theorem 3.1

Let C satisfy the hypotheses of the theorem; let $C = W^K$, for some computably enumerable set W. For every t, let $\kappa_t = \chi_{K^t} \upharpoonright k(t)$, where k is some $1-1$ computable function such that $K = \text{range}(k)$ and $K^t = \{k(s) | s \leq t\}$. Define a Σ_2^0 approximation $\{C^t\}_{t\in\omega}$ to C by letting

$$C^t = W_t^{\kappa_t},$$

As $C' \equiv_T K''$, there is an $f \leq_T C$ that dominates all Δ_2^0 total functions (see e.g. [Ler83, p. 85]). Let $f = \Psi^C$, for some Turing functional Ψ, be such a function.

We need the following lemma:

Lemma 3.3. *There exists a computable sequence $\{B_i^s\}_{i,s\in\omega}$ of finite sets such that, if*

$$B_i = \{x : (\exists t)(\forall s \geq t)[x \in B_i^s]\}$$

then

1. *for every $B \in \Delta_2^0$, there is an i such that $B =^* B_i$ and, for almost all x, $\lim_s B_i^s(x)$ exists;*
2. *the relation $x \in B_i$ (as one of x and i) is computable in C.*

Proof. Given u and X, with $X = K$, or $X = K^v$ for some $v \geq u$, we say that u is X-*true* if $\kappa_u \subseteq \chi_X$. We will use the fact that for every $B \in \Delta_2^0$ there exists some i such that $\chi_B = \varphi_i^K$. Roughly speaking, we will have $x \in B_i^s$ if there exists some K^s-true stage $t < w$, with $\varphi_{i,t}^{K_t}(x) = 1$, where w is the least K^s-true stage such that $\Psi_w^\sigma(x)\downarrow$, for some $\sigma \subset \chi_C$. Then we use the fact that Ψ^C dominates all Δ_2^0 functions to verify that, for all but finitely many x, there exists a K-true stage t such that $\varphi_{i,t}^{K_t}(x) = 1$ and $t < \Psi^C(x) < w$. The main difficulty here is that one can not find, in a computable way, the right w at s. For every i, x, s, we will therefore define the values of a finite set $B_i^s \subseteq \omega$, a finite set $L(x, s) \subseteq \omega \times 2^{<\omega}$ and a linear ordering $<_{x,s}$ on $L(x, s)$. We "assign preconditions" to elements of $\omega \times 2^{<\omega}$ subject to the following rules: $L(x, s)$ may contain only pairs $\langle r, \rho\rangle$ with preconditions which have been satisfied at some stage $u \leq s$. At stage s, we will choose the $<_{x,s}$-first element $\langle r, \rho\rangle$ of $L(x, s)$. We will argue that infinitely many times we choose the correct $\langle w, \sigma\rangle$ and we eventually choose only pairs $\langle r, \rho\rangle$ with $r \geq w$.

Let i, x be given. The formal definitions are given by induction on s.

Stage 0: Define $B_i^0 = \emptyset$ and $L(x, 0) = <_{x,0} = \emptyset$. No $\langle r, \rho\rangle$ has a precondition at 0.

Stage $s+1$: If $x \geq s+1$ then $x \notin B_i^{s+1}$; otherwise, we distinguish two cases:

- if $L(x, s) = \emptyset$, then

$$x \in B_i^{s+1} \Leftrightarrow x \in B_i^s;$$

- otherwise, let $\langle w, \sigma\rangle$ be the $<_{x,s}$-least element of $L(x, s)$. Then,

  (a) if there is no $t < w$ such that t is K^{s+1}-true, then $x \notin B_i^{s+1}$;

  (b) otherwise, for the least such t, $x \in B_i^{s+1}$ if and only if $\varphi_{i,t}^{K_t}(x) = 1$.

In the latter case, i.e. when $L(x, s) \neq \emptyset$, we extract $\langle w, \sigma\rangle$ from $L(x, s+1)$ and cancel the related precondition. Hence $\langle w, \sigma\rangle$ has no precondition at any stage

$v \geq s+1$ prior to the smallest stage $v' > s+1$ (if any) at which we again assign a precondition to $\langle w, \sigma \rangle$.

We *assign a precondition* to each pair $\langle r, \rho \rangle$ such that

1. r is K^{s+1}-true;
2. $\Psi_r^{\rho}(x)\downarrow$;
3. $\rho \subseteq \chi_{C^r}$;
4. ρ of minimal length, satisfying 2. and 3. (i.e. if $\Psi_r^{\rho'}(x)\downarrow$ and $\rho' \subseteq \chi_{C^r}$ then $\rho \subseteq \rho'$; notice that $|\rho'| < r$, for each such ρ', by the definition of the use function as in [Soa87, p. 49]);
5. $\langle r, \rho \rangle$ does not have a precondition at $s+1$.

At any $v > s+1$, we say that this precondition becomes *satisfied at* v if

$$(\forall i < |\rho|)[\rho(i) = 0 \Rightarrow (\exists t)[s+1 \leq t \leq v \,\&\, i \notin C^t]].$$

Let

$$L(x, s+1) = (L(x, s) - \{\langle w, \sigma \rangle\}) \cup \{\langle r, \rho \rangle : \langle r, \rho \rangle \text{ has a precondition that becomes satisfied at } s+1\}.$$

Finally, we order $L(x, s+1)$ as follows: if $\langle r, \rho \rangle, \langle r', \rho' \rangle \in L(x, s+1)$, then let $\langle r, \rho \rangle <_{x,s+1} \langle r', \rho' \rangle$ if either

1. $\langle r, \rho \rangle, \langle r', \rho' \rangle \in L(x, s)$ and $\langle r, \rho \rangle <_{x,s} \langle r', \rho' \rangle$, or
2. $\langle r, \rho \rangle \in L(x, s)$ and $\langle r', \rho' \rangle \notin L(x, s)$, or
3. $\langle r, \rho \rangle, \langle r', \rho' \rangle \notin L(x, s)$ and $r < r'$.

We now check that the sequence B_i^s has the desired properties.

Claim. Let x be given, let σ be the least string such that $\sigma \subset \chi_C$ and $\Psi^{\sigma}(x)\downarrow$. Let w be the least K-true stage such that $\Psi_w^{\sigma}(x)\downarrow$. Then

1. at infinitely many stages s we extract $\langle w, \sigma \rangle$ from $L(x, s)$;
2. there exists a stage t_0 such that we do not extract any pair $\langle r, \rho \rangle$ with $r < w$ from $L(x, s)$ at any stage $s \geq t_0$.

Proof. Since $\sigma \subset \chi_C$, it is clear that there are infinitely many stages at which the requirements (1-4) for assigning a precondition to $\langle w, \sigma \rangle$ are fulfilled. Moreover, once assigned at a stage s_0, there exists a stage $s_1 > s_0$ such that the precondition becomes satisfied at s_1 and so is then in $L(x, s)$ until extracted. As there are only finitely many elements of $L(x, s_1)$ before $\langle w, \sigma \rangle$ in the ordering and no new ones can later be inserted before it $\langle w, \sigma\rangle$ is eventually extracted. Hence, there exist infinitely many stages s such that $\langle w, \sigma\rangle \in L(x, s)$ and we extract $\langle w, \sigma\rangle$ from $L(x, s)$ at infinitely many stages.

Let $t < w$, and assume for a contradiction that at infinitely many stages s, we extract $\langle t, \rho_s\rangle$, for some string ρ_s. Thus $|\rho_s| < t$ by the definition of the use function, since $\Psi_t^{\rho_s}(x)\downarrow$. Then there exist a ρ, with $|\rho| < t$, and infinitely many stages u_s at which we assign a precondition to $\langle t, \rho\rangle$ which becomes satisfied at some stage $v_s \leq s$ and $\langle t, \rho\rangle \in L(x, v)$ for every v such that $v_s \leq v \leq s$. Then t is K-true. Let t_0 be a stage such that

$$(\forall s \geq t_0)(\forall i < t)[i \in C \Rightarrow i \in C^s].$$

By the minimality of σ and w, and since $C^t \subseteq C$ and t is K-true, it follows that there exists some $i < |\rho| < t$ such that $i \in C$ and $\rho(i) = 0$. But no pair $\langle r, \rho\rangle$ with $\rho(i) = 0$, for some $i < |\rho|$ such that $\chi_C(i) = 1$, can have a precondition assigned to $\langle r, \rho\rangle$ at some stage $u \geq t_0$ which becomes later satisfied. □

We now conclude the proof of the lemma. Let $B \in \Delta_2^0$, and let i be such that $\chi_B = \varphi_i^K$. Let

$$t(x) = \min\{t : t \text{ is } K\text{-true and } \varphi_{i,t}^{K_t}(x)\downarrow\}.$$

Then t is total and so a Δ_2^0 function. It follows that there exists some number x_0 such that $f(x) > t(x)$, for all $x \geq x_0$.

Given $x \geq x_0$, let w and σ be as in the previous claim (for x). Then $t(x) < f(x) < w$ (since $f(x) = \Psi_w^\sigma(x)$). Moreover, if t_0 is as in the proof of the previous claim, then, for every pair $\langle r, \rho\rangle$ such that $\langle r, \rho\rangle$ is extracted from $L(x, s)$ at any stage $s \geq t_0$, we have $t(x) \leq r$. Hence for all $x \geq x_0$, $\chi_{B_i}(x) = \varphi_i^K(x)$.

Since $f(x)$, $\psi(x)$, and w can be computed by C, we easily conclude that the relation $x \in B_i$ is computable in C. □

Remark 3.4. Note that if $t(x) \geq w$, then $\lim_s B_i^s(x)$ need not exist, but, in any case, $x \notin B_i$, since at every large enough stage at which we extract $\langle w, \sigma\rangle$ from $L(x, s)$ we have $x \notin B_i^s$.

We now go back to the proof of the theorem. We will build a Σ_2^0 set A such that, for every Δ_2^0 set B, $A \not\equiv_e B$, and $K^A \leq_T C$. This implies that $\deg_e(A)$ is properly Σ_2^0 and, by Lemma 1.1, $J_e(A) \leq_e \chi_C$.

3.2. The strategies

The properly Σ_2^0-strategy. Let $\{\Phi_e, \Psi_e\}_{e\in\omega}$ be some effective listing of all pairs of e-operators. To make A of properly Σ_2^0 e-degree, it is enough to satisfy the following requirements, for every $e, i \in \omega$:

$$\mathcal{P}_{e,i}: \qquad A = \Phi_e^{B_i} \,\&\, B_i = \Psi_e^A \Rightarrow (\exists^\infty x)[\lim_s B_i^s(x)\uparrow]$$

where $\{B_i\}_{i\in\omega}$ and $\{B_i^s\}_{i,s\in\omega}$ are as given in Lemma 3.3.

Indeed, if we satisfy these requirements for every e, i, then $\deg_e(A)$ is properly Σ^0_2. Suppose, for the sake of a contradiction, that $A \equiv_e B$ and $B \in \Delta^0_2$. Then, by the previous lemma, $A = \Phi_e^{B_i}$ and $B_i = \Psi_e^A$ for some e, i, with $B =^* B_i$ and so $\lim_s B_i^s(x)$ does not exist for infinitely many x for the desired contradiction, since $\lim_s B_i^s(x)$ exists for almost all x.

The strategy to meet $\mathcal{P}_{e,i}$ is a slight modification of the canonical properly Σ^0_2 strategy as given in [CC88], and described as follows:

(a) appoint a witness x and let $x \in A$;

(w_1) wait for finite sets D, E such that $x \in \Phi_e^D$ and $D \subseteq \Psi_e^E$;

(b) fix $E - \{x\} \subseteq A$;

(w_2) wait for $D \subseteq B_i$;

(w_3) let $x \notin A$, wait for $D \not\subseteq B_i$;

(ℓ) let $x \in A$; go back to (w_2).

A triple x, D, E as above is called a *follower* of $\mathcal{P}_{e,i}$.

As described in [CC88, Theorem 1], for a given follower x, D, E this strategy may have the following outcomes: (w_1) yields $x \in A - \Phi_e^{B_i}$ or $y \in B_i - \Psi_e^A$ for some y; (w_2) corresponds to the case $D \subseteq \Psi^A$, $D \not\subseteq B_i$; (w_3) corresponds to the case $x \in \Phi^{B_i} - A$; finally, the infinitary outcome ℓ entails that $\lim_s B_i^s(y)$ does not exist for some $y \in D$.

The subrequirements $\mathcal{P}_{e,i,j}$. It follows by the analysis of the outcomes of the previous strategy that if $B_i^s(x)$ does have limit on every $x \in D$, then $A \neq \Phi_e^{B_i}$ or $B_i \neq \Psi_e^A$. The only complication here (see Remark 3.4) is that there might exist finitely many numbers x such that $\lim_s B_i^s(x)$ does not exist, thus, for some $y \in D$, $\lim_s B_i^s(y)$ need not exist. We cope with this difficulty by attacking $\mathcal{P}_{e,i}$ through infinitely many subrequirements $\mathcal{P}_{e,i,j}$, with $j \in \omega$. The strategy for $\mathcal{P}_{e,i,j}$ consists in looking for a follower x, D, E such that $D \upharpoonright j = B_i \upharpoonright j$: thus, for almost all j, if we appoint a follower x, D, E as before, we are bound to conclude that $B_i^s(y)$ exists on every $y \in D$. Thus $\mathcal{P}_{e,i}$ is satisfied through some subrequirement $\mathcal{P}_{e,i,j}$ (in fact cofinitely many such subrequirements). Before acting, the subrequirement $\mathcal{P}_{e,i,j}$ must therefore be provided with some knowledge of what numbers $x < j$ are in fact in B_i. This information is coded in the first component, $h(\sigma, s)$, of the outcome of the node corresponding to $\mathcal{P}_{e,i,j}$ in the tree of outcomes.

The strategy for $K^A \leq_T C$. For every i, we will look for a finite set D such that $i \in \Phi_i^D$. If such a D exists then we let $D \subseteq A$. Notice that we can determine computably in $\emptyset'$ and, thus, in C, whether or not such a finite set exists.

3.3. The tree of outcomes

For notation and terminology for strings and trees, the reader is referred to [Soa87]. The tree of outcomes is the smallest set T of strings σ such that

1. if $|\sigma|$ is even then $\sigma\widehat{\ }(h, r) \in T$, for every $h \in \omega$ and $r \in \{0, 1\}$;

2. if $|\sigma|$ is odd then $\sigma\widehat{\ }r \in T$, for every $r \in \{0, 1\}$.

The strings of even length are assigned to the (sub)requirements $\mathcal{P}_{e,i,j}$, according to some fixed priority listing. The first component, $h(\sigma, s)$, of the outcome of σ at stage s will be an assessment as to which numbers $x < j$ are in fact in B_i: at stage $s+1$, $h(\sigma, s)$ will be chosen to be the first element of a list $\mathcal{L}(\sigma, s)$ of numbers. Each element h of the list is the canonical index of a finite subset of $\{x : x < j\}$. Its position in the list measures how well the set $\{x : x < j\} - B_i$ is approximated by the finite set D_h. Having decided on the first component, h, of the outcome at σ, the strategy for $\mathcal{P}_{e,i,j}$ is ready to act at $\sigma^+ = \sigma\widehat{\ }h$. The outcome 1 at σ^+ corresponds to (w_1) or (w_2); the outcome 0 corresponds to (w_3) or (ℓ).

The strings of odd length are devoted to guaranteeing that $K^A \leq_T C$: if $|\sigma| = 2i+1$ then we have outcome 0 if there exists (modulo higher priority constraints) some finite set D such that $i \in \Phi_i^D$; otherwise we have outcome 1.

Let $\hat{T} = T \cup \{\sigma\widehat{\ }h : |\sigma| \text{ even } \& h \in \omega\}$. For $\sigma \in \hat{T}$, the parameter $\alpha(\sigma, s)$ is intended to record some finite set which we want to keep in A for the sake of our actions at σ; the parameter $\epsilon(\sigma, s)$ is meant to record some finite set of elements which we want to keep out of A.

The ordering $\preceq$ of T is determined in the usual way by the ordering of the outcomes given that we define $(h, r) < (h', r')$ if

$$h > h' \text{ or } [h = h' \,\&\, r < r'].$$

We extend $\preceq$ to $\hat{T}$ in the obvious way.

Finally, let $\{\xi_\sigma\}_{\sigma \in \hat{T}}$ be a computable partition of ω into infinite computable sets.

3.4. The construction

The construction proceeds by stages. At stage s we define a finite set A^s, a string δ_s, and the values of several parameters. Unless otherwise specified, at each stage each parameter retains the same value as at the preceding stage.

Stage 0: Define $\delta_0 = \emptyset$. For every $\sigma \in \hat{T}$, let

$$\alpha(\sigma, 0) = \epsilon(\sigma, 0) = \mathcal{L}(\sigma, 0) = \emptyset.$$

Let $x(\sigma, 0)$ and $p(\sigma, h, 0)$ be undefined for every $x, h \in \omega$. Finally, let $A^0 = \emptyset$.

Stage $s+1$: Suppose that we have defined $\delta_{s+1} \upharpoonright n$, where $n < s+1$: let $\sigma = \delta_{s+1} \upharpoonright n$. Our aim is to define a string σ^{++} which we will be $\delta_{s+1} \upharpoonright n+1$.

$|\sigma|$ even. Let $\mathcal{P}_{e,i,j}$ be the requirement assigned to σ. For simplicity, drop subscripts, and let $\Phi_e = \Phi$, $\Psi_e = \Psi$ and $B_i = B$.

Our first task is to define the first component, $h(\sigma, s+1)$, of the outcome. We define $h(\sigma, s+1)$ to be the least element of $\mathcal{L}(\sigma, s)$ if $\mathcal{L}(\sigma, s) \neq \emptyset$, otherwise $h(\sigma, s+1) = 0$. Then we cancel the precondition for $h(\sigma, s+1)$ by letting $p(\sigma, h(\sigma, s+1), s+1)\uparrow$.

To every h such that $\max D_h < j$ and h does not have a precondition, we assign the precondition $p(\sigma, h, s+1)$ which becomes satisfied at some later stage $v > s+1$ if, for every $x < j$ and $x \in D_h$, there exists u such that $s+1 \leq u \leq v$ and $B^u(x) = 0$.

Define

$$\mathcal{L}(\sigma, s+1) = (\mathcal{L}(\sigma, s) - \{h(\sigma, s+1)\}) \cup \{h : h \text{ has a precondition that is satisfied at } s+1\}$$

and order $\mathcal{L}(\sigma, s+1)$ in the usual way: for every $h, h' \in \mathcal{L}(\sigma, s+1)$, define $h <_{\sigma,s+1} h'$ if either

1. $h, h' \in \mathcal{L}(\sigma, s)$ and $h <_{\sigma,s} h'$, or
2. $h \in \mathcal{L}(\sigma, s)$ and $h' \notin \mathcal{L}(\sigma, s)$, or
3. $h, h' \notin \mathcal{L}(\sigma, s)$ and $h < h'$.

Let $\sigma^+ = \sigma^\frown h(\sigma, s+1)$.

Now we are ready to activate the strategy for $\mathcal{P}$.

Let $x = x(\sigma^+, s+1)$ be the least number in ξ_{σ^+} such that $x \notin \alpha(\rho, s+1)$, for every $\rho \prec \sigma^+$.

Case 1).

$$(\exists D)(\exists E)[x \in \Phi_s^D \,\&\, D \cap D_{h(\sigma,s+1)} = \emptyset \,\&\, D \subseteq \Psi_s^E \,\&\, E \cap \bigcup\{\epsilon(\rho, s+1) : \rho \preceq \sigma\} = \emptyset].$$

Choose the least such pair D, E.

In this case, let $\alpha(\sigma^+, s+1) = E - \{x\}$:

1. if $D \subseteq B^s$, then let $\sigma^{++} = \sigma^{+\frown}0$ and $\epsilon(\sigma^{++}, s+1) = \{x\}$;
2. otherwise, let $\sigma^{++} = \sigma^{+\frown}1$ and $\alpha(\sigma^{++}, s+1) = \{x\}$.

Case 2). Otherwise, let $\sigma^{++} = \sigma^{+\frown}1$ and $\alpha(\sigma^{++}, s+1) = \{x\}$.

$|\sigma|$ odd. Let $|\sigma| = 2i + 1$. We distinguish two cases.

Case 1). $(\exists D)[i \in \Phi_i^D \;\&\; D \cap \bigcup\{\epsilon(\rho, s+1) : \rho \preceq \sigma\} = \emptyset]$.

In this case, let $\sigma^{++} = \sigma^\frown 0$, and let $\alpha(\sigma, s+1) = D$ for the least such D.

Case 2). Otherwise, let $\sigma^{++} = \sigma^\frown 1$.

Definition of A^{s+1}. At the end of stage $s+1$, let

$$A^{s+1} = (A^s \cup \bigcup\{\alpha(\rho, s+1) : \rho \preceq \delta_{s+1}\}) - \bigcup\{\epsilon(\rho, s+1) : \rho \preceq \delta_{s+1}\}.$$

3.5. Verification

The verification is based upon the following lemmas.

Lemma 3.5. *For every n, $\sigma_n = \liminf_s \delta_s \upharpoonright n$ exists.*

Proof. Assume by induction that the claim is true of n. The only nontrivial case is when $|\sigma_n|$ is even, where, say, the requirement $\mathcal{P}_{e,i,j}$ is assigned to σ_n.

Let h be the canonical index of $\overline{B_i} \upharpoonright j$. It is clear that, whenever we assign a precondition to h, then this precondition becomes satisfied at some later stage. Hence, at infinitely many stages s, $h \in \mathcal{L}(\sigma_n, s)$, and at infinitely many stages t, $h = h(\sigma_n, t)$. On the other hand, it is also clear that for almost all stages s, if $h' \in \mathcal{L}(\sigma_n, s)$, then $D_{h'} \subseteq D_h$, hence $h' \leq h$ by the usual coding of canonical sets. Therefore it follows that $\sigma_{n+1} = \sigma_n{}^\frown(h, r)$, for some $r \in \{0, 1\}$. □

Let $f = \bigcup_{n\in\omega} \sigma_n$.

Lemma 3.6. *For every $\tau \in \hat{T}$, if $\tau \subset f$, then $\alpha(\tau) = \lim_s \alpha(\tau, s)$, $\epsilon(\tau) = \lim_s \epsilon(\tau, s)$ and $x(\tau) = \lim_s x(\tau, s)$ exist. Moreover, if $\tau = \sigma_n$ for some n, then the requirement assigned to σ_n is satisfied.*

Proof. By induction on n, we show that if $\tau = \sigma_n$ or $\tau = \sigma_n^+$, where $\sigma_{n+1} = \sigma_n^{+\frown} r$ for some $r \in \{0, 1\}$ (of course $\sigma_n^+ = \sigma_n$ if n is odd), then $\lim_s \alpha(\tau, s)$, $\lim_s \epsilon(\tau, s)$ and $\lim_s x(\tau, s)$ exist, and the requirement assigned to σ_n is satisfied. The case $n = 0$ is trivial as are the existence of the required limits for all nodes to the left of the true path.

Assume that the claim is true of n. For every $\tau \in \hat{T}$ such that $\tau \preceq \sigma_n$, let $\alpha(\tau) = \lim_s \alpha(\tau, s)$, $\epsilon(\tau) = \lim_s \epsilon(\tau, s)$ and $x(\tau) = \lim_s x(\tau, s)$; and let t be a stage such that, for every $s \geq t$ and $\tau \preceq \sigma_n$, $\alpha(\tau) = \alpha(\tau, s)$, $\epsilon(\tau) = \epsilon(\tau, s)$ and $x(\tau) = x(\tau, s)$.

Suppose first that $|\sigma_n|$ is even, and let $\mathcal{P}_{e,i,j}$ be the requirement assigned to σ_n. Let $\sigma_{n+1} = \sigma_n\widehat{\ }(h, i)$, and let $\sigma_n^+ = \sigma_n\widehat{\ }h$. Then

$$x(\sigma_n^+) = \min x \in (\xi_{\sigma_n^+} - \bigcup_{\tau \preceq \sigma_n} \alpha(\tau)).$$

- If there are no finite sets D, E such that $D \cap D_h = \emptyset$, $x(\sigma_n^+) \in \Phi_e^D$, $E \cap \bigcup_{\tau \preceq \sigma_n} \epsilon(\tau) = \emptyset$ and $D \subseteq \Psi_e^E$, then $\sigma_{n+1} = \sigma_n^+\widehat{\ }1$,

  $$\lim_s \epsilon(\sigma_n^+, s) = \lim_s \epsilon(\sigma_{n+1}, s) = \emptyset,$$

  $\lim_s \alpha(\sigma_n^+, s) = \emptyset$, $\lim_s \alpha(\sigma_{n+1}, s) = \{x(\sigma_n^+)\}$ and $x(\sigma_n^+) \in A$. Moreover, either $x(\sigma_n^+) \in A - \Phi_e^{B_i}$, or $x(\sigma_n^+) \in \Phi_e^D$, for some $D \subseteq B_i$, but $D \not\subseteq \Psi_e^A$.

- If D, E exist, then we eventually choose the least such pair D, E, hence $\alpha(\sigma_n^+) = E - \{x(\sigma_n^+)\}$, $\epsilon(\sigma_n^+) = \emptyset$, and either (a) or (b) holds:

  (a) $\sigma_{n+1} = \sigma_n^+\widehat{\ }0$ and $\alpha(\sigma_{n+1}) = \emptyset$, $\epsilon(\sigma_{n+1}) = \{x(\sigma_n^+)\}$;
  (b) $\sigma_{n+1} = \sigma_n^+\widehat{\ }1$ and $\alpha(\sigma_{n+1}) = \{x(\sigma_n^+)\}$, $\epsilon(\sigma_{n+1}) = \emptyset$.

  In case (a) either

  (a_1) there exist infinitely many stages s such that $\sigma_n^+\widehat{\ }1 \subseteq \delta_s$, in which case, there exists some $y \in D$ such that $\lim_s B_i^s(y)$ does not exist; or

  (a_2) $D \subseteq B_i$ but $x(\sigma_n^+) \notin A$, giving $x(\sigma_n^+) \in \Phi_e^{B_i} - A$.

  In (b) we have $D \not\subseteq B_i$, but $E \subseteq A$, hence $D \subseteq \Psi_e^A$.

Remark 3.7. Notice that if j is such that $\lim_s B_i^s(y)$ exists for every $y \geq j$, then (a_1) does not occur, by Lemma 3.3.

If $|\sigma_n| = 2i+1$ is odd and $i \in \Phi_i^D$, for some finite set D such that $D \cap \bigcup_{\tau \preceq \sigma_n} \epsilon(\tau) = \emptyset$, then $\sigma_{n+1} = \sigma_n\widehat{\ }0$ and $\alpha(\sigma_{n+1}) = D$, for some such D; otherwise $\sigma_{n+1} = \sigma_n\widehat{\ }1$ and $\alpha(\sigma_{n+1}) = \emptyset$. In either case $\epsilon(\sigma_{n+1}) = \emptyset$.

The proof of the lemma is now complete. □

Lemma 3.8. $K^A \leq_T C$.

Proof. We will show that, for every n, one can compute σ_n recursively in C.

Now, $\sigma_0 = \emptyset$. Assume by induction that we can compute σ_n and a stage s_n such that $\tau \not\subseteq \delta_s$, for every $s \geq s_n$ and $\tau <_L \sigma_n$ and each parameter at any $\tau \preceq \sigma_n$ has reached its limit by stage s_n. Assume first that $|\sigma_n|$ is even, and let $\mathcal{P}_{e,i,j}$ be the requirement assigned to σ_n. Since C can compute $B_i \upharpoonright j$, it follows that C can compute the first component, $h = \lim_s h(\sigma_n, s)$, of the outcome of σ_n. Moreover, since $\emptyset' \leq_T C$, C can compute the least stage $s_n^+ \geq s_n$ such that $\tau \not\subseteq \delta_s$, for every $\tau <_L \sigma_n\widehat{\ }h$ (since for every $x \in B_i \upharpoonright j$ and every t, one can compute in $\emptyset'$ whether there exists some $s \geq t$ such that $x \notin B_i^s$). Again, using $\emptyset'$ as an oracle, one can compute whether Case 1 or Case 2 of the construction holds, thus computing σ_{n+1} and the corresponding s_{n+1}.

A similar argument applies in the case $|\sigma_n| = 2i + 1$, for some i, since the oracle $\emptyset'$ can compute whether or not there exists some stage $s \geq s_n$ and some finite D such that $i \in \Phi_i^D$, and $D \cap \bigcup_{\tau \preceq \sigma_n} \epsilon(\tau, s_n) = \emptyset$.

It follows that $i \in K^A$ if and only if $\sigma_{2i+2} = \sigma_{2i+1}{}^\frown 0$, thus $K^A \leq_T C$. □

Remark 3.9. We expect that, by combining the above construction of A with a variant of the coding procedure and the associated guessing at outcomes used in the tree proof of the Sacks' jump inversion theorem, one can actually guarantee that $J_e(A) \equiv_e \chi_C$.

References

[BCS97] S. Bereznyuk, R. Coles, and A. Sorbi, The distribution of properly Σ_2^0 enumeration degrees, J. Symbolic Logic, to appear.

[CC88] S. B. Cooper and C. S. Copestake, Properly Σ_2 enumeration degrees, Z. Math. Logik Grundlag. Math. 34 (1988), 491–522.

[Coo90] S. B. Cooper, Enumeration reducibility, nondeterministic computations and relative computability of partial functions, in: Recursion Theory Week, Oberwolfach 1989 (eds. K. Ambos-Spies, G. Müller, and G. E. Sacks), Lecture Notes in Math. 1432, Springer-Verlag, Heidelberg, 1990, 57–110.

[Ler83] M. Lerman, Degrees of Unsolvability, Perspect. Math. Logic, Springer-Verlag, Heidelberg, 1983.

[LS92] A. H. Lachlan and R. A. Shore, The n-rea enumeration degrees are dense, Arch. Math. Logic 31 (1992), 277–285.

[MC85] K. McEvoy and S. B. Cooper, On minimal pairs of enumeration degrees, J. Symbolic Logic 50 (1985), 983–1001.

[McE85] K. McEvoy, Jumps of quasi-minimal enumeration degrees, J. Symbolic Logic 50 (1985), 839–848.

[Med55] Y. T. Medevdev, Degrees of difficulty of the mass problems, Dokl. Nauk. SSSR 104 (1955), 501–504.

[Rog67] H. Rogers, Jr., Theory of Recursive Functions and Effective Computability, McGraw-Hill, New York, 1967.

[Soa87] R. I. Soare, Recursively Enumerable Sets and Degrees, Perspect. Math. Logic, Omega Series, Springer-Verlag, Heidelberg, 1987.

Enumeration reducibility and the problem of the nontotal property of e-degrees

Boris Solon

Abstract. The question about the characterization of nontotal e-degrees in terms of total functions is considered. It is proved that any nontotal e-degree either contains a C-quasi-minimal set for some set C or the set $C = \bigcup_{s \in N} C_s$ where $\{C_s : s \in N\}$ is a strictly increasing sequence of sets chosen from suitable total e-degrees. A number of sufficient conditions for nontotality of e-degrees is given.

We use notation and terminology similar to that of the monograph [1]. We recall those which will be used in this article. Let N denote the set of natural numbers; $A, B, C, \ldots$ (with indexes or without) are a subsets of N; D_u is a finite set with the canonical index u; $\langle x, y\rangle$ is Cantor's number of an ordered pair (x, y). Let $|D|$ be the number of elements of the finite set D.

Let $\delta\alpha$ be the domain and $\rho\alpha$ be the range of a partial function α, $\tau\alpha = \{\langle x, y\rangle : x \in \delta\alpha \wedge y = \alpha(x)\}$ be the graph of α. The set A is said to be *single-valued* if $A = \tau\alpha$ for some function α. If A is a single-valued set and $A = \tau\alpha$ then we will write $A(x)$ instead of $\alpha(x)$. We denote the fact that $x \in \delta\alpha$ by $!\alpha(x)$. We restrict the use of the symbols f and g only for notation of total functions ($\delta f = \delta g = N$).

We recall that A *is e-reducible* to B *via* n if

$$\forall x[x \in A \iff \exists u[\langle x, u\rangle \in W_n \wedge D_u \subseteq B]]$$

and A *is reducible* to B ($A \leq_e B$) if there is $n \in N$ such that A is e-reducible to B via n. Let Φ_n be an e-operator that is $\Phi_n(B) = \{x : \exists u[\langle x, u\rangle \in W_n \wedge D_u \subseteq B]\}$. So $A \leq_e B \iff \exists n[A = \Phi_n(B)]$. We write $A <_e B$ if $A \leq_e B$ and $B \not\leq_e A$; $\alpha \leq_e A$ if $\tau\alpha \leq_e A$. We define the partial order $\leq$ on set of e-degrees $\mathbf{L_e}$ in usual way: $d_e(A) \leq d_e(B) \iff A \leq_e B$. It is well known that $(\mathbf{L_e}, \leq)$ is the upper semilattice with the $\mathbf{0} = d_e(\emptyset) = \{W_z : z \in N\}$. For any e-degrees $d_e(A)$ and $d_e(B)$ the e-degree $d_e(A \oplus B)$, where $A \oplus B = \{2x : x \in A\} \cup \{2x + 1 : x \in B\}$, is the supremum of $d_e(A)$ and $d_e(B)$. Despite the fact that in $\mathbf{L_e}$ there are pairs of e-degrees which have infimum it was shown by Case [2] that there are e-degrees which have no infimum.

An e-degree is said to be *total* if it contains the graph of a total function. The first serious question about e-reducibility was solved by Medvedev [3]: *there are nontotal e-degrees*. Medvedev constructed a non-r.e. set A such that

$$\forall f[f \leq_e A \to f \text{ is recursive}].$$

Case [2] called Medvedev's sets *quasi-minimal* and their degrees *quasi-minimal e-degrees*. We note that there are nontotal e-degrees which are not quasi-minimal. For this we recall an extension of the notion of a quasi-minimal set. A set A is said *B-quasi-minimal* if

$$B <_e A \wedge \forall g[g \leq_e A \to g \leq_e B].$$

If B is a r.e. set then any B-quasi-minimal set A is quasi-minimal. It immediately follows that for any set B the e-degree of a B-quasi-minimal set is nontotal. In the following theorem we show that for any set B there is a B-quasi-minimal set A.

Theorem 1. *For any set B there is a B-quasi-minimal set A.*

Proof. We prove the theorem by obtaining the desired set A as the union of a sequence of sets $A_n = B \oplus C_n$ where C_n is a finite set which will be constructed on the step n. In the process of the construction an auxiliary set Z will be formed; Z_n is that part of Z which will be constructed on the step n. For all $m, n \in N$ we will have $C_n \subseteq C_{n+1}$, $Z_n \subseteq Z_{n+1}$, $C_n \cap Z_m = \emptyset$. Let $C = \bigcup_{n\in N} C_n$, $Z = \bigcup_{n\in N} Z_n$ and $A = B \oplus C$. Clearly, $C \cap Z = \emptyset$.

Step 0. We shall define $C_0 = \emptyset$ and $Z_0 = \emptyset$.

Step $2n+1$. See whether

$$\exists z[z \in \Phi_n(B) - A_{2n}]. \tag{1}$$

If (1) is true then two cases are possible.

Let z^* be the least integer z which satisfies (1) and if

$$z^* = 2z_1^* + 1 \tag{2}$$

we define $Z_{2n+1} = Z_{2n} \cup \{z_1^*\}$ and $C_{2n+1} = C_{2n}$ Otherwise

$$z^* = 2z_2^* \tag{3}$$

and we define $Z_{2n+1} = Z_{2n}$ and $C_{2n+1} = C_{2n}$.

If (1) is false then $\Phi_n(B) \subseteq A_{2n} = B \oplus C_{2n}$. In this case we'll define $Z_{2n+1} = Z_{2n}$ and $C_{2n+1} = C_{2n} \cup \{c_{2n}+1\}$ where $c_{2n} = \max(C_{2n} \cup Z_{2n})$.

Step $2n+2$. Let D range over the set of all finite sets. See whether

$$\exists D[D \cap Z_{2n+1} = \emptyset \wedge \Phi_n(B \oplus D) \text{ is not single-valued}]. \tag{4}$$

If so, set $Z_{2n+2} = Z_{2n+1}$ and $C_{2n+2} = C_{2n+1} \cup D^*$ where D^* is the finite set with the least canonical index which satisfies (4).

If not, we define $Z_{2n+2} = Z_{2n+1}$ and $C_{2n+2} = C_{2n+1}$.

The description of the construction is completed. We show that $A = B \oplus C$ satisfies the theorem. Clearly, $B \leq_e A$. Suppose that if $A \leq_e B$ then there is $n \in N$ such that $A = \Phi_n(B)$. Let us consider the step $2n+1$. If the condition (2) was true on this step then $z^* \in \Phi_n(B)$ and $z_1^* \in Z$. As $Z \cap C = \emptyset$ then $z_1^* \notin C$ so $z^* = 2z_1^* + 1 \notin A$. If (3) was true then $z^* \in \Phi_n(B)$ and $z_2^* \notin B$ so $z^* = 2z_2^* \notin A$.

Hence $A \neq \Phi_n(B)$ and therefore (1) was false that is, $\Phi_n(B) \subseteq A_{2n} = B \oplus C_{2n}$. In this case $z = 2(c_{2n}+1)+1 \in A$ but $z \notin \Phi_n(B)$ as $\max\{y : 2y+1 \in \Phi_n(B)\} \leq c_{2n}$. So $A \leq_e B$ is impossible, hence $B <_e A$.

Now let g be a total function such that $g \leq_e A$, that is $\tau g = \Phi_n(A)$ for some $n \in N$. Let us consider the step $2n+2$. Clearly the condition (4) cannot be true. It is easy to verify that in this case $\Phi_n(B \oplus \bar{Z}_{2n+1})$ is a single-valued set. As $A = B \oplus C \subseteq B \oplus \bar{Z}_{2n+1}$ and Φ_n is a monotone operator then $\tau g = \Phi_n(A) \subseteq \Phi_n(B \oplus \bar{Z}_{2n+1})$. So g is a total function and τg is a single-valued subset of the single-valued set $\Phi_n(B \oplus \bar{Z}_{2n+1})$. This is possible only if $\tau g = \Phi_n(B \oplus \bar{Z}_{2n+1})$. As $\bar{Z}_{2n+1}$ is a recursive set $g \leq_e B \oplus \bar{Z}_{2n+1} \leq_e B$ and the theorem is proved. □

Previously Sasso [4] proved the existence of B-quasi-minimal sets for any B. Using the proof of Theorem 1 we get two corollaries.

Corollary 1. *For any set B there is an uncountable family of B-quasi-minimal sets which are mutually incomparable.*

Proof. Let us agree to name a family of all elements from $\mathbf{Q_B}$ which are mutually incomparable as an antichain. Assume that this antichain is at most a countable family, let $A_0, A_1, \ldots, A_n, \ldots$ be all its elements. Using the notation of Theorem 1 we'll construct a set A such that A is B-quasi-minimal and the sets A and A_s are incomparable for all $s \in N$.

Step 0. Define $C_0 = Z_0 = \emptyset$ and $c_0 = 0$.

Step $4n+1$. Copy the step 2n+1 of Theorem 1.

Step $4n+2$. Copy the step 2n+2 of Theorem 1.

Step $4n+3$. Let $n = \langle k, l\rangle$. See whether

$$\exists D[D \cap Z_{4n+2} = \emptyset \wedge \Phi_k(B \oplus D) \nsubseteq A_l]. \tag{5}$$

If so, set $Z_{4n+3} = Z_{4n+2}$ and $C_{4n+3} = C_{4n+2} \cup D^*$ where D^* is a finite set with the least canonical index which satisfies (5). If not, set $Z_{4n+3} = Z_{4n+2}$ and $C_{4n+3} = C_{4n+2}$.

In the first case, it is obvious that $\Phi_k(A) \neq A_l$. In the second case we have

$$\forall D[D \cap Z_{4n+2} = \emptyset \rightarrow \Phi_k(B \oplus D) \subseteq A_l].$$

Hence $\Phi_k(B \oplus \tilde{N}) \subseteq A_l$ where $\tilde{N} = N - Z_{4n+2}$ is a co-finite set. If $\Phi_k(A) = A_l$ then $\Phi_k(A) \subseteq \Phi_k(B \oplus \tilde{N}) \subseteq A_l$ and so $\Phi_k(B \oplus \tilde{N}) = A_l$ that is $A_l \leq_e B \oplus \tilde{N} \leq_e B$ which contradicts the condition that A_l is B-quasi- minimal and in particular that $A_l \nleq_e B$.

Step $4n+4$. Let $n = \langle k, l\rangle$. See whether

$$\Phi_k(A_l) \subseteq B \oplus C_{4n+3}. \tag{6}$$

If so, set $Z_{4n+4} = Z_{4n+3}$ and $C_{4n+4} = C_{4n+3} \cup \{c_{4n+3}\}$. If not, let $x \in \Phi_k(A_l) - (B \oplus C_{4n+3})$ and x^* is the least such x. If $x^* = 2z^*$, set $Z_{4n+4} = Z_{4n+3}$ and $C_{4n+4} = C_{4n+3}$, if $x^* = 2z^*+1$, set $Z_{4n+4} = Z_{4n+3} \cup \{z^*\}$ and $C_{4n+4} = C_{4n+3}$.

The steps $4n+1$ and $4n+2$ guarantee that A is B-quasi-minimal. The steps $4n+3$ and $4n+4$ ensure A and A_s are incomparable for all $s \in N$. But A does not belong to our antichain which contradicts the premise. Hence the antichain in $\mathbf{Q_B}$ must be uncountable as required. □

Corollary 2. *For any set B there exists an uncountable strictly increasing sequence of B-quasi-minimal e-degrees.*

Proof. Let us agree to name a strictly increasing sequence by a chain. Let there be some given set B and let $\mathbf{Q_B}$ be the family of all B-quasi-minimal sets. We show that $\mathbf{Q_B}$ contains an uncountable chain. Let $\mathbf{C}$ be the set of all chains formed by elements of $\mathbf{Q_B}$ and $\mathbf{C}$ is partially ordered by the relation $\subseteq$. We show that $\mathbf{C} \neq \emptyset$ for any set B. By Theorem 1, there is the set A_0 such that $B <_e A_0$ and A_0 is B-quasi-minimal. Further, there is a set A_1 such that $A_0 <_e A_1$ and A_1 is A_0-quasi-minimal, that is $f \leq_e A_1 \rightarrow f \leq_e A_0 \rightarrow f \leq_e B$ for any total function f so A_1 is B-quasi-minimal set and so on.

Now we show that every chain from $\mathbf{C}$ has least upper bound in $\mathbf{C}$. Let $\mathbf{A} = \{\mathbf{c}_i : i \in I\}$ be a linearly ordered subset of $\mathbf{C}$ then $\mathbf{c}^* = \bigcup_{i \in I} \mathbf{c}_i$ is the chain belonging to $\mathbf{C}$ which is the the least upper bound for $\mathbf{A}$. By Zorn's Lemma $\mathbf{C}$ has at least one maximal element $\mathbf{m}$. Suppose that the chain $\mathbf{m}$ is at most countable, that is $\mathbf{m} = \{M_s : s \in N\}$ and $M_0 <_e M_1 <_e \cdots$ where M_s is B-quasi-minimal for all $s \in N$. We construct a set A such that $M_s <_e A$ for all $s \in N$ and A is a B-quasi-minimal set. This contradicts the maximality of $\mathbf{m}$. It show that $\mathbf{m}$ is an uncountable set.

To form the set A we construct step by step a sequence of sets A_s where $A_s \subseteq A_{s+1}$ for all $s \in N$. Let D and F be variables which range over finite sets and $\tilde{M}_s = \{\langle s, x\rangle : x \in M_s\}$, $\tilde{N}_s = \{\langle n, x\rangle : n > s \wedge n, x \in N\}$ for all $s \in N$.

Step 0. Set $A_0 = \tilde{M}_0$.

Step $s+1$. Suppose A_s is constructed on the step s. See whether

$$\exists D[D \subseteq \tilde{N}_s \wedge \Phi_s(A_s \cup D) \text{ is not single-valued}]. \quad (1)$$

If so, set $A_{s+1} = A_s \cup \tilde{M}_{s+1} \cup D^*$ where D^* is a finite set with the least canonical index which satisfies (1). If not, then

$$\forall D[D \subseteq \tilde{N}_s \rightarrow \Phi_s(A_s \cup D) \text{ is single-valued}]. \quad (2)$$

Set $A_{s+1} = A_s \cup \tilde{M}_{s+1}$.

Let $A = \bigcup_{s \in N} A_s$. We prove a number of statements which will show as a result that A is the desired set.

(i) *A_s is B-quasi-minimal set for all $s \in N$.* It is clear that $A_s = \bigcup_{i \leq s} \tilde{M}_i \cup F$, where F is some finite set. As $\tilde{M}_i \equiv_e M_i$ and $M_i \leq_e M_s$ for all $i \leq s$ so $A_s \equiv_e M_s$ and A_s is B-quasi-minimal together with M_s.

(ii) *$M_s <_e A$ for all $s \in N$.*

Enumerate from A the elements of the form $\langle s, x\rangle$. The set $\tilde{M}^s = \{x : \langle s, x\rangle \in A\}$ is distinguished from M_s on at most a finite set which is added to A_i on the steps i

where $i \leq s$ in case when the condition (1) was satisfied. So $M_s \leq_e \tilde{M}^s \leq_e A$. Assume that $A \leq_e M_s$ for some $s \in N$ then for example $M_{s+1} \leq_e A \leq_e M_s$ which contradicts our choice of the sequence $\{M_s : s \in N\}$.

(iii) *A is a B-quasi-minimal set.*

Let $f \leq_e A$ for some total function f, that is $\tau f = \Phi_s(A)$ for some $s \in N$. We consider step $s+1$ and show that (1) was not satisfied. If (1) was satisfied then $\Phi_s(A_s \cup D)$ is not single-valued for some $D \subseteq \tilde{N}_s$ and D^* which was selected in this case is a subset of A and hence $\Phi_s(A)$ is also not single-valued. Now we verify that $\Phi_s(A_s \cup \tilde{N}_s)$ is a single-valued set. Suppose not, then $\Phi_s(D)$ is not single-valued for some finite $D \subset A_s \cup \tilde{N}_s$ and (1) must be satisfied which is impossible. Thus $\tau f = \Phi_s(A) \subseteq \Phi_s(A_s \cup \tilde{N}_s)$. As τf is a total single-valued set, $\tau f = \Phi_s(A_s \cup \tilde{N}_s)$ and so $f \leq_e A_s \cup \tilde{N}_s \leq_e A_s$. Using (i) we have that $f \leq_e B$ and hence A is a B-quasi-minimal set as was to be proved.

We now consider whether every nontotal e-degree is a B-quasi-minimal e-degree for a suitable set B. The following theorem sheds some light on this problem.

Theorem 2. *There exists a set A such that $d_e(A)$ is a nontotal e-degree and for any total function f the set A is not τf-quasi-minimal.*

Proof. Let f, g (with or without subscripts) be variables which range over total functions and D, F be variables which range over finite sets. We construct the set A satisfying

$$f \leq_e A \rightarrow \exists g[g \leq_e A \wedge g \not\leq_e f] \tag{1}$$

for any total function f.

At first we verify that $d_e(A)$ is of nontotal e-degree. If $d_e(A)$ is total then for some $f \in d_e(A)$, $f \leq_e A$ and $A \leq_e f$. Let g be a function, whose existence follows from (1). Then $A \leq_e f \rightarrow g \leq_e A \leq_e f$ so $g \leq_e f$ which contradicts (1).

The formation of A will be fulfilled step by step. Let $g_0 <_e g_1 <_e \cdots$ be a sequence of total functions. Let $G_s = \{\langle s, n, g_s(n)\rangle : n \in N\}$ and $\tilde{N}_s = \{\langle m, x, y\rangle : m > s \wedge m, x, y \in N\}$. In the course of the construction a sequence of sets $B_0 \subseteq B_1 \subseteq \cdots$ is found.

Step 0. Set $B_0 = G_0$.

Step $s+1$. Let B_s be the set which is constructed on the step s. See whether

$$\exists D[D \subseteq \tilde{N}_s \wedge \Phi_s(B_s \cup D) \text{ is not single-valued}]. \tag{2}$$

If so, set $B_{s+1} = B_s \cup G_{s+1} \cup D^*$ where D^* is a finite set with the least canonical index which satisfies (2).

If not, then

$$\forall D[D \subseteq \tilde{N}_s \rightarrow \Phi_s(B_s \cup D) \text{ is single-valued}]. \tag{3}$$

In this case set $B_{s+1} = B_s \cup G_{s+1}$.

The description of construction is completed. Let $A = \bigcup_{s \in N} B_s$. We prove a number of the statements from which the theorem follows.

(i) *$g_s \leq_e A$ for all $s \in N$.*

Enumerate from A all elements of the form $\langle s, x, y\rangle$. The set $G^s = \{\langle x, y\rangle : \langle s, x, y\rangle \in A\}$ is distinguished from τg_s by at most a finite set of those elements which were put into B_i on steps i where $i \leq s$. In this case $G^s \equiv_e \tau g_s \equiv_e G_s$ and $g_s \leq_e G_s \leq_e G^s \leq_e A$ hence $g_s \leq_e A$.

(ii) *$B_s \leq_e g_s$ for all $s \in N$.*

As $g_0 <_e g_1 <_e \cdots <_e g_s$ then $\bigcup_{i \leq s} G_i$ is e-reducible to G_s. From the construction one can see that $B_s = \bigcup_{i \leq s} G_i \cup F$ for some finite set F so $B_s \leq_e g_s$.

To complete the proof of the theorem, let f be a total function e-reducible to A, that is $\tau f = \Phi_s(A)$ for some $s \in N$. Let us consider the step $s + 1$. We show that in this case the condition (2) was not satisfied. Otherwise $\Phi_s(B_s \cup D)$ is not single-valued for some $D \subset \tilde{N}_s$ then D^* which was selected in this case is a subset of A, and also $\Phi(B_s \cup \tilde{N}_s)$ is a single-valued set. Thus, we have $\tau f = \Phi_s(A) \subseteq \Phi_s(B_s \cup \tilde{N}_s)$. As τf is a total single-valued set this is possible only if $\tau f = \Phi_s(B_s \cup \tilde{N}_s)$ and $f \leq_e B_s \cup \tilde{N}_s \leq B_s$. Now suppose that $g_{s+1} \leq_e f$. Then $g_{s+1} \leq_e f \leq_e B_s \leq_e g_s$ which contradicts our supposition that $g_{s+1} \not\leq_e g_s$.

Finally, for any total function f

$$f \leq_e A \rightarrow \exists s[g_s \leq_e A \wedge g_s \not\leq_e f]$$

and the statement (1) is true. The proof is completed. □

Corollary 3. *There exists a set A such that $d_e(A)$ is a nontotal e-degree and A is not B-quasi-minimal for any set B.*

Proof. Let A be the set constructed in the proof of Theorem 2. It was proved that A is not τf-quasi-minimal set for any total function f. We show that indeed A is not B-quasi-minimal for any set B, in particular if B belongs to some nontotal e-degree. Assume that there is a set B such that $B <_e A$ and $\forall f[f \leq_e A \rightarrow f \leq_e B]$. Let $g_0 <_e g_1 <_e \cdots$ be total functions used for the construction of A so $g_0 <_e g_1 <_e \cdots \leq_e B$. As $G^0 <_e G^1 <_e \cdots \leq_e B$ and $G_s \equiv_e \tilde{G}^s = \{\langle s, x, y\rangle : \langle x, y\rangle \in G^s\}$ then $\tilde{G}^0 <_e \tilde{G}^1 <_e \cdots \leq_e B$. Clearly $A = \bigcup_{s \in N} \tilde{G}^s$ thus $A \leq_e B$ which contradicts the premise.

Theorem 3. *Any nontotal e-degree contains either C-quasi-minimal set for some set C or the set $C = \bigcup_{s \in N} C_s$ where $C_0 <_e C_1 <_e \cdots <_e C_s <_e \cdots$ and C_s belongs to a suitable total e-degree for all $s \in N$.*

Proof. Let $\mathbf{a}$ be a nontotal e-degree. If there is a total function f such that $A \in \mathbf{a}$ is τf-quasi-minimal set then the theorem is proved for $C = \tau f$.

Assume that for all $A \in \mathbf{a}$ the set A is not τf-quasi-minimal for all total functions f, that is

$$f \leq_e A \rightarrow \exists g[g \leq_e A \wedge g \not\leq_e f].$$

Let $B \in \mathbf{a}$, and we enumerate all total functions which are e-reducible to B. Let them be $f_0, f_1, \ldots$. We construct the sequence $g_0 <_e g_1 <_e \cdots \leq_e B$ such that

$$\forall f[f \leq_e B \rightarrow \exists n[f \leq_e g_n]]. \tag{1}$$

Set $g_0 = f_0$, let $g_0 \leq_e \cdots \leq_e g_{n-1}$ be already chosen when we dealt with $f_0, \ldots, f_{s-1}$. Let f_s be given. There are two cases:

(i) $f_s \leq_e g_{n-1} \vee g_{n-1} \leq_e f_s$ and

(ii) $f_s \not\leq_e g_{n-1} \wedge g_{n-1} \not\leq_e f_s$.

In case (i), if $g_{n-1} <_e f_s$ then set $g_n = f_s$, otherwise let $g_n = g_{n-1}$. In case (ii) set $g_n = g_{n-1} \oplus f_s$.

We prove that the constructed sequence $\{g_n : n \in N\}$ satisfies (1). Assume that it is not true and

$$\exists f[f \leq_e B \wedge \forall n[f \not\leq_e g_n]]. \tag{2}$$

Let $f = f_{s_0}$ for some $s_0 \in N$ then (i) or (ii) was satisfied for some $n = n_0$. If (i) was satisfied then necessarily $g_{n_0-1} <_e f_{s_0}$ and so $g_{n_0} = f_{s_0}$ which contradicts (2). Thus (ii) occurred and $g_{n_0} = g_{n_0-1} \oplus f_{s_0}$. In this case $f_{s_0} \leq_e g_{n_0}$ which contradicts (2) also. So, the constructed sequence $\{g_n : n \in N\}$ satisfies the condition (1).

Now we check that this sequence of total function forms an infinite strictly increasing sequence $g_0 <_e g_1 <_e \cdots \leq_e B$. Indeed, if there is s_0 such that $g_n \leq_e g_{s_0}$ for all $n > s_0$ then for any total function $f \leq_e B$ there is $n \in N$ such that $f \leq_e g_n \leq_e g_{s_0}$, hence B is τg_{s_0}-quasi-minimal which contradicts the premise.

Using the sequence $\{g_n : n \in N\}$ we form the set C with the help of the construction used in the proof of Theorem 2. Let $G^s = \{\langle x, y\rangle : \langle s, x, y\rangle \in C\}$ and $\tilde{G}_s = \{\langle s, x, y\rangle : \langle x, y\rangle \in G^s\}$. The construction ensures that $g_s \equiv_e G^s \equiv_e \tilde{G}_s \leq_e B$ for all $s \in N$ so $C = \bigcup_{s \in N} \tilde{G}_s \leq_e B$ where $\tilde{G}_0 <_e \tilde{G}_1 <_e \cdots$ and every set $\tilde{G}_s$ belongs to a total e-degree for all $s \in N$.

If $C <_e B$ then B is a C-quasi-minimal set and the theorem is proved. If $C \equiv_e B$ then the given e-degree $\mathbf{a}$ contains the set $C = \bigcup_{s \in N} C_s$ where $C_s = \tilde{G}_s$ and $\tilde{G}_s$ belongs to a total e-degree for all $s \in N$ as was to be proved. □

An important objective in the study of e-degrees is to investigate the context of the total degrees within the e-degrees, in particular to look for characterizations of the nontotal e-degrees in terms of properties of some of their elements. One of the first results of such sort belongs to Copestake [5]: *every 1-generic e-degree is quasi-minimal.*

In [6] the notion of e-hyperimmune set was introduced. We recall the necessary definitions. Let $a_0, a_1, \ldots, a_n, \ldots$ be the members of an infinite set A in strictly increasing order (or *a principal enumeration* of A). We call the set A *e-hyperimmune* (e-h.i.) if A is infinite and there is no total function $g \leq_e A$ such that $a_n \leq g(n)$ for all $n \in N$. If $a_n \leq g(n)$ for all $n \in N$ then we say that g *majorizes* A. So an infinite set A is an e-h.i. iff it is not majorized by any total function $g \leq_e A$. In [6] the existence of e-h.i. sets was proved and a question was posed: does every nontotal e-degree contain an e-h.i. set? Below the negative answer to this question is given. For this we extend the class of e-h.i. sets in such a way that every set of this class belongs to a nontotal e-degree.

We call an infinite set A an *e-immune* (e-i.) if for any total function f

$$f \leq_e A \wedge \rho f \subseteq A \rightarrow \rho f \text{ is finite.}$$

The existence and simplest properties of e-i. sets are exhibited in the following theorem.

Theorem 4.

a) *If A is an e-h.i. set then A is a e-i. set;*

b) *If A is an e-i. set then A is an immune set;*

c) *If A is an e-i. set then $d_e(A)$ is a nontotal e-degree;*

d) *If A is an immune quasi-minimal set then A is an e-immune set.*

Proof. a) Assume that an infinite set A is not e-i. Then there is a total function $f \leq_e A$ such that $\rho f \subseteq A$ and ρf is infinite. We show that there exists a total function $f \leq_e A$ which majorizes A. We define the function g as follows: $g(0) = f(0)$ and $g(n+1) = f(x_n)$ where $x_n = \mu x[f(x) > g(n)]$. As ρf is infinite then $\{x : f(x) > g(n)\}$ is infinite and so g is a total function. Moreover $\rho g \subseteq \rho f$, ρg is infinite and $g(n) < g(n+1)$ for all $n \in N$. Clearly $g \leq_e f$. We now show that g majorizes A. As $g(0) \in A$ then $a_0 \leq g(0)$. Let $a_i \leq g(i)$ for all $i < n$ and suppose that $a_n > g(n)$ then $g(n) \leq a_{i_0}$ for some $i_0 < n$. We obtained that $g(n) \leq a_{i_0} \leq g(i_0)$ what contradicts the premise.

All other items are trivial too. The proof is completed. □

It is possible to prove the existence of e-i. sets in the following stronger form.

Theorem 5. *There is an e-i. set A such that $\overline{A}$ is also an e-i. set.*

Proof. We construct step by step the sets A and B such that

(i) $\forall g[g \leq_e A \wedge \rho g \subseteq A \rightarrow \rho g \text{ is finite}]$;

(ii) $\forall g[g \leq_e B \wedge \rho g \subseteq B \rightarrow \rho g \text{ is finite}]$;

(iii) $B = \overline{A}$.

Let A_s and B_s be parts of A and B respectively which are constructed on the step s such that $A_s \cup B_s$ is an initial segment of N and $A_s \cap B_s = \emptyset$. We will ensure $A_s \subseteq A_{s+1}$ and $B_s \subseteq B_{s+1}$ for all $s \in N$. We define $A = \bigcup_{s\in N} A_s$ and $B = \bigcup_{s\in N} B_s$.

Step 0. Let $A_0 = B_0 = \emptyset$.

Step $s+1$. In turn we verify the truth of the following conditions and after a definition of A_{s+1} and B_{s+1} move to step $s+2$. Let $z_s = \max(A_s \cup B_s)$ then $A_s \cup B_s = \{0, \dots, z_s\}$.

Let $s = 4n$. If W_n is finite then set $A_{s+1} = A_s$ and $B_{s+1} = B_s$. If W_n is infinite then $\exists z[z \in W_n - A_s \cup B_s]$. Let z^* is the least such z, set $A_{s+1} = A_s$ and $B_{s+1} = B_s \cup \{z_s + 1, \dots, z^*\}$.

Let $s = 4n+1$. Fulfill the above-mentioned part of step $s+1$ where B_s replaces A_s and vice-versa.

Let $s = 4n+2$. See whether

$$\exists D[D \cap B_s = \emptyset \wedge \Phi_n(A_s \cup D) \text{ is not single-valued}]. \tag{1}$$

If so, set $A_{s+1} = A_s \cup D^*$ and $B_{s+1} = \{0, \dots, z_{s+1}\} - A_{s+1}$ where D^* has the least canonical index among finite sets D which satisfied the condition (1) and $z_{s+1} = \max(A_{s+1} \cup B_s)$. If (1) is false then

$$\forall D[D \cap B_s = \emptyset \to \Phi_n(A_s \cup D) \text{ is single-valued}].$$

In this case $\Phi_n(A)$ is a single-valued set for any set $A \subseteq \overline{B_s}$ that is $\Phi_n(A) = \tau g$ for some function g. We write $!\Phi_n(A)(m)$ if $!g(m)$ and $\Phi_n(A)(m) = g(m)$.

See whether

$$\exists D[D \cap B_s = \emptyset \wedge \rho\Phi_n(A_s \cup D) \not\subseteq A_s \cup D]. \tag{2}$$

If so, then

$$\exists D \exists m[D \cap B_s = \emptyset \wedge !\Phi_n(A_s \cup D)(m) \wedge \Phi_n(A_s \cup D)(m) \in \rho\Phi_n(A_s \cup D) - (A_s \cup D)]. \tag{3}$$

Set $A_{s+1} = A_s \cup D^*$ and $B_{s+1} = \{0, \dots, z_{s+1}\} - A_{s+1}$ where D^* has the least canonical index among finite sets D satisfying the condition (2) and $z_{s+1} = \max(A_{s+1} \cup B_s \cup \{\Phi_n(A_s \cup D^*)(m^*)\})$ where m^* is the least m which satisfied (3) if $D = D^*$. Particularly note that if $\tau g = \Phi_n(A)$ then $g(m^*) \in B_{s+1}$ hence $g(m^*) \in B$. If (2) is false then set $A_{s+1} = A_s$ and $B_{s+1} = B_s$.

Let $s = 4n+3$. Fulfill the above-mentioned part of the step $s+1$ where B_s replaces A_s and vice-versa.

The description is completed. We prove that the constructed sets A and B satisfy conditions (i)–(iii). The step $s+1$ where $s = 4n$ and $s = 4n+1$, $n \in N$ guarantee an immunity of sets A and B. Let g be a total function $\leq_e A$ then $\tau g = \Phi_n(A)$ for some $n \in N$. Consider the step $s+1$ where $s = 4n+2$. The case (1) cannot be satisfied for otherwise $\Phi_n(A)$ is not a single-valued set. If the case (2) was satisfied then $g(m^*) \in B$ and on the next steps $g(m^*)$ was not included into A hence in this case $\rho g \not\subseteq A$.

Suppose that (2) is not satisfied then

$$\forall D[D \cap B_s = \emptyset \to \rho\Phi_n(A_s \cup D) \subseteq A_s \cup D]. \tag{4}$$

In this case $\rho g = \rho\Phi_n(A) \subseteq A$. Otherwise, $\exists D[D \subseteq A \wedge \rho\Phi_n(D) \not\subseteq A]$ that is

$$\exists D \exists m[D \cap B_s = \emptyset \wedge !\Phi_n(A_s \cup D)(m) \wedge \Phi_n(A_s \cup D)(m) \in \rho\Phi_n(A_s \cup D) - A]$$

and so $\Phi_n(A_s \cup D)(m) \in \rho\Phi_n(A_s \cup D) - (A_s \cup D)$. This contradicts our supposition that (3) is false. As $A \subseteq \overline{B_s}$ so $\tau g = \Phi_n(A) \subseteq \Phi_n(\overline{B_s})$. It is easy see that $\Phi_n(\overline{B_s})$ is single-valued so $\tau g = \Phi_n(\overline{B_s})$ hence g is recursive function. As a result $\rho g \subseteq A$ where A is an immune set so ρg is finite. Thus A is an e-i. set and the condition (i) is satisfied.

Similarly it is possible to verify the satisfiability of the condition (ii). From the construction $B = \overline{A}$ and the theorem is proved. □

We denote by **I** the class of all immune sets, by **EI** the class of all e-i. sets, by **EHI** the class of all e-h.i. sets and by **Q** the class of all quasi-minimal sets. Clearly **EHI** $\subseteq$ **EI** $\subseteq$ **I** and **I** $\cap$ **Q** $\subseteq$ **EI**. We prove with the help of the following theorem that the first inclusion is strict.

Theorem 6. *There is an immune quasi-minimal but not e-h.i. set.*

Proof. We construct step by step the set A as the union of a sequence of sets A_s where A_s is a finite set which will be constructed on the step s such that

(i) $\forall n[W_n$ is infinite $\to W_n \cap \overline{A} \neq \emptyset]$;

(ii) $\forall f[f \leq_e A \to f$ is recursive function];

(iii) $\forall k[|A \cap \{0, \dots, 2k\}| \geq k]$.

Note that if the condition (iii) is satisfied then A is an infinite non hyperimmune set. In fact if (iii) is satisfied then $a_k \leq 2k$ for all $k \in N$ where $a_0, a_1, \dots, a_k, \dots$ is the principal enumeration of A. Suppose that $a_0 \leq 0,\ a_1 \leq 2, \dots, a_{k-1} \leq 2(k-1)$ where $k > 1$ and $|A \cap \{0, \dots, 2k\}| \geq k$. If $a_k > 2k$ then $|A \cap \{0, \dots, 2k\}| = k - 1$ which would contradict the condition (iii). Hence $a_k \leq 2k$.

Step 0. Define $a_0 = 0$ and $A_0 = \{a_0\}$.

Step $2n+1$. Let $2n = s$ and $a_0 < a_1 < \cdots < a_{k_s}$ be the members of A_s. If W_n is a finite set then go to the next step. If W_n an infinite r.e. set then there exists $x \in W_n$ and $x > a_{k_s}$. Let x^* be the least such x. We define $A_{s+1} = A_s \cup \{a_{k_s}, \dots, x^* - 1, x^* + 1\}$.

We verify that if $|A_s \cap \{0, \dots, 2k_s\}| \geq k_s$ then $|A_{s+1} \cap \{0, \dots, 2k\}| \geq k$ for all k such that $k_s < k \leq k_{s+1}$. In fact even if $a_{k_s} = 2k_s$ then $a_{k_s+1} = a_{k_s} + 1$ under $x \neq a_{k_s} + 1$ and $a_{k_s+1} = a_{k_s} + 2$ under $x = a_{k_s} + 1$. In either case $a_{k_s+1} \leq 2k_s + 2$. Moreover the steps $2n + 1$, $n \in N$ guarantee the satisfaction of the condition (i) and so A is an immune set.

Step $2n+2$. Let $s = 2n+1$. Denote by $Z_s = \{0, \dots, a_{k_s}\} - A_s$. We define F to range over finite sets such that $F \subseteq A_s \vee [F \nsubseteq A_s \to F \cap Z_s = \emptyset]$. If in the second case $d = |F - A_s|$ then we must have $|(A_s \cup F) \cap \{0, \dots, 2k\}| \geq k$ for all $k \leq k_s + d$.

See whether

$$\exists F[\Phi_n(A_s \cup F) \text{ is not single-valued}]. \qquad (1)$$

If so, set $A_{s+1} = A_s \cup F^*$ where F^* has the least canonical index among finite sets F which satisfied the condition (1). If not, set $A_{s+1} = A_s$.

We prove that the construction gives as a result the quasi-minimal set A. Let $f \leq_e A$ via n that is $\tau f = \Phi_n(A)$. In this case condition (1) is not satisfied and we can show that $\Phi_n(\overline{Z_s})$ is a single-valued set. Suppose that $\Phi_n(\overline{Z_s})$ is not single-valued and so there exists $D \subseteq \overline{Z_s}$ such that $\Phi_n(D)$ is not a single-valued set. Clearly $D \nsubseteq A_s$ and $D \cap Z_s = \emptyset$. Let $m = \max D$ and so $m > a_{k_s}$. We denote $d = |D - A_s|$ and let us consider $F = D \cup \{a_{k_s} + 1, \dots, m\}$. As $|A_s \cap \{0, \dots, 2k\}| \geq k$ for all $k \leq k_s$ then $|(A_s \cup F) \cap \{0, \dots, 2k\}| \geq k$ for all $k \leq k_s + d$. Hence the condition (1) must be satisfied contradicting our supposition.

Thus $\Phi_n(\overline{Z_s})$ is a single-valued set and $A \subseteq \overline{Z_s}$ so $\tau f = \Phi_n(A) \subseteq \Phi_n(\overline{Z_s})$. Hence $\tau f = \Phi_n(\overline{Z_s})$ and f is a recursive function. Hence condition (ii) is satisfied.

Finally, the set A is formed on every step of the construction in such a way that it satisfies (iii). The proof is completed. □

Corollary 4. *There is an e-immune but not e-hyperimmune set.*

Proof. Let A be an immune quasi-minimal set that is not e-h.i. which is constructed in the course of proof of Theorem 6. So by Theorem 4.d A is an e-i. set but not an e-h.i. set. □

So, **EHI** $\subset$ **EI**. As any nonzero total e-degree contains an immune set and by Theorem 4.c an e-i. set cannot belong to a total e-degree, we have **EI** $\subset$ **I**. With the aid of the following theorem we give a less trivial example.

Theorem 7. *There is an immune but not e-immune set which belongs to a nontotal e-degree.*

Proof. At first we note that there is a total nonrecursive function f such that τf is an immune set and ρf is an infinite set. Indeed suppose that $f(0), \dots, f(x_n)$ is defined by step n. If W_{n+1} is an infinite r.e. set then let m be the least x such that

$$x > x_n \wedge \exists y[\langle x, y\rangle \in W_{n+1} - \{\langle 0, f(0)\rangle, \dots, \langle x_n, f(x_n)\rangle\}].$$

Let y_0 be the least y which satisfies this condition. Set $f(m) = y_0 + a + 1$ where $a = \max\{f(0), \dots, f(x_n)\}$ and $f(x) = 0$ for all x between x_n and m. So, if W_{n+1} is an infinite r.e. set then $W_{n+1} \nsubseteq \tau f$ as $\langle m, y_0\rangle \in W_{n+1} - \tau f$. Clearly f is a nonrecursive function.

If $C = \tau f$ then by Theorem 1 there is a set B such that $C <_e B$ and for any total function $g \leq_e B$ we have $g \leq_e C$. As $d_e(C)$ is a total e-degree and $d_e(C) < d_e(B)$ there is an immune set $B_1 \in d_e(B)$. Let $A = \tau f \oplus B_1$. We verify that the set A satisfies all the requirement of the theorem. First we check that A is an immune set. Let $W \subseteq A = \tau f \oplus B_1$ be an r.e. set and $V_1 = \{x : 2x \in W\}$ and $V_2 = \{x : 2x+1 \in W\}$. Clearly $W = V_1 \oplus V_2$ and $V_1 \subseteq \tau f$ and $V_2 \subseteq B$. As τf and B_1 are both immune sets V_1 and V_2 are both finite sets. Hence W is a finite set and A is an immune set. Now we check that A is not an e-i. set. If $f_1(x) = 2\langle x, f(x)\rangle$ then $f_1 \equiv_e f \leq_e A$ and $\rho f_1 \subseteq A$. Moreover ρf_1 and τf are both infinite. Finally we check that $d_e(A)$ is a nontotal e-degree. Assume that there is a total function g such that $g \in d_e(A)$. As $f <_e B_1$ and $g \leq_e A = \tau f \oplus B_1$ then $g \leq_e B_1$ thus $g \leq_e B$ and $g \leq_e f$ so $B_1 \not\leq_e g$. If $A \leq_e g$ then $\tau f \oplus B_1 \leq_e g$ and $B_1 \leq_e g$ what contradicts the foregoing fact. The proof is completed. □

As was shown by Rozinas [7] the family of quasi-minimal immune sets is uncountable and so **EI** is uncountable too.

The following theorem shows that $\mathbf{I} \cap \mathbf{Q} \subset \mathbf{EI}$. In other words, there are e-i. sets which do not belong to quasi-minimal e-degrees.

Theorem 8. *For any retraceable set B there exists a countable family of e-i. sets $A_0, A_1, \ldots, A_k, \ldots$ such that $B <_e A_k$ for all $k \in N$ and $A_k \not\leq_e A_l$ for all $k \neq l$.*

Proof. Let B be an infinite retraceable set. We construct step by step the sets $A_0, A_1, \ldots$ such that

(i) $\forall k[A_k \subseteq B \wedge A_k$ is infinite];

(ii) $\forall k \forall g[g \leq_e A_k \wedge \rho g \subseteq A_k \rightarrow \rho g$ is finite];

(iii) $\forall k \forall l[k \neq l \rightarrow A_k \not\leq_e A_l]$.

Note that if A_k is an infinite set and $A_k \subseteq B$ then $B \leq_e A_k$ and $A_k \not\leq_e B$ because otherwise $A_k \leq_e B \leq_e A_l$ for some $k \neq l$ which contradicts (iii). If (ii) is satisfied then A_k is an e-i. set.

We denote by A_{k_t} the part of A_k which is formed on the step t. Set $A_k = \bigcup_{t \in N} A_{k_t}$, $k \in N$. In the course of the construction the auxiliary set Z will be formed, Z_t be a finite part of Z which is formed on the step t. Let $g_0, g_1, \ldots, g_n, \ldots$ be all total function which are reducible to B.

Step 0. Set $A_{k_0} = \emptyset$ for all $k \in N$ and $Z_0 = \emptyset$.

Step $t+1$. Let $t = \langle k, l, s \rangle$. We check the conditions stated below one after another and move on to the step $t+2$ after the definition of $A_{k_{t+1}}$.

If $s = 3n$ and ρg_n is finite then set $A_{k_{t+1}} = A_{k_t} \cup \{x^*\}$ and $Z_{t+1} = Z_t$ where $x^* = \min(B - Z_t)$. If ρg_n is infinite then

$$\exists z[z \in \rho g_n - A_{k_t}]. \tag{1}$$

Let z^* be the least z which satisfies (1), set $Z_{t+1} = Z_t \cup \{z^*\}$ and $A_{k_{t+1}} = A_{k_t} \cup \{x^*\}$ where $x^* = \min(B - Z_{t+1})$.

If $s = 3n + 1$ see whether

$$\exists D[D \subseteq B \wedge D \cap Z_t = \emptyset \wedge \Phi_n(A_{k_t} \cup D) \text{ is not single-valued}]. \tag{2}$$

If so, set $A_{k_{t+1}} = A_{k_t} \cup D^*$ where D^* has the least canonical index among finite sets D which satisfy the condition (2) and $Z_{t+1} = Z_t$. If not, then

$$\forall D[D \subseteq B \wedge D \cap Z_t = \emptyset \rightarrow \Phi_n(A_{k_t} \cup D) \text{ is single-valued}]$$

that is $\Phi_n(A_{k_t})$ is a single-valued set for all $t \in N$. In this case $\Phi_n(A_k) = \tau g$ for some function g.

See whether

$$(\exists D)(\exists m)[D \subseteq B \wedge D \cap Z_t = \emptyset \wedge !\Phi_n(A_{k_t} \cup D)(m) \wedge \Phi_n(A_{k_t} \cup D)(m) \notin A_{k_t} \cup D]. \tag{3}$$

If so, set $A_{k_{t+1}} = A_{k_t} \cup D^*$ and $Z_{t+1} = Z_t \cup \{\Phi_n(A_{k_t} \cup D^*)(m^*)\}$ where D^* has the least canonical index among finite sets which satisfy (3) and m^* is the least m which satisfies (3) for $D = D^*$. If not, set $A_{k_{t+1}} = A_{k_t}$ and $Z_{t+1} = Z_t$.

If $s = 3n + 2$ see whether

$$\exists x \exists D[x \in \overline{A_{k_t}} \wedge D \subseteq B \wedge D \cap Z_t = \emptyset \wedge x \in \Phi_n(A_{l_t} \cup D) \cap B]]. \tag{4}$$

If so, set $A_{k_{t+1}} = A_{k_t}$ and $Z_{t+1} = Z_t \cup \{x^*\}$ where x^* is the least number which satisfies (4) and $A_{l_{t+1}} = A_{l_t} \cup D^*$ where D^* has the least canonical index among finite sets which satisfy (4) for $x = x^*$. If not, then

$$\forall x \forall D[x \in \overline{A_{k_t}} \wedge D \subseteq B \wedge D \cap Z_t = \emptyset \rightarrow x \notin \Phi_n(A_{l_t} \cup D) \cap B]]. \tag{5}$$

Let $x^* = \min(B - Z_t)$ set $A_{k_{t+1}} = A_{k_t} \cup \{x^*\}$, $A_{l_{t+1}} = A_{l_t}$ and $Z_{t+1} = Z_t$.

The description of the construction is completed. We now show that A_k satisfies the conditions (i)–(iii) for all $k \in N$. One can see from the construction that $A_{k_t} \subseteq A_{k_{t+1}} \subseteq B$ for all $k, t \in N$ so $A_k \subseteq B$ for all $k \in N$ and the condition (i) is satisfied.

The step $t + 1$ where $t = \langle k, l, s \rangle$ and $s = 3n$ guarantee the satisfiability of the following condition for all $k \in N$

$$\forall n[\rho g_n \subseteq A_k \rightarrow |\rho g_n| < \infty].$$

Let g be a total function $\leq_e A_k$, that is $\tau g = \Phi_n(A_k)$ for some $n \in N$. Consider the step $t + 1$ where $t = \langle k, 0, s >$ and $s = 3n + 1$. If (2) was satisfied then $\Phi_n(A_k)$ would not be single-valued so

$$\forall D[D \subseteq B \wedge D \cap Z_t = \emptyset \rightarrow \Phi_n(A_{k_t} \cup D) \text{ is single-valued}].$$

Clearly in this case $\Phi_n(B - Z_t)$ is single-valued. Since $A_k \subseteq B - Z_t$ then $\tau g = \Phi_n(A_k) \subseteq \Phi_n(B - Z_t)$ and hence $\tau g = \Phi_n(B - Z_t)$. In other words $g \leq_e B - Z_t \leq_e B$

so $g = g_n$ for some $n \in N$. As noted above, for any total function g

$$\rho g \subseteq A_k \to \rho g \text{ is finite}$$

and the condition (ii) is satisfied.

Finally we show that $A_k \not\leq_e A_l$ for all $k \neq l$. Suppose that it is not true and there is $n \in N$ such that $A_k = \Phi_n(A_l)$. Consider the step $t+1$ where $t = \langle k, l, s\rangle$ and $s = 3n+2$. If the condition (4) is satisfied then $x^* \notin A_k$ and $x^* \in \Phi_n(A_{l_t} \cup D^*)$, therefore $x^* \in \Phi_n(A_l)$. If the condition (5) is satisfied then $x^* \in A_{k_{t+1}} \subseteq A_k \subseteq B$ and

$$\forall D[D \subseteq B \wedge D \cap Z_t = \emptyset \to x^* \notin \Phi_n(A_{l_t} \cup D) \cap B]$$

so $\forall D[D \subseteq B \wedge D \cap Z_t = \emptyset \to x^* \notin \Phi_n(A_{l_t} \cup D)]$. This means that $x^* \notin \Phi_n(A_l)$. In any case $A_k \neq \Phi_n(A_l)$ and the condition (iii) is satisfied. The proof is completed. □

Corollary 5. *There is an uncountable increasing chain of e-i. sets.*

Proof. Let A_0 be an e-i. set and $\mathbf{J}(A_0)$ is the jump of A_0. It is well known that there exists a retraceable set $B_1 \equiv_e \mathbf{J}(A_0)$. So by Theorem 8 there is an e-i. set A_1 such that $B_1 <_e A_1$. Thus $A_0 <_e A_1$ and so on. Let $\mathbf{C}$ be the family of all chains of e-i. sets which is partially ordered by the relation $\subseteq$. Every chain of elements of $\mathbf{C}$ has a least upper bound in $\mathbf{C}$, that is the degree which is the union of all sequences forming the given chain from $\mathbf{C}$. By Zorn's Lemma $\mathbf{C}$ has at least one maximal element $\mathbf{m} = \{M_s : s \in N\}$ where $M_0 <_e M_1 <_e \cdots$ and M_s is an e-i. set for all $s \in N$. Let $M = \{\langle s, x\rangle : x \in M_s\}$. Clearly $M_s \leq_e M$ for all $s \in N$. Let $B \in \mathbf{J}(M)$ be a retraceable set. As there is an e-i. set A such that $B <_e A$ then $M_s <_e A$ for all $s \in N$ what contradicts the maximality of $\mathbf{m}$. This shows that $\mathbf{m}$ is an uncountable set. □

Corollary 6. *For any retraceable set B there is an uncountable family of e-i. sets which are over B in the semilattice of e-degrees and are mutually incomparable.*

Proof. Reconstruct the proof of Theorem 8 as was made in Corollary 1. Let $\mathbf{EI_B}$ be the family of all e-i. sets over B. Assume that the antichain in $\mathbf{EI_B}$ is a countable set, let $A_0, A_1, \ldots$ be all elements of this antichain. We construct a set C such that it is e-i. set, $B <_e C$ and C and A_k are incomparable for all $k \in N$. Let C_t be the part of C formed on the step t. The set Z will play an auxiliary role and Z_t denotes that finite part of Z which is formed on the step t.

Step 0. Define $C_0 = Z_0 = \emptyset$.

Step $t+1$. Let $t = \langle k, s\rangle$.

If $s = 4n$ and ρg_n is finite then set $C_{t+1} = C_t \cup \{x^*\}$ and $Z_{t+1} = Z_t$ where $x^* = \min(B - Z_t)$. If ρg_n is infinite then

$$\exists z[z \in \rho g_n - C_t]. \tag{1}$$

Let z^* is the least z which satisfies (1), set $Z_{t+1} = Z_t \cup \{z^*\}$ and $C_{t+1} = C_t \cup \{x^*\}$ where $x^* = \min(B - Z_{t+1})$.

If $s = 4n + 1$ see whether

$$\exists D[D \subseteq B \wedge D \cap Z_t = \emptyset \wedge \Phi_n(C_t \cup D) \text{ is not single-valued}]. \qquad (2)$$

If so, set $C_{t+1} = C_t \cup D^*$ where D^* has the least canonical index among finite sets D which satisfy (2) and $Z_{t+1} = Z_t$. If not, then

$$\forall D[D \subseteq B \wedge D \cap Z_t = \emptyset \rightarrow \Phi_n(C_t \cup D) \text{ is single-valued}]$$

that is $\Phi_n(C_t)$ is a single-valued for all $t \in N$. In this case $\Phi_n(C) = \tau g$ for some function g. See whether

$$(\exists D)(\exists m)[D \subseteq B \wedge D \cap Z_t = \emptyset \wedge !\Phi_n(C_t \cup D)(m) \wedge \Phi_n(C_t \cup D)(m) \notin C_t \cup D]. \qquad (3)$$

If so, set $C_{t+1} = C_t \cup D^*$ and $Z_{t+1} = Z_t \cup \{\Phi_n(C_t \cup D^*)(m^*)\}$ where D^* has the least canonical index among finite sets which satisfy (3) and m^* is the least m which satisfies (3) for $D = D^*$. If not, set $C_{t+1} = C_t \cup \{x^*\}$ and $Z_{t+1} = Z_t$.

If $s = 4n + 2$, see whether

$$\exists x[x \in \overline{Z_t} \wedge x \in \Phi_n(A_k) \cap B]. \qquad (4)$$

If so, set $C_{t+1} = C_t$ and $Z_{t+1} = Z_t \cup \{x^*\}$ where x^* is the least number which satisfies (4). If not, then

$$\forall x[x \in \overline{Z_t} \rightarrow x \notin \Phi_n(A_k) \cap B].$$

Let $x^* = \min\{x : x \in B - Z_t\}$, set $C_{t+1} = C_t \cup \{x^*\}$ and $Z_{t+1} = Z_t$.

If $s = 4n + 3$, see whether

$$\exists D[D \subseteq B \wedge D \cap Z_t = \emptyset \wedge \Phi_n(C_t \cup D) \not\subseteq A_k]. \qquad (5)$$

If so, set $C_{t+1} = C_t \cup D^*$ and $Z_{t+1} = Z_t$ where D^* has the least canonical index among finite sets D which satisfy (5). If not, set $C_{t+1} = C_t$ and $Z_{t+1} = Z_t$. In this case

$$\forall D[D \subseteq B \wedge D \cap Z_t = \emptyset \rightarrow \Phi_n(C_t \cup D) \subseteq A_k],$$

and hence $\Phi_n(\tilde{B}) \subseteq A_k$ where $\tilde{B} = B - Z_t$. If $\Phi_n(C) = A_k$ then $\Phi_n(C) \subseteq \Phi_n(C_t \cup \tilde{B}) \subseteq A_k$ so $\Phi_n(C_t \cup \tilde{B}) = A_k$ that is $A_k \leq_e C_t \cup \tilde{B} \leq_e B$ which contradicts the assumption that $B <_e A_k$ for all $k \in N$.

From the construction it follows that C is an infinite subset of the retraceable set B so $B \leq_e C$. The steps $t + 1$ where $t = \langle k, s\rangle$ and $s = 4n$ and $s = 4n + 1$ ensure that C is an e-i. set. The steps $t + 1$ where $t = \langle k, s\rangle$ and $s = 4n + 2$ and $s = 4n + 3$ ensure that $C \not\leq_e A_k$ and $A_k \not\leq_e C$ for all $k \in N$. Note that $C \not\leq_e B$ as otherwise $C \leq_e B \leq_e A_k$ as was to be proved. □

Now we answer the question of whether a nontotal e-degree contains an e-i. set. We give negative answer to this question. With that goal in mind we define the following class of sets. Let g be a recursive function and $V_{n,g} = \{x : \exists u[\langle x, g(u)\rangle \in W_n]\}$. We call the set A *strongly nontotal* if A is not r.e. and

$$\forall n[\Phi_n(A) \text{ is infinite} \rightarrow (\exists g)(\forall u)[D_{g(u)} \subseteq A \wedge V_{n,g} \text{ is infinite}]].$$

First we note an important property of strongly nontotal sets and then we will prove the existence of such sets.

Theorem 9. a). *If A is strongly nontotal set and $B \leq_e A$ then the set B is not immune;*
b). *If A is a strongly nontotal set then A is the quasi-minimal set.*

Proof. a). Let $B \leq_e A$ via n that is

$$B = \Phi_n(A) = \{x : \exists u[\langle x, u\rangle \in W_n \wedge D_u \subseteq A]\}.$$

If $\Phi_n(A)$ is an infinite set then there exists a recursive function g such that $D_{g(u)} \subseteq A$ for all $u \in N$ and $V_{n,g} = \{x : \exists u[\langle x, g(u)\rangle \in W_n]\}$ is infinite so $V_{n,g} \subseteq B$ and B is not immune.

b). Suppose that A is not a quasi-minimal set, then there exists a total function such that $f \leq_e A$ and f is not recursive. We consider $d_e(\tau f)$, and by Case's theorem [2] there exists a retraceable set, $B \in d_e(\tau f)$. As any nonrecursive retraceable set is immune, B is immune. This contradicts the fact that $B \leq_e A$ and so B is not immune by item a). The proof is completed. □

Corollary 7. *There are quasi-minimal e-degrees which do not contain e-i. sets.*

Proof. As an e-i. set is immune and e-degree of a strongly nontotal set does not contain an immune set, then the e-degree of a strongly nontotal set does not contain e-i. sets. □

Theorem 10. *There is an uncountable collection of strongly nontotal sets.*

Proof. Suppose that the cardinal number of the collection **SNT** of strongly nontotal sets is at most countable. We enumerate the sets from **SNT**$\cup$**W** where **W** is the collection of all r.e. sets. Let them be $C_0, C_1, \ldots, C_n, \ldots$. We define two nondecreasing sequences of sets $\{A_s\}$ and $\{B_s\}$, $s \in N$ such that $A_s \cap B_s = \emptyset$, A_s is a r.e., co-infinite set, $\bar{A}_s$ is not immune and B_s is finite for all $s \in N$.

Step 0. Set $A_0 = B_0 = \emptyset$.

Step $2n + 1$. Let $s = 2n$ and $z = \min \overline{A_s \cup B_s}$. See whether

$$z \in C_n. \tag{1}$$

If so, set $A_{s+1} = A_s$ and $B_{s+1} = B_s \cup \{z\}$. If not, set $A_{s+1} = A_s \cup \{z\}$ and $B_{s+1} = B_s$.

Step $2n+2$. Let $s = 2n+1$. See whether

$$\exists D[B_s \subseteq D \wedge A_s \cap D = \emptyset \wedge |\Phi_n(\bar{D})| < \infty]. \tag{2}$$

If so, set $A_{s+1} = A_s$ and $B_{s+1} = D^*$ where D^* has the least canonical index among finite sets which satisfy the condition (2). If not, then

$$\forall D[B_s \subseteq D \wedge A_s \cap D = \emptyset \rightarrow |\Phi_n(\bar{D})| = \infty]. \tag{3}$$

As $\bar{A}_s$ is an infinite not immune set on all previous steps, $\bar{A}_s$ contains some infinite r.e. subset U. We form, with the help of the following construction, step by step a recursive function g and an auxiliary r.e. set Q such that

$$\forall u[D_{g(u)} \subseteq A_{s+1} \wedge |V_{n,g}| = \infty] \wedge B_{s+1} \subseteq Q \wedge Q \cap A_{s+1} = \emptyset.$$

We denote by Q^k the part of Q and by g^k the initial segment of g which are constructed on the substep k.

Substep 0. Set $g^0 = \emptyset$ and $Q^0 = B_s$.

Substep $k+1$. Let $g^k = \{\langle 0, g(0)\rangle, \ldots, \langle x_k, g(x_k)\rangle\}$. We enumerate W_n in standard order until a pair $\langle x, y\rangle$ is found such that $x > x_k$ and $D_y \subseteq \overline{Q^k}$. Such pair $\langle x, y\rangle$ must appear in the standard enumeration of W_n as the condition (3) implies

$$\Phi_n(\overline{Q^k}) = \{x : \exists y[\langle x, y\rangle \in W_n \wedge D_y \subseteq \overline{Q^k}]\} \text{ is infinite.}$$

We define $g^{k+1} = \{\langle 0, g(0)\rangle, \ldots, \langle x_k, g(x_k)\rangle, \langle x_k+1, y\rangle, \ldots, \langle x, y\rangle\}$. (So, the function g will have in the points $x_k+1, \ldots, x$ the same value $y = g(x)$ which is different from any of previous values of the function g). Next we enumerate U in the standard order until such $z \in U$ is found that $z \notin \bigcup_{i \leq x} D_{g(i)} \cup Q^k$. This z exists necessarily as U is an infinite set and $\bigcup_{i \leq x} D_{g(i)} \cup Q^k$ is finite.

Set $Q^{k+1} = Q^k \cup \{z\}$. From our construction it is implied that g is a recursive function and Q is an infinite r.e. set. Let $A_{s+1} = \bigcup_{n \in N} D_{g(n)} \cup A_s$ and $B_{s+1} = B_s$. Move on to the next step. It is obvious that A_{s+1} is an r.e. co-infinite set and $\bar{A}_{s+1}$ is not immune.

Let $A = \bigcup_{s \in N} A_s$. We show that A is a strongly nontotal set. Assume that $\Phi_n(A)$ is an infinite set then consider the step $2n+2$. The condition (3) must be satisfied in this case. The construction of this step permits us to form the recursive function g such that

$$\forall u[D_{g(u)} \subseteq A_{s+1} \wedge |V_{n,g}| = \infty].$$

As $A_{s+1} \subseteq A$ then A is a strongly nontotal set.

The steps $2n+1, n \in N$ guarantee that $A \notin \mathbf{SNT} \cup \mathbf{W}$ so A is not r.e. and strongly nontotal. This contradicts the fact that the formed set A is a strongly nontotal set. Thus $\mathbf{SNT} \cup \mathbf{W}$ is an uncountable set and $\mathbf{SNT}$ is an uncountable set too. The proof is completed. □

One traditional question in reducibility theory is the question about the interior structure of degrees in terms of strong reducibilities. We now consider the strengthening of e-reducibility introduced by Skordev [8]. The set A is *pc-reducible to B* via p.r.f.ϕ if

$$\forall x[x \in A \iff !\phi(x) \wedge D_{\phi(x)} \subseteq B].$$

We write $A \leq_{pc} B$ if there is a p.r.function ϕ for which A is pc-reducible to B via ϕ. As usual $d_{pc}(A) = \{B : B \leq_{pc} A \wedge A \leq_{pc} B\}$ is the *pc-degree* of A. It is obvious for all sets A and B, $A \leq_{pc} B$ implies $A \leq_e B$ but there are A and B such that $A \leq_e B$ and $A \not\leq_{pc} B$. So the pc-reducibility is a *strengthening* of e-reducibility. In this case any e-degree consists of some set of pc-degrees. If e-degree $d_e(A)$ contains more than one pc-degree then we will say that $d_e(A)$ is *decomposable*. Clearly $\mathbf{0_e}$ is not decomposable, that is it consists of only one pc-degree. Zaharov [9] showed that there are nonzero e-degrees which consist of a single pc-degree. On the other hand as it was shown in [10] any nonzero total e-degree is decomposable as pc-degrees. From this it follows

Theorem 11. *If a nonzero e-degree consists of a single pc-degree then it is nontotal.*

But as the following theorem shows the non-decomposition condition is sufficient but not necessary for non-totality of e-degrees.

Theorem 12. *No quasi-minimal e-degree which contains an e-i. set is decomposable.*

Proof. Let A be an e-i. set which belongs to no quasi-minimal e-degree. Then there is a nonrecursive total function $f \leq_e A$ (in fact, $f <_e A$). Assume that $\tau f \leq_{pc} A$ then for some p.r.function ϕ

$$\forall z[z \in \tau f \iff !\phi(z) \wedge D_{\phi(z)} \subseteq A].$$

Let $D \subseteq A$ be a finite set. We describe an effective process in an enumeration of A which permits us to construct a finite nonempty set $D^* \subseteq A$ such that $D^* \cap D = \emptyset$. In this case A cannot be an e-i. set since there is a total function $g \leq_e A$ such that $\rho g \subset A$ and ρg is infinite.

As $f \leq_e A$ then effectively enumerate the members of τf in the following order: $\langle 0, f(0)\rangle, \langle 1, f(1)\rangle, \ldots$ relative to any listing of A. We compute $\phi(\langle 0, f(0)\rangle)$, $\phi(\langle 1, f(1)\rangle), \ldots$ where the values are defined as $\tau f \subseteq \delta\phi$. We show that

$$\exists n[D_{\phi(\langle n, f(n)\rangle)} \not\subseteq D].$$

Suppose this is not true and $\forall n[D_{\phi(\langle n, f(n)\rangle)} \subseteq D \subseteq A]$.

Then

$$\forall z[z \in \tau f \iff !\phi(z) \wedge D_{\phi(z)} \subseteq D \subseteq A]$$

from which it follows that τf is an r.e. set which contradicts the non-recursiveness of f. Set $D^* = D_{\phi(<n^*, f(n^*)>)} - D$ where n^* is the least integer n for which $D_{\phi(\langle n, f(n)\rangle)} \nsubseteq D$. Clearly $D^* \neq \emptyset$ and $D^* \subseteq D_{\phi(\langle n^*, f(n^*)\rangle)} \subseteq A$. We define the total function g. We take at first $D = \emptyset$ then find D^* with the help of the process described above. Let $D^* = \{y_0, \ldots, y_k\}$, set $g(0) = y_0, \ldots, g(k) = y_k$. Assume that the values $g(0), \ldots, g(m)$ are defined already. Now take $D = \{g(0), \ldots, g(m)\}$ and find D^*. Let $D^* = \{y_{m+1}, \ldots, y_l\}$ where $m \leq l$, set $g(m+1) = y_{m+1}, \ldots, g(l) = y_l$. Clearly $g \leq_e A$, $\rho g \subseteq A$ and ρg is infinite. Thus A is not an e-i. set which contradicts our hypothesis.

So, $\tau f \nleq_{pc} A$. Let $B = \tau f \times A$ and so as $\tau f \leq_e A$ then $B \equiv_e A$. It is obvious that $B \nleq_{pc} A$. At the same time $A \leq_1 B$ and therefore $A <_{pc} B$. So $d_e(A)$ contains at least two pc-degrees $d_{pc}(A)$ and $d_{pc}(B)$. The proof is completed. □

References

[1] H. Rogers, Jr,, Theory of Recursive Functions and Effective Computability, McGraw-Hill, New York, 1967.

[2] J. Case, Enumeration reducibility and partial degrees, Ann. Math. Logic 4 (1971), 419–439.

[3] Yu. T. Medvedev, Degrees of difficulty of the mass problem, Dokl. Acad. Nauk SSSR 104 (1955), 501–504 (in Russian).

[4] L. P. Sasso, A survey of partial degrees, J. Symbolic Logic 40 (1975), 130–140.

[5] C. S. Copestake, 1-genericity in the e-degrees, J. Symbolic Logic 53 (1988), 878–887.

[6] B. Solon, E-hyperimmune sets, Sibirsk. Mat. Zh. 33 (1992), 211–214; Siberian Math. J. 33 (1992), 364–366.

[7] M. Rozinas, Partial degrees of immune and hyperimmune sets, Sibirsk. Mat. Zh. 19 (4) (1978), 866–870, 955.

[8] D. Skordev, About the partial conjunctive reducibility, II All Union Conf. of Math. Log., Abstract, Moscow, 1972, 43–44 (in Russian).

[9] S. D. Zaharov, About the inside structure of e-degrees, IX All Union Conf. of Math. Log., Abstract 61, Leningrad, 1988 (in Russian).

[10] B. Solon, The decomposition theorem for total degrees, Anniversary Conf. of University of Ivanovo, Abstract 185, Ivanovo, 1994 (in Russian).

Algebras of recursive functions

V. Solovyev *

Almost since its creation, the mainstream development of the theory of algorithms has spilled over from the study of both numeric functions computable by algorithms and the sets of natural numbers assocaited with these to the general ideas of computability. At the same time, the algorithms — their structure, inductive methods for their generation via more simple ones, etc. — were only somewhat investigated within the classical theory of algorithms. Within the last few decades, algorithms have been actively studied in complexity theory, but only to a certain degree. At present, we cannot refer the reader to a structural theory of algorithms, though there do exist the structural theory of finite-state automata ([Krohn-Rhodes, 1]) and the structural theory of databases ([Plotkin, 2]); one should also mention the structural theory of algebraic objects, e.g., rings, fields, etc. This situation can probably be explained by the fact that Turing's formalization of algorithms — the first one in a string of such formalizations, which thus turned to be classical — breaks up the algorithm into too small elements (e.g., the position of the read/write head, and so on) and only somewhat helps to construct a theory which could study the structure of algorithms as a whole. A certain negative influence goes back perhaps to an imaginary equivalence between different formalizations of the notion of an algorithm.

However, let us consider more closely one of these formalizations — the concept of a "partial recursive function" (p.r.f.). A p.r.f. is the result of applying composition (S), primitive recursion (R), and minimization (μ) to the basic functions (o, s, I_n^m). Obviously, the set of p.r.f.'s (denoted in what follows by Φ_r) can be considered as an algebra $(\Phi_r, \mathcal{L})$ with the signature $\mathcal{L} = \langle S, R, \mu \rangle$. This allows us to raise different algebraic problems. Their source is universal algebra, which has posed the problems of completeness, the description of the structure of the set of subalgebras, and so on. By solving these problems, we start to understand some new aspects of algorithms: in what way some algorithms are constructed from others, etc.

Universal algebra is not the only framework within which the objects of this kind are studied. They are also discussed by means of the theory of functional systems (see [Kudriavtsev, 3]), which was developed by S. Yablonsky, the Moscow mathematical cybernetics school. By a "functional system" one can also understand an algebra of the form $(\mathcal{A}, \mathcal{L})$, but its components are supposed to be non-abstract, possessing content. This approach enables one to ask some additional questions and derive substantial

*The author received support from RFBR Grant no. 96–01–00830 during the preparation of this article

answers. Before then, the functional systems of two-valued logic, k-valued logic, and finite-state automata were thoroughly investigated. In what follows we shall use the terminology of universal algebra.

The number of papers on algebras of recursive functions exceeds several hundred, but they are poorly known among the specialists in the field of recursion functions theory (e.g., they even were not mentioned in the fundamental monographs on the theory of algorithms: see [Rogers, 4], [Odifreddi, 5]).

In defining algebras of recursive functions, we have freedom in the choice of both the set of basic functions, and the set of means for creating the closure. The latter is of greater importance because the signature does indeed determine the properties of an algebra.

Three basic types of algebras of recursive functions can be distinguished.

1. The signature consists solely of composition. Algebras of this type were thoroughly investigated first by S. Marchenkov [Marchenkov, 6–9, 15, 38].

2. The signature consists of algebraic operations; for instance, addition and inversion of functions. These algebras are called Robinson algebras (named after their founders R. Robinson [Robinson, 10] and J. Robinson [Robinson, 11]). Later, they were thoroughly investigated by A. Maltsev's Novosibirsk school of mathematical logic.

3. The signature consists of programmable operations, e.g., consecutive execution, conditional branching, looping, recursive procedures, and so on. The basic functions used in the assignment operations are of the form $y := f(x_1, \dots, x_n)$. Algebras of this class attracted attention not too long ago, and recently some interesting results were obtained in this field. At the same time, these algebras are potentially of major interest since they are directly connected with the actual problems of programming: description of computability for abstract datatypes, equivalent transformations of programs, etc.

Together with the more specific features of these algebras we should emphasize that along with the recursive functions, the recursive predicates are also considered at the same level. This is due to the presence of branching by predicates. Another interesting feature is the fact that this set of algebras is very extensive. Various classes of program schemes are considered here using closure tools: standard schemes (flow charts), those with counters, arrays, etc.

Let us formulate the principal questions which must be studied for algebras of recursive functions of all kinds. Below, we shall use the notation $\langle \mathcal{A}, \mathcal{L} \rangle$ for any algebra of the types cited above. For the set A of recursive functions (and predicates), $[A]_{\mathcal{L}}$ will stand for the closure of the set A with respect to operators from $\mathcal{L}$. We denote by Φ_r, Φ_o, and Φ_n the p.r.f.'s, the recursive functions (r.f.), and the primitive recursive functions, respectively. By changing the symbol Φ to $\mathcal{P}$, we obtain the corresponding predicate sets.

I. Bases. A set A of recursive functions (and predicates) is said to be "complete in the algebra $\langle \mathcal{A}, \mathcal{L} \rangle$" if $[A]_{\mathcal{L}} = \mathcal{A}$. The set A is called a "basis of the algebra $\langle \mathcal{A}, \mathcal{L} \rangle$" if it is complete and no proper subset of A is complete.

Problems. Does a certain algebra have bases? If the answer is "yes", when does it have finite bases? If "yes" again, then the problem arises of finding the simplest bases.

II. Completeness. The problem of determining bases can be continued to its logical end: describe all complete sets. The most popular way to state this problem is as follows: describe all finite complete sets of the algebra $\langle \mathcal{A}, \mathcal{L} \rangle$.

III. The structure of subalgebras. A set A of recursive functions (and predicates) is called a "subalgebra of $\langle \mathcal{A}, \mathcal{L} \rangle$" if $[A]_{\mathcal{L}} = A$.

The problem is: describe the structure of the set (partially ordered by inclusion) of all subalgebras for a given algebra $\langle \mathcal{A}, \mathcal{L} \rangle$.

We state the main cases of this problem as follows:

III a. Describe the maximal subalgebras of the algebra $\langle \mathcal{A}, \mathcal{L} \rangle$.

This problem is of particular importance because its solution enables us to derive a completeness criterion of the Post type: A set is complete if and only if it is not contained in any maximal subalgebra.

III b. Describe the automorphisms which act on the structure of subalgebras.

IV. Algebraic properties of the subalgebras. Simplicity is the property studied most often among the algebraic properties. A subalgebra is said to be "simple" if on this algebra there are no nontrivial congruencies (in the general algebraic sense).

Problem. Describe the set of all simple subalgebras.

Let us cite the basic results which were obtained for all questions mentioned above and which concern all types of algebras.

I. Bases

1. Algebras based on composition.

1 a. The class K of Calmar's elementary functions (see [Grzegorczyk, 12]) is the minimal class of functions that includes $\{x + 1,\ x + y,\ x \dot{-} y\}$ and is closed under composition, bounded summation, and bounded multiplication: $\sum_{i \le x} f(i, y)$ and $\prod_{i \le x} f(i, y)$.

In [12], Grzegorczyk posed the following question: Does the class K have a finite basis under composition? A positive (but very complicated) answer was given by D. Rödding ([Rödding, 13]). In the paper [Jones and Matijasevic, 14]) one can find simple examples of bases, which allow one to realize all of Calmar's elementary functions that have a finite number of values. Simple bases for the whole class K were described in [Marchenkov, 6, 9]). One of them is the function set $\left\{x+1, \left[\frac{x}{y}\right], x^y, \phi(x, y)\right\}$ with the function $\phi(x, y)$ equaling 0 as $x < 1$, and equaling the least index of a zero-bit in the x-based notation for y as $x > 1$. The function $\phi(x, y)$ can be replaced by a function $\tau(x)$ which equals 0 as $x = 0$ and equals the power of 2 in the prime factor decomposition of x as $x > 0$.

1 b. Classes E^n by Grzegorczyk.

In [Grzegorczyk, 12] , the question of the existence of finite bases in the classes E^n was posed. A positive answer was given in [Rödding, 13] for $n > 3$, and in [Marchenkov, 38] for $n = 2$. A more general method for the construction of bases was given in [Muchnik, 39].

1 c. We say that a function class "is closed under primitive recursion" if it contains the functions $\{0, x + 1, I_n^m(x_1, \ldots, x_n) = x_m\}$ and is closed under composition and primitive recursion. Rather general results for the bases under composition in these classes were described by S. Marchenkov. In particular, the following two statements were proved:

– every countable class of total functions, which is closed under primitive recursion, has a basis under composition [Marchenkov, 7];

– the set of one-place functions contained in any of these classes has a countable basis under composition [Marchenkov, 8].

1d. We denote by F_k^{ψ} the minimal class of functions which includes $\{1, x \dot{-} y, x \cdot y, I_n^m, \psi\}$ and is closed under composition and limited recursion repeated k times. In [Marchenkov, 15], it was proved that the classes F_k^{ψ} have no finite bases under composition, but that the sets of all one-place functions from F_k^{ψ} do have countable bases [Marchenkov, 8]. All the results mentioned above were proved by applying different techniques.

2. Algebras defined by J. Robinson.

2 a. In [11], Robinson started considering the algebra $\mathcal{F}_o = \langle \Phi_o^1, +, \cdot, ^{-1} \rangle$, where the symbol "$^{-1}$" stands for the operation of function inversion. The algebra $\mathcal{F}_r = \langle \Phi_r^1, +, \cdot, ^{-1} \rangle$ also possesses similar properties.

For these algebras, the problem of searching for bases turned out to be simpler than for algebras under composition. It was shown in [Robinson, 10] that the r.f.'s are generated by a pair of functions $\{s(x) = x + 1,\ q(x) = x - [\sqrt{x}]^2\}$. A more general formulation of this result was given in [Zakharov, 16]: For any function f with $f(0) = 0$, every pair from $(f, l), (f, r), (f, q)$, where l and r mean the inverse functions for the standard pairing function $\langle x, y \rangle : l(\langle x, y \rangle) = x, r(\langle x, y \rangle) = y$, generates the set of all one-place r.f.'s. A similar result holds for the algebra $\mathcal{F}_r$ ([Zakharov, 16]). In addition, the algebras $\mathcal{F}_o$ and $\mathcal{F}_r$ have one-element bases. Here we cite the example by Kuznetsov, [17] of a one-element basis:

$$f_m(x) = \begin{cases} m^2 + 3m + 5, & \text{if } x \text{ is odd,} \\ l(t), & \text{if } x = 2t + 1. \end{cases}$$

A detailed study of the bases of the algebras $\mathcal{F}_o$ and $\mathcal{F}_r$ can be found in works in [Nasibullov, 18] and [Polyakov, 19]. In particular, the existence of a countable number of n-element bases was proved for every n.

2 b. In [10], Robinson introduced the algebra $\mathcal{F}_n = \langle \Phi_n^1, +, \cdot, i_a \rangle$, where i_a means iteration via the scheme $i_a f = g \Longleftrightarrow g(0) = a \;\&\; \forall x \geq 0\ (g(x+1) = fg(x))$. An example of a simple basis (namely, (s, l)) was indicated as well. Later, in [Polyakov, 20], one-element bases were found. For example,

$$f_m(x) = \begin{cases} t + 1, & \text{if } x = mt, \\ l(t), & \text{if } x = mt + 1 \text{ and } m \geq 2, \\ 0, & \text{else.} \end{cases}$$

For all n, n-element bases were found by Polyakov [21].

3. Program type algebras.

All algebras in which classes of program schemes are used to define closure can be divided into two types, depending on the number of memory cells (variables) used in these schemes: the first type consists of those with a single cell used, and the other of those with several cells used.

3 a. The case of a single cell.

The example from this group investigated most deeply is Glushkov's system of algorithmic algebras [Glushkov, 22]. We denote by $FDS(1)$ this class of program schemes. Minsky's program machines with one register are also related to this group. M. Minsky demonstrated in [Minsky, 23] the computability of any p.r.f. by applying the set $\{2x, 3x, 2^x, \left[\frac{x}{6}\right], [\log_2 x], x \equiv 0 \pmod 6\}$ to a program machine with one register. (Here "$x \equiv 0 \pmod 6$" is a one-place predicate.) As was proved in [Barzdin, 24], the functions 2^x and $[\log_2 x]$ also cannot be excluded from this set. All the above remains valid for Glushkov's system of algorithmic algebras (Golunkov, [25]).

3 b. The case of several cells.

We shall select the three most interesting algebras in this group, depending on the class of program schemes used for closure. We shall cite here a non-formal description of the basic classes of program schemes.

The standard program schemes (see [Kfoury, Urzyczyn, 26]) admit only branching (if P then Q_1 else Q_2, where P is a predicate and Q_1 and Q_2 are operators) and consecutive execution of operations.

Following [Kfoury, Urzyczyn, 26], the program schemes with equality and arrays include also a variable M to designate an infinite array with cells $M(0), M(1), \ldots$ All the variables used in these program schemes can be divided into the following two groups: subject variables and counters. The possible operations here are $y := M[i]$

and $M[i] := y$; where y is a subject variable, and i is a counter. Working with counters, we can use the following operations: $i := i + 1$, $i := i \dot{-} 1$, $i := j$, $i = 0$. The last operation is a predicate that can be used in branching instructions. The equality predicate may also be applied to the subject variables: $x = y$. The program schemes with arrays, equality, and indirect addressing (or RAM-machines in accordance with [Aho, Hopcroft, Ullman, 27]) allow the use of commands such as $M[x] := y$ and $y := M[x]$ with a subject variable x.

These classes and algebras of recursive functions are denoted, respectively, by FD, $FDA_=$, $FDAA_=$ and $\langle \mathcal{A}, FD\rangle$, $\langle \mathcal{A}, FDA_=\rangle$, $\langle \mathcal{A}, FDAA_=\rangle$, where the character "$\mathcal{A}$" means either $\Phi_r \cup \mathcal{P}_r$, or $\Phi_o \cup \mathcal{P}_o$, or $\Phi_o^1 \cup \mathcal{P}_o^1$, or $\Phi_r^1 \cup \mathcal{P}_r^1$. First simple examples of the bases of the algebra $\langle \Phi_r \cup \mathcal{P}_r, FD\rangle$ were given in [Ershov, 28], [Shepherdson, Sturgis, 29]: They are $\{0, x + 1, x = y\}$ and $\{x + 1, x \dot{-} 1, x = 0\}$. Both these sets produce also bases of the algebra $\langle \Phi_o \cup \mathcal{P}_o, FDA_=\rangle$. The bases of the algebra $\langle \Phi_o \cup \mathcal{P}_o, FDAA_=\rangle$: $\{0, x + 1\}$ and $\{x + 1, x \dot{-} 1\}$ are simpler.

The general trend is the simplification of the bases along with a strengthening of the closure tools.

II. Completeness

The completeness problem seems to be very complicated for algebras of p.r.f. It is more difficult than the same problem for either two- or k-valued logics. For algebras under composition and the Robinson algebras, the problem has not even been stated yet. For the algebras of program type, the completeness problem was stated in [Ershov, Lyapunov, 31] and was referred to there as one of the most important problems of theoretical programming. However, the problem has not been solved for algebras in the one-memory-cell case, though much effort was applied in this direction (see [Golunkov, 32]). The main trend in these investigations is a description of all maximal subalgebras. This would enable one to obtain a completeness criterion of the Post type (mentioned above).

The situation is more promising in the case of several cells. For the algebras $\langle \Phi_o \cup \mathcal{P}_o, FDA_=\rangle$ and $\langle \Phi_o \cup \mathcal{P}_o, FDAA_=\rangle$, the problem of completeness has been solved by means of the following results:

A completeness criterion for $\langle \Phi_o \cup \mathcal{P}_o, FDA_=\rangle$ ([Solovyev, 33]). *A set $\mathcal{L} = \{f_1, \ldots, f_n, p_1, \ldots, p_m\}$ is complete if and only if*

1) $\forall x, y \exists \tau (\tau$ *is a term in* $\mathcal{L}$ & $\tau(x) = y)$;

2) $\exists a \exists \Phi (\Phi(x)$ *is formula without quantifiers in the signature* $\mathcal{L} \cup \{=\}$ & $\Phi(a)$ & $\forall y \neq a\ \neg\Phi(y))$.

A completeness criterion for $\langle \Phi_o \cup \mathcal{P}_o, FDAA_= \rangle$ ([Solovyev, 34]. *A set* $\mathcal{L} = \{f_1, \ldots, f_n, p_1, \ldots, p_m\}$ *is complete if and only if*

$$\forall x, y \exists \tau (\tau \text{ is a term in } \mathcal{L} \,\&\, \tau(x) = y).$$

A completeness criterion for $\langle \Phi_o \cup \mathcal{P}_o, FD_= \rangle$ was also obtained [Solovyev, 52], but we do not cite it here due to its cumbersome form.

We give below a simple example of a set of functions, which is complete for one algebra and incomplete for another. The set $\{0, g\}$, where

$$g(i, j) = \begin{cases} j + 1, & \text{if } i = \left[\frac{j}{2}\right], \\ 0, & \text{otherwise}, \end{cases}$$

is complete for $\langle \Phi_r \cup \mathcal{P}_r, FDA_= \rangle$ and incomplete for $\langle \Phi_r \cup \mathcal{P}_r, FD_= \rangle$ [Kfoury, 46]. In [45], Nepomnyashchii proposed a program of investigation based on a comparison of different classes of program schemes by the complexity of solving completeness problem for appropriate algebras. Since different classes of program schemes formalize different concepts of computability, we obtain a working tool to compare various concepts of computability. Thus, by comparing the above-mentioned completeness criterion for $\langle \Phi_o \cup \mathcal{P}_o, FDA_= \rangle$ and $\langle \Phi_o \cup \mathcal{P}_o, FDAA_= \rangle$, we can note that indirect addressing in the latter expression is equivalent to the possibility of distinguishing a certain number among others ("equivalent" in the sense of the completeness problem).

In general, we can observe the same tendency, as we saw for bases — stronger closure tools imply a simpler completeness criterion, and thus the completeness problem can be solved more easily. The same situation appears in the comparison of the T-degrees of the completeness problems.

For the algebras $\langle \Phi_o \cup \mathcal{P}_o, FDA_= \rangle$ and $\langle \Phi_o \cup \mathcal{P}_o, FDAA_= \rangle$, it is $0''$ [Solovyev, 35]; for the algebra $\langle \Phi_r^1 \cup \mathcal{P}_o^1, FDS(1) \rangle$ with weaker closure tools, it is $0'''$ [Golunkov, 30]. Then, although the index sets of the completeness problem for the algebras $\langle \Phi_o \cup \mathcal{P}_o, FDA_= \rangle$ and $\langle \Phi_o \cup \mathcal{P}_o, FDAA_= \rangle$ belong to the same T-degree, they occupy different places in the arithmetical hierarchy. The index set of the completeness problem for the algebra $\langle \Phi_o \cup \mathcal{P}_o, FDA_= \rangle$ belongs to the level $(\sum_3 \cap \prod_3) \setminus (\sum_2 \cup \prod_2)$ in the Kleene-Mostowski hierarchy; it is $\prod_2$-complete for the algebra $\langle \Phi_o \cup \mathcal{P}_o, FDAA_= \rangle$ (see [Solovyev, 35]).

III. The structure of subalgebras

For algebras with composition, the structure of subalgebras has not yet been investigated. For Robinson algebras, only maximal subalgebras have been considered. A complete description of the maximal subalgebras was not obtained, and up to now we have only some examples of maximal subalgebras and their families.

The algebra $\mathcal{F}_r$.

1. [Shcheglov, 36]. $\{f \mid f(0) = 0 \vee f(0)$ *is not defined*$\}$ *is a maximal subalgebra.*

2. [Shcheglov, 36]. $\{f \mid pf$ *is finite* $\vee\ \delta f \neq N\}$ *is a maximal subalgebra.*

3. [Shcheglov, 37]. *There exists a continuum of maximal subalgebras.*

4. [Shcheglov, 36]. *The Frattini subalgebra (the intersection of all maximal subalgebras) is infinite.*

The algebra $\mathcal{F}_o$.

1. [Polyakov, 38]. $\{f \mid f(0) = 0\}$ *is a maximal subalgebra.*

2. [Polyakov, 39]. *Assume that the set M differs from N, includes 0, is closed under addition, and $\bar{M}$ is finite. Then* $\{f \mid \forall a(a \in M \rightarrow f(a) \in M)\}$ *is a maximal subalgebra.*

3. [Polyakov, 40]. $\{f \mid pf$ *is finite* $\vee\ \forall x(x \geq n \rightarrow f(x) \geq n)\}$ *is a maximal subalgebra for all* $n \geq 1$.

The algebra $\mathcal{F}_n$.

1. [Polyakov, 39]. *Let $M \neq N$, $0 \in M$, and assume that the set M is be closed under addition. Then* $\{f \mid \forall x(x \in M \rightarrow f(x) \in M)\}$ *is a maximal subalgebra.*

2. [Kozminych, 41]. *There exists a continuum of maximal subalgebras.*

3. [Kozminych, 42]. *The Frattini subalgebra is infinite.*

Additional information on the maximal subalgebras of Robinson algebras can be found in [Zakharov, 16].

As for program type algebras, they were investigated more closely.

The algebra $\langle \Phi_r^1 \cup \mathcal{P}_r^1, FDS(1)\rangle$.

Some classes of maximal subalgebras were described in the literature, but a complete description has not been accomplished. Below, we describe three classes of this kind. Such a description seems to be of interest in view of the use of some terms of the classical theory of recursive functions.

1. [Golunkov, 43]. *A subalgebra* $\mathcal{U}_A = \{f \mid f(A) \cap \bar{A} = \emptyset\} \cup \mathcal{P}_r^1$ *is maximal if and only if A is either finite or cofinite.* The subalgebra $\mathcal{U}_A$ is called the "algebra of preserving membership to a set".

A cohesive set F is maximal [Golunkov, 44] if and only if $\forall W(W$ is recursively enumerable & $F \not\subseteq W \rightarrow \exists R(R$ is a recursive set & $F \not\subseteq R \not\subseteq W))$.

2. [Golunkov, 44]. *If a set $A \subseteq N$ is either a) infinite recursive with an infinite complement, or b) the largest cohesive set, then the subalgebra* $\mathcal{U}'_A = \{f \mid f(\bar{A}) \cap A$ *is finite*$\} \cup \mathcal{P}_r^1$ *is maximal.*

Cohesive sets A and B are said to be f-equivalent [Golunkov, 44] if a p.r.f. ϕ exists such that $\phi(A)$ almost coincides with B. A set is quasicohesive (see [Rogers, 4]) if it is a finite union of cohesive sets.

3. [Golunkov, 44]. *If A is a quasicohesive set and $A = \cup_{i=1}^{k} A_i$ (where A_i is a cohesive set for $i = 1, \ldots, k$), then the subalgebra $\mathcal{U}'_A = \{f \mid f(\bar{A}) \cap A$ is finite$\} \cup \mathcal{P}_r^1$ is maximal if and only if all A_i are maximal cohesive sets and pairwise f-equivalent for $i = 1, \ldots, k$.*

Since the completeness problem for the algebra $\langle \Phi_r^1 \cup \mathcal{P}_r^1, FDS(1) \rangle$ has not been solved, a modified completeness problem was considered.

A function and predicate set M is called "A-complete" [Golunkov, 47] if $\forall t \in \Phi_r^1 \exists z \exists h \in [M]_{FDS(1)} \forall x \leq z(t(x) = h(x))$. A set M is said to be "A-precomplete" if it is not A-complete, but will turn out to be A-complete after addition of an arbitrary element from $\Phi_r^1 \setminus M$. We write $\mathcal{B} = \{\mathcal{U}_A \mid A \neq N \ \& \ A \neq \emptyset\}$ (here $\mathcal{U}_A$ is the subalgebra from item 1 of this section). Then w is a non-diagonal, binary, symmetric, and reflexive relation which is defined on the set N, and ξ is a family of subalgebras of the form:

$\mathcal{U}(w) = \{f \mid \forall a, b((a, b) \in w \ \& \ f(a)! \ \& \ f(b)! \to (f(a), f(b)) \in w))\} \cup \{p \mid \forall a, b((a, b) \in w \ \& \ p(a)! \ \& \ p(b)! \to p(a) \equiv p(b))\}$ for all w as above.

As one can see in the following result, A-precomplete classes have a structure which is simpler than that of maximal subalgebras.

[Golunkov, 47]. *The set of A-precomplete classes is $\mathcal{B} \cup \xi$.*

In spite of its rather simple description, the problem of A-completeness remains unsolved (in its complete form). As before, the set of A-precomplete classes remains of size the continuum. This obstacle does not allow us to obtain an A-completeness criterion similar to that by Post.

In connection with this, we provide below a new definition. A set T of subalgebras will be called an "α-criterion system" (for A-completeness) if for all M whose cardinality is less than α, M is A-complete if and only if M is not contained in any subalgebra in T. It can be easily shown (using non-constructive methods) that there exist countable $\aleph_0$-criterion systems. The first constructive example of such a system was announced in [47].

[Golunkov, 47]. *The set $\{\mathcal{U}_A \mid A$-recursively enumerable, non-simple$\} \cup \{\mathcal{U}(w) \mid w$-recursively enumerable, non-simple$\}$ is a countable $\aleph_0$-criterion system for A-completeness.*

The algebra $\langle \Phi_o^1 \cup \mathcal{P}_o^1, FDA_{=} \rangle$.

It appears to be the only algebra with a continuum of maximal subalgebras and a complete description of these subalgebras. Within this description, all maximal subalgebras are divided into 4 classes. The description can be found in [Solovyev, 48]. Here, we shall give a description of one of these classes, given in terms of the classical theory of algorithms.

Let R be a finite set of r.f. Let us define a quasiorder $\leq_R$ on N as follows: $x \leq_R y \leftrightarrow \exists \tau(\tau$ is a term of $R \ \& \ x = \tau(y))$.

By factorizing this quasiorder with respect to the equivalence relation $x \equiv_R y \leftrightarrow x \leq_R y \;\&\; y \leq_R x$, we obtain a partial order written as $\leq_R$ and called effectively finite.

[Solovyev, 48]. *The subalgebras* $\mathcal{U}_A = \{f \mid f(A) \cap \bar{A} = \emptyset\} \cup \mathcal{P}_o^1$ *are maximal if and only if A is a non-trivial initial segment of a certain effectively finite order with largest number.*

It can be proved that A is not hyperimmune in this case, and that there is a continuum of such maximal subalgebras. A more detailed investigation of the effectively finite orders can be found in [Solovyev, 49].

[Solovyev, 52]. *The Frattini algebra is finite and is given via the expression* $\mathcal{U}_0 = \{\text{TRUE}, \text{FALSE}, \ \text{id}(x) = x\}$.

In [Solovyev, 50] one can find a complete description of all finitely-generated subalgebras of finite rank, i.e. when there are only a finite number of intermediate subalgebras between this subalgebra and the whole algebra $\langle \Phi_o^1 \cup \mathcal{P}_o^1, FDA_= \rangle$. The description is written in the language of group theory and category theory.

Below, we give one more result concerning subalgebras which lie near $\mathcal{U}_0$.

[Solovyev, 52]. *The set* $\mathcal{U}$ *of subalgebras such that* $\neg\exists \mathcal{U}_1, \ldots, \mathcal{U}_n, \ldots \forall i(\mathcal{U}_i < \mathcal{U}_{i+1} \;\&\; \mathcal{U}_i < \mathcal{U})$ *is not an upper semilattice.*

The most interesting part of the result lies in its proof, which is carried out with the use of the existence of Burnside groups having the following property: a group is infinite, generated by a and b — two elements of prime order, semiregular, and obeying the following fact: any of its non-trivial subgroups containing an element c coincides with $[c]$. The existence of these groups was proved in [Olshansky, 53].

Let $\mathcal{G}$ be the set of all subalgebras of the algebra $\langle \Phi_o^1 \cup \mathcal{P}_o^1, FDA_= \rangle$, which is partially ordered by inclusion. Among other problems which were investigated in this direction, we can refer to the problem of describing the automorphisms of $\mathcal{G}$. Let us denote by α a recursive permutation of N. It induces the following mapping over functions and predicates: $\phi_\alpha : \Phi_o^1 \cup \mathcal{P}_o^1 \longrightarrow \Phi_o^1 \cup \mathcal{P}_o^1$, $\phi_\alpha(f) = \alpha f \alpha^{-1}$, $\phi_\alpha(p) = p\alpha^{-1}$. In turn, this induces an automorphism (called internal) of $\mathcal{G}$: $\phi_\alpha(\mathcal{A}) = \{\phi_\alpha(f) \mid f \in \mathcal{A}\} \cup \{\phi_\alpha(p) \mid p \in \mathcal{A}\}$.

[Solovyev, 35]. *The set of the automorphisms of* $\mathcal{G}$ *consists only of internal automorphisms.*

To prove this proposition, we applied a new method. Its principal idea consists in the representation of the functions in the form of unions of chains $f = \bigcup_x \{x, f(x), \ldots, f^k(x), \ldots\}$ and the analysis of the structure of this representation for all functions included in the algebra.

In order to demonstrate the method we give here the proof of this result.

A function f with the property $\exists n \forall i \forall x((x \notin A \rightarrow f(x) = x) \;\&\; (x \in A \;\&\; 0 < i < n \rightarrow f^n(x) = x \;\&\; f^i(x) \neq x))$ is said to be "cyclic on A", the number n is called the "cycle length".

Since any subalgebra of $\langle \Phi_0, FDA_= \rangle$ contains the elements id, TRUE, and FALSE, in what follows we shall not mention their existence, and by the words "elements of

$[L]$" we shall mean elements which differ from id, TRUE, and FALSE. The algebra $\langle$id, TRUE, FALSE$\rangle$ is the least element of the structure G. $\mathcal{A}$ is an atom of $\mathcal{G}$ if and only if $\forall \mathcal{B}(\mathcal{B} \subseteq \mathcal{A} \rightarrow \mathcal{B} = \mathcal{A} \vee \mathcal{B} = \mathcal{U}_0)$.

Lemma 1. *A subalgebra a is an atom of $\mathcal{G}$ if and only if $a = [p]$, where $p \neq$ TRUE, $p \neq$ FALSE, or $a = [f]$, where f is a cyclic function on N with its cycle length being a prime number.*

Proof. Assume that $a \in \mathcal{G}$ is an atom. Clearly, a cannot contain two predicates if these are not negations of each other — in this case a would have a proper subalgebra generated by one of these predicates. If a contains one predicate, then this leads to $[p_M] = \{p_M, \neg p_M\}$ for a set M, where $M \neq \emptyset$, $M \neq N$.

Now let us assume that a does not contain any predicate but a function f. If none of the relations of the form $f^n(x) = f^m(x)$ holds for any possible x, then all the functions $f, f^2, f^3, \ldots$ are distinct, and a has a proper subalgebra generated by f^2, which contradicts the statement that a is an atom. Thus, some relation of the form $f^n(x) = f^m(x)$ must hold for a certain x_0. But then it must hold for all possible x, otherwise a would contain a predicate defined via the equality $f^n(x) = f^m(x)$. Also, the equality $f^k(x) = x$ holds with $k =| n - m |$. Let us take the least among those appropriate k. If k had a divisor l, the function $f^l(x)$ then would generate a subalgebra which is strictly contained in a, contradicting the statement that a is an atom. All this gives us the description of atoms which is contained in the formulation of Lemma 1. □

Lemma 2. $\forall x \exists y\ (\gamma([p_{\{x\}}]) = [p_{\{y\}}])$.

Proof. From Lemma 1, it follows that either $\gamma([p_{\{x\}}]) = [p_M]$, or $\gamma([p_{\{x\}}]) = [f]$ since $\gamma([p_{\{x\}}])$ must be an atom. Let us prove that the second case is impossible. Assume that an equality $\gamma([p_{\{x\}}]) = [f]$ holds for a certain x and a certain function f. Let $B(\neq \emptyset, \neq N)$ be a set such that $f(B) = B$. The subalgebra $[f, p_B]$ contains the following elements: $p_B, \neg p_B, f, g_1$, and g_2, where both g_1 and g_2 are functions with the properties $g_1/B = \text{id}$, $g_1/\bar{B} = f$, and $g_2/B = f$, $g_2/\bar{B} = \text{id}$; it also contains the iterations of f, g_1, and g_2. This subalgebra has the following proper subalgebras: $[f]$, $[p_B]$, $[p_B, g_1]$, and $[p_B, g_2]$. Let us clarify what $\gamma^{-1}([p_B])$ is.

If $\gamma^{-1}([p_B]) = [p_C]$, then $[p_{\{x\}}, p_C]$ must have four proper subalgebras, but it has only three: $[p_{\{x\}}]$, $[p_C]$, and either $[p_{C \setminus \{x\}}]$ or $[p_{C \cup \{x\}}]$ depending on whether $x \in C$ or $x \notin C$. It may happen that $\gamma^{-1}([p_B]) = [h]$. But the algebra $[h, p_{\{x\}}]$ has more than four subalgebras. For example, $[h]$, $[p_{\{x\}}]$, $[p_{\{h(x)\}}]$, $[p_{\{x,h(x)\}}]$, and $[p_{\{x\}}, g]$, where g is a function coinciding with h on x and with id on other sets. Thus, we arrive at a contradiction which proves that $\gamma[p_{\{x\}}]) = [p_M]$.

This statement can be slightly generalized. We prove that $\forall C \exists M(\gamma([p_C]) = [p_M])$. Indeed, assume $C = \{x_0, x_1, \ldots\}$. Then $[p_C]$ is a subalgebra of the algebra $[p_C, p_{\{x_0\}}]$. If the equality $\gamma([p_C]) = [f]$ were to hold, then $\gamma([p_C, p_{\{x_0\}}]) =$

$[f, p_{M_0}]$, where $p_{M_0} = \gamma([p_{\{x_0\}}])$. But, as we have shown above, the algebra $[f, p_{M_0}]$ contains at least four proper subalgebras, which contradicts the assertion that $[p_C, p_{\{x_0\}}]$ has only three. This proves that the equality $\gamma([p_C]) = [f]$ cannot hold. Therefore, $\gamma([p_C]) = [p_M]$ for a certain M.

The reverse statement $\forall C \exists M\ (\gamma^{-1}([p_C]) = [p_M])$ is also true, because γ^{-1} is an automorphism.

Thus, having $\gamma([p_{\{x\}}]) = [p_M]$ for a certain M, we shall continue the proof of Lemma 2. We have to show that M (or $\bar{M}$) has a single element.

Assume, for the sake of a contradiction, that M and $\bar{M}$ each contain at least two numbers, and let $x_0 \in M$. Take C such that $\gamma([p_C]) = [p_{\{x_0\}}]$ and $x_0 \notin C$ (one can always pass to complements if necessary). The algebra $[p_M, p_{\{x_0\}}]$ has the proper subalgebras $[p_{\{x_0\}}]$, $[p_M]$, and $[p_{M\setminus\{x_0\}}]$. As $[p_M, p_{\{x_0\}}]$ is a least upper bound for both $[p_M]$ and $[p_{\{x_0\}}]$, we have $[p_M, p_{\{x_0\}}] = \gamma([p_{\{x\}}, p_C])$ and $[p_{M\setminus\{x_0\}}] = \gamma([p_{\{x\}\cup C}])$ in accordance with the properties of automorphisms.

Let us choose $z \in \bar{M}$ and assume that $\gamma([p_D]) = [p_{\{z\}}]$, where $x \notin D$ and $D \neq \{x\}$. By comparing the least upper bounds for the pairs $[p_{M\setminus\{x_0\}}]$, $[p_z]$ and $[p_{\{x\}\cup C}]$, $[p_D]$, we can see that $[p_{(M\setminus\{x_0\})\cup\{z\}}] = \gamma([p_{\{x\}\cup C\cup D}])$.

Finally, let us consider the least upper bound for the elements $[p_{\{x\}}]$ and $[p_{\{x\}\cup C\cup D}]$. It contains three proper subalgebras — $[p_{C\cup D}]$ besides the above-mentioned. But an automorphic image — the least upper bound for $[p_M]$ and $[p_{(M\setminus\{x_0\})\cup\{z\}}]$ — contains more than three subalgebras, namely: $[p_{\{x_0\}}]$, $[p_{\{z\}}]$, $[p_{M\setminus\{x_0\}}]$, $[p_M]$, and others. This contradiction proves the lemma. □

This proof holds true for every automorphism, in particular for γ^{-1}. Thus, the statement "$\forall x \exists y\ (\gamma^{-1}([p_{\{x\}}]) = [p_{\{y\}}])$" is also true. For every number x, we fix a corresponding y in accordance with the assertion of Lemma 2. It will constitute the permutation "α".

Lemma 3. $\forall M\ (\gamma([p_M]) = [p_{\alpha(M)}] = [p_M \alpha^{-1}])$.

Proof. Suppose that $|M| > 1\ \&\ |\bar{M}| > 1$; otherwise, the assertion was already proved. Let $a \in M$. As has been shown in the proof of Lemma 2, $\exists C\ (\gamma([p_M]) = [p_C])$. Assume $\alpha(a) \in C$ ($\bar{C}$, respectively).

Let us take a $b \in M$, an arbitrary element which is different from a. We have $\gamma([p_{\{a,b\}}]) = [p_{\{\alpha(a),\alpha(b)\}}]$, which immediately follows from Lemma 2. The algebra $[p_M, p_{\{a,b\}}]$ does not contain subalgebras of the form $[p_{\{d\}}]$. Consequently, the same holds for the algebra $[p_C, p_{\{\alpha(a),\alpha(b)\}}]$, which implies that $\{\alpha(a), \alpha(b)\} \in C$. Since $b \in M$ is arbitrary, we see that $\alpha(M) \subseteq C$. One can show similarly that $\alpha(\bar{M}) \subseteq \bar{C}$. As α is a permutation of N, $\alpha(M) = C$. □

Lemma 4. *If D is a subalgebra consisting only of predicates, then* $\gamma(D) = \{\gamma([p]) \mid p \in D\} = \{p\alpha^{-1} \mid p \in D\}$.

Proof. $p \in D \iff [p] \subseteq D \iff \gamma([p]) \subseteq \gamma(D) \iff p\alpha^{-1} \in \gamma(D)$. □

Now, let us consider elements of $\mathcal{G}$ which contain functions. Here, we shall describe a method which will often be of use in what follows.

Problem. Prove that $\gamma([f]) = [\alpha f \alpha^{-1}]$ for a function f.

Solution. Assuming $\gamma([f]) = [g_1, \ldots, g_n, p_1, \ldots, p_m]$, where g_i is a function and p_i is a predicate, we create a program which computes the function $\alpha f \alpha^{-1}$ in the signature $\{g_1, \ldots g_n, p_1, \ldots, p_m\}$. If this is successful, we shall have that $[\alpha f \alpha^{-1}] \subseteq \gamma([f])$. Applying the automorphism γ^{-1} to the function $\alpha f \alpha^{-1}$, we similarly derive that $[f] \subseteq \gamma^{-1}([\alpha f \alpha^{-1}])$. These two inclusions result in the equality $\gamma([f]) = [\alpha f \alpha^{-1}]$.

Lemma 5. *Assume that $k_{a,b}$ is subject to the relation $\{k(a) = b, k(x) = x$ where $x \neq a\}$ for $a \neq b$. Then $\gamma([k_{a,b}]) = [k_{\alpha(a),\alpha(b)}]$.*

Proof. As $[k_{a,b}]$ has only the proper subalgebra $[p_{\{a\}}]$, $\gamma([k_{a,b}])$ has only the proper subalgebra $[p_{\{\alpha(a)\}}]$. Therefore, $\gamma([k_{a,b}])$ does not have other predicates except $p_{\alpha(a)}$ and $\neg p_{\alpha(a)}$ and still contains some function f'. Assume that $\exists x_0 \neq \alpha(a)(f'(x_0) \neq x_0)$. Then $[k_{a,b}, p_{\{\alpha^{-1}(f'(x_0))\}}]$ does not contain other atom subalgebras except $[p_{\{a\}}]$, $[p_{\{\alpha^{-1}(f'(x_0))\}}]$ and $[p_{\{a,\alpha^{-1}(f'(x_0))\}}]$.

At the same time $[f', p_{\{f'(x)\}}, p_{\{\alpha(a)\}}]$ contains $[p_{\{\alpha(a)\}}]$, $[p_{\{f'(x_0)\}}]$, $[p_{\{\alpha(a), f'(x_0)\}}]$ and also an atom subalgebra $[p_{\{y \mid f'(y) = f'(x_0)\}}]$ which is different from the three mentioned above, which contradicts the existence of an isomorphism between them. Therefore, $\forall x \neq \alpha(a)(f'(x) = x)$. Assume $f'(\alpha(a)) = c$, (i.e. $f' = k_{\alpha(a),c}$). Let us introduce one more algebra $[k_{b,d}]$ for some element $d \neq a, \neq b$. Now let $\gamma([k_{b,d}]) = [k_{\alpha(b),e}]$. Then the fact that $[k_{a,b}, k_{b,d}]$ contains the subalgebra $[k_{a,d}]$ implies that $[k_{\alpha(a),c}, k_{\alpha(b),e}]$ must contain the subalgebra $[k_{\alpha(a),j}]$ for some element $j \neq c$. But this holds only if $c = \alpha(b)$. □

Lemma 6. *Let $A = \{a_0, a_1, \ldots\}$ be an infinite set. Let the function j be subject to $\{j(a_i) = a_{i+1}$ for $i \geq 0$, $j(x) = x$ for $x \not\subseteq A\}$. Then $\gamma([j]) = [\alpha j \alpha^{-1}]$.*

Proof. The algebra $[j]$ contains the predicates p_A, $\neg p_A$ and the functions $\{j^n \mid n \geq 1\}$. Therefore, the algebra $\gamma([j])$ contains only the predicates $p_{\alpha(A)}$, $\neg p_{\alpha(A)}$ and some functions. Let $\gamma([j]) = [p_{\alpha(A)}, g_1, \ldots, g_n]$. Assume $x \notin A$. Then $[j, p_{\{x\}}]$ does not contain any function of the form $k_{a,b}$, and therefore $[p_{\alpha(A)}, g_1, \ldots, g_n, p_{\{\alpha(x)\}}]$ also does not contain the functions of the form $k_{a,b}$. Each of the functions $g_1, \ldots, g_n$ takes the value $\alpha(x)$ on $\alpha(x)$. (If $g_i(\alpha(x)) \neq \alpha(x)$, then $[p_{\alpha(A)}, g_1, \ldots, g_n, p_{\{\alpha(x)\}}]$ contains the function $k_{\alpha(x),g_i(\alpha(x))}$, which yields a contradiction).

Now let $x \in A$. The algebra $[j, p_{\{x\}}]$ has exactly subalgebras of the form $k_{a,b}$: $[k_{j^k(x),j^n(x)}]$ where $k \geq 0$, $n > k$. Consequently, $[p_{\alpha(A)}, g_1, \ldots, g_n, p_{\{\alpha(x)\}}]$ contains exactly subalgebras of the form $k_{a,b}$: $[k_{\alpha(j^k(x)),\alpha(j^n(x))}]$, where $k \geq 0,\ n > k$. Hence, any term τ of the signature $g_1, \ldots, g_n$ takes its value from the set $\{\alpha(j^n(x)) \mid n > k\}$ on the numbers of the form $\alpha(j^k(x))$.

Since the algebra $[k_{\alpha(x),\alpha(j(x))}]$ is contained in $[p_{\alpha(A)}, g_1, \ldots, g_n, p_{\{\alpha(x)\}}]$, a certain term τ must take the value $\alpha(j(x))$ on $\alpha(x)$. Let us show that one of the functions $g_1, \ldots, g_n$ may be taken as that term. Indeed, if none of the functions mentioned above were appropriate (i.e., $\forall i \leq n\ (g_i(\alpha(x)) = \alpha(j^{n_i}(x)))$ with some $n_i > 1$), then the relation $\tau(\alpha(x)) = \alpha(j^n(x))$, where $n > 1$, would hold for all terms τ by what was proved in the previous section, which leads to a contradiction.

Let us show that one can efficiently find the number $\alpha(j(x))$ via a number $\alpha(x)$ (in the algebraic system with signature $V = \{g_1, \ldots, g_n\}$).

Let $\leq_V$ be an effectively finite order defined on the set V. Since the number n is non-negative, we have that $\tau(\alpha(j^k(x))) = \alpha(j^{k+n}(z))$ exists for any term τ of the signature V and any number k. Then $\cdots \leq_V \alpha(j^2(x)) \leq_V \alpha(j(x)) \leq \alpha(x)$, and $\alpha(j(x))$ is maximal in the set $\{z \mid z < \alpha(z)\}$. Due to this fact, we obtain its determination.

Applying any possible term of the signature $\{g_1, \ldots, g_n\}$ to all elements of the set $\{g_k(\alpha(x)) \mid 1 \leq k \leq n\}$, we proceed as far as we have fixed an element of it by which all other elements (of the set) are generated. In fact, $\alpha(j(x))$ will be this element. In accordance with the remark made above, such an element is sure to occur.

Therefore, the function $\alpha j \alpha^{-1}$ is computable in the signature $\{g_1, \ldots, g_n, p_{\alpha(A)}\}$. As has been shown before Lemma 5, the stated proof of this fact is sufficient for obtaining the equality $\gamma([j]) = [\alpha j \alpha^{-1}]$. □

The assertion of Lemma 6 is valid as well for the set $A = \{\ldots, a_{-1}, a_0, a_1, \ldots\}$ with the function j subject to $\{j(a_i) = a_{i+1}$ for all integer i, $j(x) = x$ for $x \notin A\}$. The proof is quite similar.

Lemma 7. *Let $A \cap B = \emptyset$, $h(A) = B$ and $h(x) = x$ if $x \notin A$. Then $\gamma([h]) = [\alpha h \alpha^{-1}]$.*

Proof. As has been established in the previous lemma, $\gamma([h]) = [p_{\alpha(A)}, g_1, \ldots, g_n]$, where all functions $g_1, \ldots, g_n$ are exactly determined on $\alpha(A)$. From the fact that $[h, p_{\{a\}}]$ contains only subalgebras of the form $[k_{c,d}]$ for $a \in A$ (namely, $[k_{a,h(a)}]$) it follows that $[p_{\alpha(A)}, g_1, \ldots, g_n]$ contains only subalgebras of the same form $[k_{\alpha(a),\alpha(h(a))}]$.

Consequently, $g(\alpha(a)) = \alpha(h(a))$ for each function $g \in [p_{\alpha(A)}, g_1, \ldots, g_n]$ and any $a \in A$, i.e., only one function g is contained in $[p_{\alpha(A)}, g_1, \ldots, g_n]$, and this function is $\{g(\alpha(a)) = \alpha(h(a))$ for $a \in A$, $g(\alpha(a)) = \alpha(a)$ for $a \notin A\}$. □

Lemma 8. *Let l be a function subject to the following relation:*

$$\exists x \exists n (l^n(x) = x)\ \&\ \forall y \in \{x, \ldots, l^{n-1}(x)\}(l(y) = y).$$

Then $\gamma([l]) = [\alpha l \alpha^{-1}]$.

Proof. Let $\gamma([l]) = [g_1, \ldots, g_n, p_1, \ldots, p_m]$. We have to prove that $\alpha l \alpha^{-1} \in [g_1, \ldots, g_n, p_1, \ldots, p_m]$. Let x be the element from the assertion. Let us consider the algebra $[l, p_{\{x\}}]$. It contains a subalgebra $[k_{x,l(x)}]$. By Lemma 5, we have that $[g_1, \ldots, g_n, p_1, \ldots, p_m, p_{\{\alpha(x)\}}]$ contains $[k_{\alpha(x),\alpha(l(x))}]$. This implies the existence of the composition τ of the functions $g_1, \ldots, g_n$, where $\tau(\alpha(x)) = \alpha(l(x))$.

Further, let us take an infinite set $A = \{a_0, a_1, \ldots\}$ which has no intersection with $\{x, \ldots, l^{n-1}(x)\}$. We also take the functions j and h as follows: $\forall i \geq 0\ (j(a_i) = a_{i+1})$ and $j(y) = y$ for $y \notin A$, $\forall i \geq 0\ (h(a_i) = l^i(x))$ and $h(y) = y$ for $y \notin A$.

It can be easily seen that within the algebra $[l, j, h]$, every equality of the terms $\tau_1 = \tau_2$ which holds true for one element of A is also true for all other elements of A. Therefore, this algebra contains only the predicates p_A, $p_{\{x,\ldots,l^{n-1}(x)\}}$ as well as those generated by them. So, the algebra $[g_1, \ldots, g_n, p_1, \ldots, p_m, \alpha(j), \alpha(h)]$ (by Lemmas 6 and 7) contains only the predicates $p_{\alpha(\{x_1,\ldots,l^{n-1}(x)\})}$, $p_{\alpha(A)}$ as well as those generated by them. Since $\alpha(h)(\alpha(a_0)) = \alpha(x), \ldots$, the equality $\tau(\alpha(h)(y)) = \alpha(h)(\alpha(j)(y))$ which is true for $\alpha(a_0)$ must also be true for all $y \in \alpha(A)$. Otherwise, it would define another predicate. Proceeding in the same manner, we arrive at the equality $\tau(\alpha(z)) = \alpha(l(z))$ which must be true for all $z \in \{x, \ldots, l^{n-1}(x)\}$, i.e. $\tau = \alpha l \alpha^{-1}$ for $\alpha(\{x, \ldots, l^{n-1}(x)\})$. □

Lemma 9. *Let l be a function with the property $\forall x \exists n (l^n(x) = x)$. Then $\gamma([l]) = [\alpha l \alpha^{-1}]$.*

Proof. Contrary to the situation in the previous lemma, l can contain several cycles with different length. Assume that $\gamma([l]) = [g_1, \ldots, g_n, p_1, \ldots, p_m]$.

The fact $\forall x \forall i \exists k (g_i(x) = (\alpha l \alpha^{-1})^k(x))$ implies that, for any x, the algebra $[l, p_{\{x\}}]$ contains exactly the subalgebras $[k_{x,l^m(x)}]$ of the form $[k_{a,b}]$. Consequently, $[g_1, \ldots, g_n, p_1, \ldots, p_m, p_{\{x\}}]$ contains only the subalgebras $[k_{x,(\alpha l \alpha^{-1})^t(x)}]$ of the form $[k_{a,b}]$. Furthermore, if M is one of the cycles of the function l, then (in accordance with the previous lemma) $[g_1, \ldots, g_n, p_1, \ldots, p_m, p_{\alpha(M)}]$ contains the subalgebra $\gamma([l/M]) = [(\alpha l \alpha^{-1})/\alpha(M)]$; for $[l, p_M]$ it contains the subalgebra $[l/M]$. The latter means that $\forall i \exists s (g_i/\alpha(M) = (\alpha l \alpha^{-1})^s/\alpha(M))$.

Now we shall prove that $\alpha l \alpha^{-1} \in [g_1, \ldots, g_n, p_1, \ldots, p_m]$. To this end, we shall introduce a set B which includes one element from every cycle. The algebra $[l, p_B]$ contains the subalgebra $[l/B]$. By Lemmas 2 and 7, we have $\gamma([l/B]) = [(\alpha l \alpha^{-1})/\alpha(B)]$. Let us consider a program which uses the functions $g_1, \ldots, g_n$ and the predicates $p_1, \ldots, p_m$, $p_{\alpha(B)}$ and which computes the function $(\alpha l \alpha^{-1})/\alpha(B)$.

Let a number x from $\alpha(B)$ be the input of the program. Obviously, if we change all tests of $p_{\alpha(B)}$ to the test of $z = x$, then the program with $x \in \alpha(B)$ as input will give the same result and compute a certain function g' with the property $g'/\alpha(B) = (\alpha l \alpha^{-1})/\alpha(B)$; moreover, $g' \in [g_1, \ldots, g_n, p_1, \ldots, p_m]$. One can easily notice the following two facts. Firstly, the value $g'(x)$ is equal to the value of the term τ composed by the functions $g_1, \ldots, g_n$; secondly, the value of g' for another element from the same cycle (to which x belongs) is equal to the proper value of the same term τ. But this implies that $g' = \alpha l \alpha^{-1}$ since $\alpha(B)$ includes one element from every cycle. □

Now, let us turn to the study of the structure of the algebra $[f]$ for an arbitrary function f.

Write $R_{n,m} = \{x \mid f^n(x) = f^m(x)\}$ and $R'_{n,m} = R_{n,m} \setminus \bigcup_{p<n,q\leq m} R_{p,q}$ for $n > m \geq 0$. By Lemma 9, if $[f]$ has the subalgebra $[f/R'_{n,0}]$, then $\gamma([f])$ has the subalgebra $[(\alpha f \alpha^{-1})/\alpha(R'_{n,0})]$. By Lemma 7, if $[f]$ has the subalgebra $[f/R'_{n,m}]$, $m > 0$, then $\gamma([f])$ has the subalgebra $[(\alpha f \alpha^{-1})/\alpha(R'_{n,m})]$.

Let us suppose that $\gamma([f]) = [g_1, \ldots, g_n, p_1, \ldots, p_m]$. One can note that $\forall x \forall i \exists k (g_i(x) = (\alpha f \alpha^{-1})^k(x))$. This fact implies that the algebra $[f, p_{\{x\}}]$ contains exactly the subalgebras $[k_{x,f^i(x)}]$ of the form $[k_{a,b}]$ for any x. Therefore, $[g_1, \ldots, g_n, p_1, \ldots, p_m, p_{\{x\}}]$ contains only the subalgebras $[k_{x,(\alpha f \alpha^{-1})^i(x)}]$ of the form $[k_{a,b}]$. We have to prove that $\alpha f \alpha^{-1} \in [g_1, \ldots, g_n, p_1, \ldots, p_m]$.

The statement will be proved separately for the case where $\bigcup_{n,m} R_{n,m} = N$ and the case where $\bigcup_{n,m} R_{n,m} \neq N$.

Lemma 10. *If f obeys the condition $\bigcup_{n,m} R_{n,m} = N$, then $\gamma([f]) = [\alpha f \alpha^{-1}]$.*

Proof. Let us write $R_1 = \bigcup_{n>0} R'_{n,0}$, $R_2 = \bigcup_{n>1} R'_{n,1}$, $R_3 = \bigcup_{n>k>1} R'_{n,k}$. By Lemma 9, we have $(\alpha f \alpha^{-1})/\alpha(R_1) \in \gamma([f])$ since $[f]$ contains the subalgebra $[f/R_1]$. In accordance with Lemma 7, $(\alpha f \alpha^{-1})/\alpha(R_2) \in \gamma([f])$ since $[f]$ contains the subalgebra $[f/R_2]$. Similarly (using again Lemma 6), we deduce the inclusion $(\alpha f \alpha^{-1})/\alpha(R_3) \in \gamma([f])$ for the subalgebra $[f/R_3]$. Since $R_1, R_2, R_3 \in [f]$, $\alpha(R_1), \alpha(R_2), \alpha(R_3) \in \gamma([f])$, we have $\alpha f \alpha^{-1} \in \gamma([f])$ ($\alpha f \alpha^{-1}$ is computed by case analysis: $x \in \alpha(R_1)$, $x \in \alpha(R_2)$, $x \in \alpha(R_3)$). □

Lemma 11. *If f is defined by the relation $\bigcup_{n,m} R_{n,m} \neq N$, then $\gamma([f]) = [\alpha f \alpha^{-1}]$.*

Proof. Let $\gamma([f]) = [g_1, \ldots, g_n, p_1, \ldots, p_m]$.

Sublemma 1. *For any predicate $p \in [f]$, its domain of truth or falsehood is the union of a finite number of sets $R_{n,m}$ (and, therefore, $R'_{n,m}$).*

Proof. Let a predicate p be computable by a program S in the signature $\{f\}$. Let the conditions $Q_1(x), \ldots, Q_n(x)$ be tested within the program S proceeding for the argument $x \in \overline{\bigcup_{n,m} R_{n,m}}$. In view of the membership of x, all these conditions are of the form $f^{s_i}(x) \neq f^{t_i}(x)$, where $i = 1, \ldots, s_i \neq t_i$ (the trivial testing $f^s(x) = f^s(x)$ can be eliminated from the program). But this means that the predicate p may possess values different from $p(x)$ only for the elements of the sets R_{s_i,t_i}, where $i = 1, \ldots, k$. But the number of those elements is finite. □

Let us take an arbitrary $x \in \alpha(\overline{\bigcup_{n,m} R_{n,m}})$. As was shown in Lemma 6, at least one function g_i must take the value $\alpha f \alpha^{-1}(x)$ for x. Let us denote this function by

g_1. Now it is necessary to verify that $\forall y \in \alpha(\overline{\cup_{n,m} R_{n,m}})$ $(g_1(y) = \alpha f \alpha^{-1}(y))$. Let us assume the contrary: $\exists y \in \alpha(\overline{\cup_{n,m} R_{n,m}})$ $(g_1(y) \neq \alpha f \alpha^{-1}(y))$. In this case we shall have a proper function g_i which is subject to the relation $g_i(y) = \alpha f \alpha^{-1}(y)$.

Since we have proved before that $g_1(y) = (\alpha f \alpha^{-1})^k(y)$ holds for some k, we have $g_1(y) = \tau(g_i(y))$ for some term τ of the signature $g_1, \ldots, g_n$. But, in that case, the equality $g_1(z) = \tau(g_i(z))$ is *true* for y and *false* for x. Consequently, if $g_i(x) = g_1(x)$ and $\tau(g_i(x)) = g_i(x)$, then the equality $g_1(z) = g_i(z)$ is *true* for x and *false* for y. In both these cases, a predicate, which is not isomorphic to any predicate of the form $[f]$, is computable in $\gamma([f])$.

Thus, $\forall y \in \alpha(\overline{\cup_{n,m} R_{n,m}})$ $(g_1(y) = \alpha f \alpha^{-1}(y))$. Let us assume that $g_i(x) = g_1^k(x)$ holds for some $x \in \alpha(\overline{\cup_{n,m} R_{n,m}})$ and a function g_i. Then the latter equality defines a predicate $\overline{p_i}$, and its domain of truth (in accordance with the Sublemma 1) coincides with a finite union of the sets $R'_{n,m}$.

In order to simplify our description, we change the functions $g_i (i > 1)$ to the functions g'_i as follows: $g'_i(x) = g_1(x)$ as $\overline{p_i(x)}$ is *true* and $g'_i(x) = g_i(x)$ as $\overline{p_i(x)}$ is *false*. Obviously, $[g_1, \ldots, g_n, p_1, \ldots, p_m] = [g_1, g'_2 \ldots, g'_n, p_1, \ldots, p_m]$.

Now, we shall prove that the inequality $g_1(x) \not\models \alpha f \alpha^{-1}(x)$ can be true only for a finite number of sets $\alpha(R'_{n,m})$. So let us assume the contrary.

The cases of the sets R_1, R_2, R_3 introduced in Lemma 10 will be considered separately. Let $g_1(x) \neq \alpha f \alpha^{-1}(x)$ on $A = \{a_0, a_1, \ldots\}$ which contains elements from an infinite number of different sets of the form $\alpha(R'_{n,0})$. One can suppose $\forall x \in A$ $(\alpha f \alpha^{-1}(x) \notin A)$; otherwise, we can exclude some elements from A.

Assume that $A' = A \cup \{x\}$ for some $x \in \alpha(\overline{\cup_{n,m} R_{n,m}})$. We shall consider the algebra $[g_1, \ldots, g'_n, p_1, \ldots, p_m, p_{A'}]$. Being isomorphic with the previous one, the algebra $[f, p_{\alpha^{-1}(A')}]$ contains the subalgebra $[f/\alpha^{-1}(A')]$.

By Lemma 7, $[g_1, \ldots, g'_n, p_1, \ldots, p_m, p_{A'}]$ contains $[\alpha f \alpha^{-1}/A']$. Let us assume that a program S, which uses the functions and predicates from $[g_1, \ldots, g'_n, p_1, \ldots, p_m, p_{A'}]$, computes the function $(\alpha f \alpha^{-1})/A'$. The conditions $Q_1, \ldots, Q_k$ are being tested while the program S proceeds for the arguments x. The program results in $g_1(x) = \alpha f \alpha^{-1}(x)$.

As was shown in Sublemma 1, we can assume that the conditions Q_j are of the form $g_1^{s_j} \neq g_1^{t_j}$, $p_{A'}$ or p_i. But then $\exists n \forall n' > n \forall y \in \alpha(R'_{n',0}) \cap A$ (the computing scheme for y is just the same as for x in the program S). For y, the program results in the same term g_1 as for x. Thus, $g_1(y) = \alpha f \alpha^{-1}(y)$, contradicting the assumption.

Next, let $g_1(x) \neq \alpha f \alpha^{-1}(x)$ on A which contains elements from an infinite number of different sets of the form $\alpha(R'_{n,1})$. As has been supposed in the previous case, $A' = A \cup \{x\}$ for some $x \in \alpha(\overline{\cup_{n,m} R_{n,m}})$. Obviously, $\alpha f \alpha^{-1}(A) \cap A = \emptyset$ for $A \subset \alpha(\cup_n R'_{n,1})$.

The algebra $[f, p_{\alpha^{-1}(A')}]$ contains the subalgebra $[f/\alpha^{-1}(A')]$. By Lemma 7, $[g_1, \ldots, g'_n, p_1, \ldots, p_m]$ contains the subalgebra $[(\alpha f \alpha^{-1})/A']$. The remainder of the proof is the same as in the previous case.

Finally, let us consider the set $\alpha(R_3)$. According to Lemma 6, one function from $g_1, \ldots, g_n'$ must take the value $\alpha f \alpha^{-1}(x)$ for every x. But all these function coincide for all $\alpha(R_{n,m}')$ $(m > 1)$ except possibly for a finite number. Thus, we have proved the fact that $g_1(x) \neq \alpha f \alpha^{-1}(x)$ is true only for a finite number of sets of the form $\alpha(R_{n,m}')$. Assume that these sets are $\alpha(R_{n_1,m_1}'), \ldots, \alpha(R_{n_p,m_p}')$. As $[f]$ contains the subalgebras $[f/R_{n_i,m_i}']$ for all $p \geq i \geq 1$, by Lemmas 7 and 9, $\gamma([f])$ contains the functions $(\alpha f \alpha^{-1})/\alpha(R_{n_i,m_i}')$ for all $p \geq i \geq 1$.

Now the function $\alpha f \alpha^{-1}$ can be computed as g_1 in the signature $[g_1, \ldots, g_n, p_1, \ldots, p_m]$ by the following case consideration, namely consider separately the case $\alpha(R_{n_i,m_i}')$, and separately the arguments which are not contained in $\cup_{1 \leq i \leq p} \alpha(R_{n_i,m_i}')$. This completes the proof of Lemma 11. □

The proof of Theorem 2.1 can be obtained similarly to that of Lemma 4. If D is an arbitrary element of $\mathcal{G}$, then $p, f \in D \Longleftrightarrow [p], [f] \subseteq D \Longleftrightarrow \gamma([p]), \gamma([f]) \subseteq \gamma(D) \Longleftrightarrow p\alpha^{-1}, \alpha f \alpha^{-1} \in \gamma(D)$.

Let us prove that the permutation α is recursive.

By Lemma 6, we have $\phi_\alpha([s]) = [\alpha f \alpha^{-1}]$ for the function $s(x) = x + 1$, and the function $g = \alpha s \alpha^{-1}$ is recursive. But we then have $g^n = (\alpha s \alpha^{-1})^n = \alpha s^n \alpha^{-1}$ and $\alpha s^n = g^n \alpha$ for all n. Substituting 0, we obtain that $\alpha(n) = g^n(\alpha(0))$, which represents the algorithm to compute the function α. This concludes the proof of the theorem.

Further, we can investigate the structure of the subalgebras more closely and describe all finitely generated subalgebras for the algebra $\langle \Phi_o^1 \cup \mathcal{P}_o^1, FDAA_= \rangle$.

[Solovyev, 34]. *$\mathcal{U}$ is a finitely generated subalgebra of the algebra*

$$\langle \Phi_o^1 \cup \mathcal{P}_o^1, FDAA_= \rangle$$

iff $\exists n \exists f_1, \ldots, f_n \exists m \exists p_1, \ldots, p_m (\mathcal{U} = \{f \mid \forall x, y(y = f(x) \rightarrow \exists l \exists k_1, \ldots, k_l(y = f_{k_1} \cdots f_{k_l}(x)))\} \cup \{P \mid P/M$ *is a Boolean combination of predicates* $p_1, \ldots, p_m\})$, *where* $M = \{x \mid \forall f \in \mathcal{U}(f(x) = x)\}$.

IV. Algebraic properties of subalgebras

We should note that we cannot refer the reader to any investigation of the algebraic properties of the algebras with composition.

The Robinson algebras.

A binary relation $\sim$ on the algebra $A = \langle \mathcal{A}, +, \cdot, ^{-1} \rangle$ is a congruence if $f_1 \sim g_1 \,\&\, f_2 \sim g_2 \rightarrow f_1 + f_2 \sim g_1 + g_2 \,\&\, f_1 \cdot f_2 \sim g_1 \cdot g_2 \,\&\, f_1^{-1} \sim g_1^{-1}$.

An algebra A is simple if there are no non-trivial congruence relations on it. Sufficient conditions for the simplicity of a subalgebra were obtained for the Robinson algebras. Below, we shall mention typical results in this field. Additional information can be found in [Zakharov, 16].

[Polyakov, 20]. *A maximal subalgebra of the algebra $\mathcal{F}_r$ which contains a constant function is simple only when it contains the constant* 1.

[Polyakov, 20]. *A subalgebra of the algebra $\mathcal{F}_n$ is simple if it contains the constant* 0 *and the function* $x \dot{-} 1$.

The algebra $\langle \Phi_o^1 \cup \mathcal{P}_o^1, FDA_= \rangle$.

We can obtain stronger results for the program type algebras. In particular, for these algebras, a complete description of all finitely generated simple subalgebras was obtained.

[Solovyev, 35]. *A finitely generated subalgebra $\mathcal{U}$ is simple iff* $\mathcal{U} = \{\text{TRUE}, \text{FALSE}, \text{id}(x) = x\}$ *or* $\mathcal{U}$ *contains some functions different from* id *and* $\forall d(\exists e \neq d \exists f \in \mathcal{U}(f(d) = e) \longrightarrow \forall f_1, f_2 \in \mathcal{U}(f_1 \neq f_2 \longrightarrow \exists g \in \mathcal{U}(f_1(g(d)) \neq f_2(g(d))))$.

As a simple consequence, we have that the algebra $\langle \Phi_o \cup \mathcal{P}_o, FDA_= \rangle$ is simple, and the maximal subalgebras $\mathcal{U}_A = \{f \mid f(A) \cap \bar{A} = \emptyset\} \cup \mathcal{P}_o$ are simple iff $\mid A \mid = 1$.

Comparing the results obtained for the program type algebras and the Robinson algebras, we can see that the former have a more regular structure, and thus algebraic results are more complete for those algebras.

Problems related to algebras of recursive functions are not totally removed and do not differ completely from other questions in the theory of algorithms. They are directly related to the generalized theory of computability.

Let us assume that we have a countable set A and a set L of functions and predicates, which is defined on A and computable in a certain intuitive sense. The question arises as to what the functions and predicates are which must be considered to be computable on A. One of the possible answers was proposed in [Friedman, 54], namely, that the functions and predicates from the closure $[\mathcal{L}]_{FDA_=}$ are computable.

Later, this approach was developed in the papers [Shepherdson, 55], [Moldestad, Stoltenberg-Hansen and Tucker, 40]. As concerns the algebras of recursive functions, $[\mathcal{L}]_{FDA_=}$ is just the subalgebra of $\langle \Phi_o \cup \mathcal{P}_o, FDA_= \rangle$.

Taking advantage of this approach, one can obtain the following results, which describe computability in some fields.

[Solovyev, 51]. *Let $Q(\sqrt{2})$ be the extension field of the rational numbers and $\mathcal{L} = \langle +, \times, -, ()^{-1}, \langle, 0, 1 \rangle$ be the signature of the ordered field. Set $0^{-1} = 0$. Then $[\mathcal{L}]_{FDA_=}$ will be isomorphic to a maximal subalgebra of the algebra $\langle \Phi_o \cup \mathcal{P}_o, FDA_= \rangle$.*

[Solovyev, 51]. *Let $Q(\alpha)$ be an algebraic extension of a prime power and $\mathcal{L} = \langle +, \times, -()^{-1}, 0, 1 \rangle$. Then $[\mathcal{L}]_{FDA_=}$ is isomorphic to a subalgebra of rank 2, i.e., "is maximal in maximal".*

The proofs of the results for the algebras of recursive functions use a wide range of methods from the classical theory of algorithms, mathematical logic, algebra (categories and groups theory), discrete mathematics, and computer science. In addition, some original specific methods have been developed in this field. Among them we can refer to the methods for describing the automorphisms of the structure of subalgebras.

References

[1] Krohn, K., Rhodes, J., Algebraic theory of machines. I. Prime decomposition theorem for finite semigroups and machines, Trans. Amer. Math. Soc. 116 (1965), 450–464.

[2] Plotkin, B. I., Galois theory of databases, in: Algebra, some current trends, Springer-Verlag, Berlin, 1988, 147–162.

[3] Kudriavtsev, V. B., Functional systems, Moscow University, Moscow, 1982.

[4] Rogers, H., Theory of recursive functions and effective computability, McGraw-Hill, New York, 1967.

[5] Odifreddi, P., Classical Recursion Theory, North-Holland, Amsterdam, 1989.

[6] Marchenkov, S. S., The simple examples of bases for superposition in the class of Kalmar's elementary functions, Combinatorics and Graph theory, Banach Center Publications 25, 1989.

[7] Marchenkov, S. S., The existence of bases for superposition in countable primitive-recursive closed classes, Mat. Zametki 39 (2) (1986), 268–276.

[8] Marchenkov, S. S., On existence of infinite bases for superposition in some classes of one place functions, Mat. Zametki 47 (4) (1990), 55–62.

[9] Marchenkov, S. S., On a certain basis for superposition in the class of Kalmar's elementary functions, Mat. Zametki 27 (3) (1980), 321–332.

[10] Robinson, R. M., Primitive recursive functions, Bull. Amer. Math. Soc. 53 (1947), 925–942.

[11] Robinson, J., General recursive functions, Proc. Amer. Math. Soc. (1950), 703–708.

[12] Grzegorczyk, A., Some classes of recursive functions, Rozprawy Matematyczne IV, Warszawa, 1953.

[13] Rödding, D., Über die Eliminierbarkeit von Definitionsschemata in der Theorie der rekursiven Funktionen, Z. Math. Logik Grund. Math. 10 (1964), 315–330.

[14] Jones, J. P., Matijasevic, Ju. V., A new representation for the symmetric binomial coefficient and its applications, Ann. Sci. Math. Quebec 6 (1982), 81–97.

[15] Marchenkov, S. S., On bounded recursions, Mathematica Balkanica 2 (1972), 124–142.

[16] Zakharov, D. A., Recursive functions, Novosibirsk University, Novosibirsk, 1970.

[17] Kuznetsov, A. V., The addition to Goodstein's book "Mathematical Logic", Inostran. lit., Moscow, 1961.

[18] Nasibullov, N. N., On recursive functions with large amplitude, Dokl. Akad. Nauk SSSR 178 (3) (1968), 539–541.

[19] Polyakov, E. A., Some properties of algebras of recursive functions, Kibernetica 2 (1967), 20–23.

[20] Polyakov, E. A., Algebras of recursive functions, Algebra i Logika 3 (1) (1964), 41–56.

[21] Polyakov, E. A., On some properties of algebras of recursive functions, Algebra i Logika 3 (3) (1964), 39–57.

[22] Glushkov, V. M., The theory of automata and formal transformations of programs, Kibernetica 5 (1965), 1–9.

[23] Minsky, M., Finite and infinite machines, Prentice-Hill, 1967.

[24] Barzdin, Ia. M., On one class of Turing machine (Minsky machine), Algebra i Logika 1 (1962), 42–51.

[25] Golunkov, Iu. V., On completeness of operations in the systems of algorithmic algebras, in: Algoritmi i avtomati, Kazan Univ., Kazan, 1978, 11–53.

[26] Kfoury, A. J., Urzyczyn, P., Necessary and sufficient conditions for the universality of programming formalisms, Acta Inform. 22 (1985), 347–377.

[27] Aho, A., Hopcroft, J., Ullman, J., The design and analysis of computer algorithms, Addison-Wesley, London, 1976.

[28] Ershov, A. P., Operator algorithms, 1, Problemi Kibernetiki 3, Fizmatgiz, Moscow, 1960.

[29] Shepherdson, J. C., Sturgis, H.,E., Computability of recursive functions, J. Assoc. Comput. Mach. 10 (1963), 217–255.

[30] Golunkov, Iu. V., The degree of unsolvability of completeness problem in algebras of recursive functions, Mat. Zametki 44 (5) (1988), 620–627.

[31] Ershov, A. P., Liapunov, A. A., On formalization of the concept ⟨program⟩, Kibernetika 5 (1967), 40–57.

[32] Golunkov, Iu. V., Completeness problem in functional systems of program type, Ph. dissertation, Moscow Univ., Moscow, 1992.

[33] Solovyev, V. D., Total recursive functions and predicates system completeness, Recursive Function Theory: Newsletter 39, 1992.

[34] Solovyev, V. D., Completeness problem in the algebras of recursive functions with additional closure tools, Izv. Vuzov. Mat. 6 (1993), 46–52

[35] Solovyev, V. D., Algebraic aspects of the abstract theory of computability, in: Matematicheskie problemi kibernetiki 3, Nauka, Moscow, 1991, 233–256.

[36] Shcheglov, A. I., The algebra of partial recursive functions, Algebra i Logika 6 (6) (1967), 33–48.

[37] Shcheglov A. I., On cardinality of the set of maximal subalgebras of algebra of partial recursive functions, Algebra i Logika 7 (3) (1968), 119–121.

[38] Marchenkov, S. S., Deletion of the recursion schemes in the Grzegorczyk's class E^2. Mat. Zametki 5 (5) (1969), 561–568.

[39] Muchnik, A. A., On two approach to classification of recursive functions, in; Problemi matematicheskoi logiki, Mir, Moscow, 1970, 123–138.

[40] Moldestad, L., Stoltenberg-Hansen, V., Tucker, J., Finite algorithmic procedures and inductive definability, Math. Scand. 46 (1980), 62–76

[41] Kozminych, V. V., On one-place primitive recursive functions, Algebra i Logika 7 (1) (1968), 75–90.

[42] Kozminych, V. V., On subalgebras of the R. Robinson's algebra, Algebra i Logika 9 (6) (1970), 672–690.

[43] Golunkov, Iu. V., Towards theory of systems of algorithmic algebras, Dokl. Akad. Nauk SSSR 232 (4) (1977), 749–752.

[44] Golunkov, Iu. V., On precomplete classes of algorithms, 1, Veroiatnostnie metodi i kibernetika 16, Kazan Univ., Kazan, 1980, 21–40.

[45] Nepomniashchy, V. A., On operation completeness in operator algorithms, Dokl. Akad. Nauk SSSR 199 (4) (1971), 780–782.

[46] Kfoury, A. J., Definability by programs in first-order structures, Theoret. Comp. Sci. 25 (1983), 1–66.

[47] Golunkov, Iu. V., Approximation completeness in algebras of partial recursive functions and predicates, Kybernetika 6 (1987), 26–30.

[48] Solovyev, V. D., The completeness of systems of recursive functions and predicates, I, II, Izv. Vuzov. Mat. 8 (1989), 56–63; 9 (1989), 60–66.

[49] Solovyev, V. D., Initial segments of effectively-finite orders, Mat. Zametki 61 (6) (1997), 950–953.

[50] Solovyev, V. D., Program closed classes of recursive functions and predicates of finite rank, Izv. Vuzov. Mat. 9 (1993), 51–63.

[51] Solovyev, V. D., The structure of program closed classes of many-place functions and predicates, Izv. Vuzov. Mat. 7 (1993), 47–50.

[52] Solovyev, V. D., Functional systems of recursive functions and predicates with strong program tools for closure, Ph dissertation, Moscow, VC RAN, 1996.

[53] Olshansky, A. Iu., The limited order groups with subgroups of prime order, Algebra i Logika 21 (5) (1982), 553–618.

[54] Friedman, H., Algorithmic procedures, generalized Turing algorithms and elementary recursion theory, Logic Coll. '69, North-Holland, Amsterdam, 1971, 361–389.

[55] Shepherdson, L. S., Computation over abstract structures: serial and parallel procedures and Friedman's definitional schemes, Logic Coll. '73., North-Holland, Amsterdam, 1975, 445–513.

[56] Moldestad, L., Stoltenberg-Hansen, V., Tucker, J., Finite algorithmic procedures and inductive definability, Math. Scand. 46 (1980), 62–76

Σ_2 Induction and cuppable degrees

Yue Yang

1. Introduction and preliminaries

The study of recursion theory on fragments of Peano arithmetic was started in the late 70's and the early 80's. One of the main tasks is to investigate the proof-theoretic strength of theorems in recursion theory. This study is parallel to the "reverse mathematics" project pioneered by H. Friedman and S. Simpson. In reverse mathematics, the proof-theoretic strength is measured by the subsystems of second-order Peano arithmetic, whereas in the study of recursion theory on fragments of Peano arithmetic, the strength is measured by the amount of (first-order) induction involved. To be more precise, over the base theory $P^- + I\Sigma_0$, the first-order theory of Peano arithmetic PA is split into an infinite hierarchy of fragments,

$$P^- + B\Sigma_1 \Leftarrow P^- + I\Sigma_1 \Leftarrow P^- + B\Sigma_2 \Leftarrow P^- + I\Sigma_2 \cdots,$$

where P^- denotes the basic axioms, and $I\Sigma_n$ ($B\Sigma_n$) denotes the induction (collection) scheme for Σ_n^0 formulas. By Kirby and Paris [10], the above arrows cannot be reversed. (From now on, we always assume $P^- + I\Sigma_0$.) One of the earliest results in this area was due to S. Simpson (unpublished), which says that Σ_1 induction is sufficient to prove the Friedberg–Muchnik theorem, (see Mytilinaios [13]).

Over the past two decades, many more results are known. A survey paper [5] by Chong and Yang lists most of the major ones. In particular, there seems to be a natural correspondence between the syntactical form of the theorem and the amount of induction required in the proof. For example, Groszek and Slaman [7] formulated an abstract version of Σ_n and Π_n construction, and established that Σ_n induction is the necessary and sufficient condition for carrying out all Π_n construction. In particular, they give the following general heuristic: $P^- + I\Sigma_1$ is the fragment for carrying out all Friedberg–Muchnik type finite injury arguments and $P^- + I\Sigma_2$ for most typical infinite injury arguments. Using the iterated trees framework by Lempp and Lerman [9], Yang [17] also established a one-to-one correspondence between Σ_n induction and constructions involving n iterated trees. However, when we look at each individual theorem in recursion theory, the general machineries do not produce the best estimates for sufficient conditions and hardly give any information on necessary conditions. It is often the case that if the conflict between requirements is not too serious (even though the syntactical form of the theorem is complicated), then the amount of induction involved is actually lower than it appears. For example, Sacks Splitting Theorem seems to belong to the Σ_2 level, but Mytilinaios [13] showed that $I\Sigma_1$ is sufficient to prove it. Moreover, Mourad [12] showed that over the base theory $P^- + B\Sigma_1$, Sacks Splitting Theorem is in fact equivalent to $I\Sigma_1$. Sacks Density Theorem provides another example. It looks like a Σ_3 construction, but Groszek, Mytilinaios and Slaman [15] have shown that $B\Sigma_2$ is sufficient to prove it, the necessary condition is still unknown. In the proof of the above two Sacks theorems, blocking method plays an important role. There are other examples, where the techniques involved are completely different from the ones in classical recursion theory (e.g., the proof of Friedberg–Muchnik Theorem under $P^- + B\Sigma_1$ by Chong and Mourad [1]).

Therefore, to pinpoint the exact level of a theorem in recursion theory in the hierarchy of fragments of Peano arithmetic is still a challenging problem. One interesting aspect is that working on different fragments requires completely different tools. For example, without Σ_1 induction, the lack of regularity of r. e. sets is often the hardest obstacle. Without $B\Sigma_2$, the usual pointwise Turing reducibility between r. e. sets may not be transitive (see [8]), so that the concept of r. e. degree is different from the one in classical recursion theory.

In our studies, we concentrate on models $\mathcal{M}$ satisfying $P^- + B\Sigma_2$, because it is a natural place to talk about r. e. degrees and to look at infinite injury arguments. If $\mathcal{M}$ satisfies $I\Sigma_2$, then many infinite injury argument can be carried through (though the noncuppable degree construction is slightly different). If a model $\mathcal{M}$ of $P^- + I\Sigma_0$ does not satisfy $I\Sigma_2$, we shall call such an $\mathcal{M}$ a $B\Sigma_2$ *model*. Then there is a Σ_2 cut I and an associated Σ_2 function $f : I \to \mathcal{M}$, which is unbounded. Recall that a cut I is a proper subset of $\mathcal{M}$ which is downward closed and closed under successor. In the study of fragments of arithmetic, the cut I plays a crucial role. More discussions can be found in the survey paper [5] by Chong and Yang. In $B\Sigma_2$ models, the cofinal map $f : I \to \mathcal{M}$ can be recursively approximated (true in I only), which offers us a tool to make various nonuniform constructions, an approach which usually fails in classical recursion theory.

We now briefly recall some earlier results about infinite injury argument and $I\Sigma_2$. Groszek and Mytilinaios [6] showed that $I\Sigma_2$ is sufficient to prove the Shoenfield Thickness Lemma and thus the existence of an incomplete high r. e. degree. Mytilinaios and Slaman [14] showed that $P^- + B\Sigma_2$ is not sufficient to prove the existence of incomplete high r. e. degrees, by constructing a $B\Sigma_2$ model in which every r. e. set is either low or complete. In [3] and [4], Chong and Yang studied the jump operator in $B\Sigma_2$ models, and established the equivalence between the existence of an incomplete high r. e. degree and $I\Sigma_2$. Recently, Chong, Qian, Slaman and Yang [2] studied the meet operators in $B\Sigma_2$ models, and showed that the existence of an r. e. minimal pair is also equivalent to Σ_2 induction.

In this paper, we study the relationship between Σ_2 induction and noncuppable degrees. Recall that an r. e. degree $\boldsymbol{a}$ is called *cuppable* if there exists an r. e. degree $\boldsymbol{b} < \boldsymbol{0}'$ such that $\boldsymbol{a} \cup \boldsymbol{b} = \boldsymbol{0}'$. An r. e. degree is *noncuppable* if it is not cuppable. The typical construction of a noncuppable degree is a pinball machine construction, first carried out by S. B. Cooper and later generalized to high r. e. degrees by Harrington, (see D. Miller [11]).

The main result of this paper is

Theorem 1. *Over the base theory $P^- + B\Sigma_2$, the existence of noncuppable r. e. degrees is equivalent to $I\Sigma_2$.*

We split the theorem into two parts. In Section 2, we show that in any $B\Sigma_2$ model, every nonrecursive r. e. degree is low cuppable. In Section 3, we show that show that $I\Sigma_2$ is sufficient to show the existence of a noncuppable r. e. degree. The proof is a tree construction of such a degree.

We now briefly recall some basic terminologies in fragments of arithmetic. More information can be found in many early references, for instance Chong and Yang [5]; or Mytilinaios [13]. Let $\mathcal{M}$ be a model of $P^- + I\Sigma_0$. A subset F of $\mathcal{M}$ is called $\mathcal{M}$-finite, if it is coded in $\mathcal{M}$. An $\mathcal{M}$-finite set can be viewed as a simple rearrangement of a number in $\mathcal{M}$. In $\mathcal{M}$, a bounded set may or may not be $\mathcal{M}$-finite. For example, in $B\Sigma_2$ models, the Σ_2 cut I is bounded but not $\mathcal{M}$-finite. However, under $I\Sigma_n$, any bounded Σ_n subset of $\mathcal{M}$ is $\mathcal{M}$-finite. In other words, any Σ_n set is regular. Recall that we say a subset X of $\mathcal{M}$ is *regular* if for all $a \in \mathcal{M}$, $X \upharpoonright a$ is $\mathcal{M}$-finite.

We use upper case Greek letters, such as Γ or Δ to denote Turing functionals in $\mathcal{M}$, which is a $\Sigma_1(\mathcal{M})$ set of consistent quadruples. $\Gamma(A; x)$ ($\Gamma(AB; x)$) is the value of Γ applied to x relative to A ($A \oplus B$). We let $\gamma(A; x)$ to denote the use of the computation $\Gamma(A; x)$. When the underlying set is clear, we also write $\gamma(x)$ instead of $\gamma(A; x)$. As we work in $B\Sigma_2$ models, we can make further conventions on Turing functional Γ applying to an r. e. set A. We assume that when $x < y$ and $\Gamma(A; y)$ is defined then $\Gamma(A; x)$ is also defined, and $\gamma(A; x) < \gamma(A; y)$. However, we often add this convention to Turing functionals given to us. When we build a Turing functional, we may or may not follow this convention.

Acknowledgment. The project was started when Ted Slaman visited National University of Singapore as a visiting professor. The author wishes to thank him for his many suggestions on various constructions involving the join operator. The author also wishes to thank C. T. Chong and L. Qian for many discussions and suggestions on details.

2. $B\Sigma_2$ and cuppable degrees

In [2], Chong, Qian, Slaman and Yang showed the following promptness result, which they used to solve the minimal pair problem in $B\Sigma_2$ models.

Definition 1. Let p be a total recursive function and W be an r.e. set. We say that an r.e. set A is *p-prompt for W*, if

$$(\exists s)(\exists x)(x \text{ enters } W \text{ at stage } s \text{ and } A \upharpoonright x[s] \neq A \upharpoonright x[p(s)]).$$

We say that A is *p-prompt* if A is p-prompt for all infinite r.e. set W. We say that A is *prompt*, if there is a total recursive function p such that A is p-prompt.

Lemma 1 ([2]). *Let $\mathcal{M}$ be a $B\Sigma_2$ model. Then any r. e. set A in $\mathcal{M}$ is either recursive or prompt.*

In this section, we apply the promptness result to show the following theorem.

Theorem 2. *Let $\mathcal{M}$ be a $B\Sigma_2$ model. Then any nonrecursive r. e. degree* $\boldsymbol{b}$ *in $\mathcal{M}$ is low cuppable, that is, there exists a low r. e. degree* $\boldsymbol{a}$ *such that* $\boldsymbol{a} \cup \boldsymbol{b} = \boldsymbol{0}'$.

Proof. We modify the classical proof as presented in Soare [16] together with blocking method. Let B be an r. e. set in degree $\boldsymbol{b}$. By Lemma 1, B is prompt. Let $p(s)$ be a recursive function such that for all s $p(s) \geq s$ and for all e, if W_e is infinite, then

$$(\exists x)(\exists s)[x \in W_{e,s} \setminus W_{e,s-1} \wedge B_s \upharpoonright x \neq B_{p(s)} \upharpoonright x].$$

We build a low r. e. set A and a recursive functional $\Gamma(AB; x)$ such that $\Gamma(AB) = K$ where K is a complete r. e. set. We may assume that A is a subset of even numbers and B is a subset of odd numbers, so that A and B are disjoint. Fix an enumeration $\{K_s\}_{s \in M}$. We use $\gamma_s(x)$ denote the use of $\Gamma(AB; x)$ at stage s. When we say "reset $\gamma_s(x)$", we mean to choose $\gamma_{s+1}(x)$ to be the least number y which is not in A_s and greater than all $\gamma_t(z)$ for $t \leq s$ and $z \leq x$.

The positive strategy P_x is to make $\Gamma(AB; x) = K(x)$. Initially we set $\Gamma(AB; x) = K(x)[s]$, which is 0, with use $\gamma_s(x)$. If x enters K, then we put $\gamma_s(x)$ in A and reset $\Gamma(AB; x) = 1 = K(x)$ with empty use. If no injury happens then P_x only acts once and remains satisfied forever.

To make A low, we meet the following negative requirements N_e:

$$N_e : (\exists^\infty s)[\Phi_e(A; e)\downarrow [s]] \rightarrow \Phi_e(A; e)\downarrow$$

where Φ_e is a fixed enumeration of Turing functionals. Let $\varphi_s(e)$ denote the use of $\Phi(A; e)$ at stage s. To overcome the conflict between the positive and negative requirements, we try to clear all uses $\gamma(x)$, $x \geq e$ from the A-use $\varphi_s(e)$, by using the promptness of B to force $B \upharpoonright \gamma_s(x)$ to change. During the construction, we define r. e. sets U_e for $e \in \mathcal{M}$. Let g be the corresponding recursive function obtained by the following Slowdown Lemma whose proof can be found in Soare [16] and can be carried over in any model satisfying $I\Sigma_1$:

Lemma 2 (Slowdown Lemma). *Let $\{U_{e,s}\}_{e,s \in \mathcal{M}}$ be a strong array of $\mathcal{M}$-finite sets such that $U_{e,s} \subseteq U_{e,s+1}$ and $U_e = \bigcup_s U_{e,s}$. Then there is a recursive function g such* *that for all e, $W_{g(e)} = U_e$ and $W_{g(e),s} \cap U_{e,at\ s} = \emptyset$ (namely any element enumerated in U_e appears strictly later in $W_{g(e)}$).*

We modify the construction of A in classical recursion theory as follows. We will explain the motivation for blocking after the construction.

We organize the requirements in blocks following the natural priority

$$T_0 < T_1 < \cdots,$$

where T_i is the set of numbers $\{n : b_i \leq n < b_{i+1}\}$ for some numbers b_i and b_{i+1} in $\mathcal{M}$. For a fixed block T_i, we give negative requirement N_e, $e \in T_i$ higher priority than P_x, $x \in T_i$. We will approximate b_i recursively by $b_{i,s}$.

Construction

Stage $s = 0$: Set $b_{i,0} = i$ for all $i \in \mathcal{M}$. Set $\Gamma(A_0 B_0; x) = 0 = K_0(x)$ with use $\gamma(x) = 2x$ for all $x \in \mathcal{M}$.

Stage $s + 1$:

Step 1. Find the least pair e and x such that $\Phi_e(A; e)\downarrow [s]$, $\gamma_s(x) \leq \varphi_s(e)$ and 'x has lower priority than e' (that is, $e \in T_{i,s}$, $x \in T_{j,s}$, and $j \geq i$). (If no such e, x exists, then go to step 2.) Enumerate $\gamma_s(x)$ in U_e. Find the least stage t such that $\gamma_s(x) \in W_{g(e),t}$.

Case 1 (Free clear). $B_s \upharpoonright \gamma_s(x) \neq B_{p(t)} \upharpoonright \gamma_s(x)$. Then for each y in block $T_{j,s}$, $j \geq i$, if $\Gamma(AB, y)$ has not been set to 1 then set $\gamma_{s+1}(y)$ to be the least number not in A_s and greater than $\varphi_s(e)$ and all $\gamma_t(z)$ for $t \leq s$ and $z \leq x$; set $\Gamma(AB; y) = 0$ with use $\gamma_{s+1}(y)$.

Case 2 (Capricious destruction). $B_s \upharpoonright \gamma_s(x) = B_{p(t)} \upharpoonright \gamma_s(x)$. Enumerate $\gamma_s(x)$ in A. For each y in block $T_{j,s}$, $j \geq i$, set $\gamma_{s+1}(y)$ and $\Gamma(AB; y)$ as in Case 1.

Step 2. If $z \in K_{s+1} - K_s$, then enumerate $\gamma_s(z)$ in to A and define $\Gamma(AB; z) = 1 = K(z)$ with empty use.

In either case, if there is an action in block T_i, then set $b_{j,s+1} = b_{j,s}$ for $j \leq i + 1$; set $b_{j,s+1} = s + j - i$ for $j > i + 1$.

End of construction

The reason for using blocking is as follows. In classical recursion theory, we have sufficient induction to show that $\gamma(x) = \lim_s \gamma_s(x)$ is defined for all x. In may not be the case in a $B\Sigma_2$ model. Fix an $x \in \mathcal{M}$. Without blocking, $\gamma(x)$ is reset by either a free clear or a capricious destruction caused by higher priority requirements. Although the free clear can happen at most 2^{x+1} times, there is no recursive upper bound on the number of capricious destructions. It is conceivable that there is a Σ_2 cut I such that the stage s, at which a capricious destruction occurs due to some N_i for $i \in I$, is cofinal in $\mathcal{M}$. Thus for all $x \notin I$, $\gamma(x)$ is undefined. The same phenomenon happens in the proof for Sacks Splitting Theorem in $I\Sigma_1$, see [13], and was overcome by blocking method too.

We now verify that the construction works. Let $I = \{i : \lim_s b_{i,s} \text{ exists}\}$ and $T_i = \{n : b_i \leq n < b_{i+1}\}$ for $i \in I$. Then I is a Σ_2 downward closed set.

Lemma 3. (1) *For any e in $\mathcal{M}$ there is an i in I such that e is in block T_i.*

(2) *I is closed under successor.*

(3) *For any x in $\mathcal{M}$, $\gamma(x) = \lim_s \gamma_s(x)$ exists and $\Gamma(AB; x) = K(x)$.*

(4) *A is low.*

Proof. (1) Initially e is in $T_{e,0}$. Notice that $b_{i,s}$ is nondecreasing with respect to s, at any stage s, the number e can only be in block $T_{i,s}$ for $i \leq e$. On the other hand, the set

$$\{i \leq e : (\exists s)[e \in T_{i,s}]\}$$

is a bounded Σ_1 set, thus $\mathcal{M}$-finite. Hence it has a least element, say i_0. Hence e will eventually be in the permanent block T_{i_0}.

(2) Suppose that i is in I, we want to show that $i+1$ is also in I, i.e., $\lim_s b_{i+1,s}$ exists. As i is in I, there is a stage s_0 such that for all $s \geq s_0$, $b_{i,s} = b_i$ and $T_{i-1,s} = T_{i-1}$. By construction, after stage s_0, b_{i+1} changes its value if and only if there is an action caused by requirements in T_{i-1}. By regularity of K, there is a stage $s_1 \geq s_0$ such that $K \upharpoonright b_i = K_{s_1} \upharpoonright b_i$. Hence after stage s_1, there is no action due to Step 2. It remains to show that after certain stage, no action is taken due to Step 1 action. Fix an e in block T_{i-1}. If there is a stage at which we see a free clear for e, then N_e will not be injured any more. On the other hand, if there are unboundedly many capricious destruction and no free clear due to N_e, then $W_{g(e)}$ defined in the construction will be a infinite r. e. set violating the choice of prompt function $p(s)$. Hence for any e in T_{i-1} there is a stage s such that for all $t > s$, no action will be caused due to N_e. By $B\Sigma_2$, there is a uniform bound $s_2 > s_1$ for every $e \in T_{i-1}$. For any stage $t > s_2$, $b_{i+1,t} = b_{i+1,s_2}$.

(3) and (4) follows from (1) and (2), which is the same as the argument in classical recursion theory. □

3. $I\Sigma_2$ and noncuppable degrees

In this section, we show that Σ_2 induction is sufficient to prove the existence of a noncuppable r. e. degree. Thus over the base theory $P^- + B\Sigma_2$, the existence of noncuppable r. e. degrees is equivalent to $I\Sigma_2$.

Theorem 3. *Let $\mathcal{M}$ be a model of $P^- + I\Sigma_2$. Then there is an r. e. degree $\mathbf{a} \neq \mathbf{0}$ such that for all r. e. degree $\mathbf{w} \neq \mathbf{0}'$, $\mathbf{a} \cup \mathbf{w} \neq \mathbf{0}'$.*

Proof. We present a tree construction of a nonrecursive r.e. set A such that for all incomplete r.e. set W, $A \oplus W$ is still incomplete. The most essential use of Σ_2 induction is to justify the existence of true path. In the construction, we also need to set links for delaying purposes, which adds some $0'''$ flavor on the construction. However it is not a true $0'''$ priority argument because there is no injury on the true path. All links on the true path will stay only for $\mathcal{M}$-finite amount of time and then be gone forever.

We wish to build an r. e. set A and for each e in $\mathcal{M}$ a Turing functional Δ_e satisfying the following requirements:

$$P_e : A \neq \Phi_e;$$

$$N_e : \text{If } \Gamma_e(AW_e) \text{ is total and } \Gamma_e(AW_e) = K, \text{ then } \Delta_e(W_e) = K,$$

where $\{\Phi_e\}_{e \in \mathcal{M}}$ and $\{\Gamma_e\}_{e \in \mathcal{M}}$ are fixed enumerations of Turing functionals in $\mathcal{M}$, and the set K is a complete r. e. set chosen as follows. In the construction, we need to use an r.e. set D to help us making K change. By Recursion Theorem, we may assume that we know the index of D. Thus we may pick $K = D \oplus K_0$, where K_0 is a fixed complete r. e. set. By doing so, we may put odd numbers directly in K, although we cannot keep even numbers out of K.

Naive version of strategies. The first attempt to satisfy P_e is to use Friedberg–Muchnik diagonalization strategy. We pick a witness x not yet in A, wait until x is realized, i.e., $\Phi_e(x)\downarrow = 0$, then put x in A. Then P_e is satisfied forever.

The noncuppable strategy N_e works as follows. We omit the indices during the discussion. N measures the length of agreement function $l(e, s)$ between $\Gamma(AW)$ and K that is,

$$l(e, s) = \mu y(\Gamma(AW_e; y) \neq K(y)[s] \wedge \forall z < y \Gamma(AW_e; z) = K(z)[s]).$$

If $l(e, s)$ increases, then for all $y < l(e, s)$, N sets $\Delta(W; y) = \Gamma(AW; y)$ with proper use $\delta(W; y)$ and preserves A up to $\gamma(AW; y)$. The choice of $\delta(W; y)$ plays an important role in the construction, which we will specify later. At this moment, we only require that $\delta(W; y) > \gamma(AW; y)$. The purpose of restraining A is as follows. Suppose that there is a number $z < l(e, s)$ entering K. If W does not change below $\gamma(AW; z)$ then the restraint on A ensures that $\Gamma(AW) \neq K$, so that N is satisfied by the failure of premise. On the other hand if W changes below $\gamma(AW; z)$, then $\Delta(W; z)$ can be reset too, since $\gamma(AW; z) < \delta(W; z)$.

Conflicts between requirements. We now look at the conflicts between P and N. Notice that when $\Gamma(AW) = K$, the restraint set by N is truly infinite, thus it is difficult for lower priority P to act. The modified strategy P will try to force W to change by attacking the D side of K. If W does not respond then $\Gamma(AW) \neq K$. If W responds then P can be satisfied without injuring N.

Suppose that P puts a number x in A. Then $\Gamma(AW)$ has a chance to change value on p for all p such that $x < \gamma(AW; p)$, which we wish to avoid since W may not change below $\delta(W; p)$. Let $p(x)$ denote the least such p. Notice that $p(x)$ depends not only on x, but also on stage s, since the use of Γ may move due to W changes. (When necessary we write $p_s(x)$ instead of $p(x)$.) For example, it is possible to have an $i < p_s(x)$, $\gamma(AW; i) < x$ at stage s and $\gamma(AW; i) > x$ at stage $t > s$, so $p_t(x) \leq i$.

Clearing the δ uses. To avoid any injury for N, before we put x in A, we must wait until $\delta(W; p(x))$ is undefined (thus for all $y > p(x)$, $\delta(W; y)$ is undefined). If we view the δ uses as movable markers on W, then we can restate the N-restraint on P as follows: Before P puts x in A, the use $\delta(W; p(x))$ on W must be cleared. (We will say that *N stops x entering* A, if there is some δ use to be cleared.) Hence we need to force W to change below $\delta(W; p(x))$. We will select a number (called a flip point) d not yet in D, and force W to change by putting d in D. The necessary condition on d is $\gamma(AW; d) < \delta(W; p(x))$. If we are able to select such a d, then when we put d in D, $A \oplus W$ must change below $\gamma(AW; d)$. If we restrain A up to $\gamma(AW; d)$, then W must change below $\gamma(AW; d)$, so that $\delta(W; p(x))$ is cleared.

The selection of witness x and the flip point d depends on P's environment. We will make it precise after we introduce some notations for the tree construction.

Description of priority tree T. Fix a priority list

$$N_0 < P_0 < N_1 < P_1 < \cdots$$

Let the priority tree T be a full binary tree in $\mathcal{M}$. Consider a node α on T. If $|\alpha| = 2e$, then we assign to α the requirement N_α, which is a version of N_e. If $|\alpha| = 2e + 1$, then we assign to α the requirement P_α, which is a version of P_e. We assume that the tree T grows upward.

The possible outcomes for P_α and N_α are 0 and 1. For a P strategy, the outcome 0 means that it is waiting for its witness x_α being realized and the outcome 1 means that it is satisfied because there is an x (not necessarily x_α) in A and $\Phi_e(x) = 0$. For an N strategy, the outcome 0 means that there are infinitely many α-expansionary stages (which includes the case where $\Gamma(AW_e) = K$ and sometimes the case where $\Gamma(AW_e)$ is partial) and the outcome 1 means that there are only $\mathcal{M}$-finitely α-expansionary stages.

Environment of α. Let α be a node on T. We use the word "environment" of α to describe a collection of informations about α, which we will use as a guide in the construction.

If $|\alpha| = 2e$ then α is responsible for building a Turing functional Δ_α. The Environment of α consists of a list of $\mathcal{M}$-finite sets:

$$E_y^\alpha = \{d_\alpha^\tau : \alpha \subseteq \tau, x_\tau < \gamma_e(AW_e; y)\}.$$

Intuitively, E_y^α is the set of flip points which may push $\delta_\alpha(W_e; y)$ forward. We will give more explanations after we specify the choice of δ.

If $|\alpha| = 2e + 1$, then α's environment is

$$E_\alpha = \{\beta : \ |\beta| \text{ is even and } \beta^\wedge \langle 0\rangle \subseteq \alpha\}.$$

Intuitively, E_α is the collection of β such that before α puts its witness in A, α must ask for permissions from β. For $|\alpha| = 2e+1$, α also has a set of parameters consisting of a witness x_α and for each β in E_α a flip point d^α_β such that for any stage s

$$\gamma_i(AW_i; d^\alpha_\beta) < \delta_\beta(W_i; (p(x))[s]$$

provided they are defined, where $i = |\beta|$.

Selection of x_α and d^α_β. Fix an α on T, $|\alpha| = 2e+1$, we now specify the selection of x_α and d^α_β. First let us deal with one negative requirement N below P_α. Assume that N is assigned to a $\beta \subseteq \alpha$ such that $|\beta| = i < e$. If $\beta \notin E_\alpha$, i.e., $\beta^\wedge \langle 1\rangle \subseteq \alpha$, then P_α does not need any flip point, it can just pick an x greater than the restraint set by nodes to the left or below α, which in particular includes β. If $\beta \in E_\alpha$, i.e., $\beta^\wedge \langle 0\rangle \subseteq \alpha$. Then α first selects a $d^\alpha_\beta > \max\{d : d \in E^\beta_y, \wedge y \leq s$, as its flip point, waits until $\gamma_i(AW_i; d^\alpha_\beta)$ is defined, then picks x_α greater than the necessary restraint and $\gamma_i(AW_i; d^\alpha_\beta)$. (There is a possibility that such a γ is never defined, then α would not be on the true path. Therefore we can ignore it.) By the selection of x_α and d^α_β, we have $d^\alpha_\beta < p_s(x)$, where s is the stage of selection. Thus

$$\gamma_i(AW_i; d^\alpha_\beta) < \gamma_i(AW_i; p_s(x)) < \delta_\beta(W_e; p_s(x)).$$

Selection of δ. There are two aspects on selection of δ. First, we need to ensure that for all $t > s$,

$$\gamma_i(AW_i; d^\alpha_\beta) < \delta_\beta(W_i; p_t(x)),$$

which could make $\delta_\beta(W_i; y) > \gamma_i(AW_i; z)$ for $y < z$. Secondly, we also need to ensure $\Delta(W_i)$ is total, provided $\Gamma(AW_i)$ is total. We now focus on the first aspect. Later we will see that the choice of δ also respect the second aspect.

Suppose that

$$\gamma_i(AW_i; d^\alpha_\beta) < \delta_\beta(W_i; p_s(x)),$$

we want to keep it at any stage $t > s$ until x_α is realized. We first preserve A up to $\gamma_i(AW_i; d^\alpha_\beta)$.

Suppose that there exist stages at which $p(x)$ moves. We consider only the first move, the rest are similar. Let t be first such stage and $y = p_t(x)$. As we have restrained A, there must be a number $z < \gamma_i(AW_i; y)$ entering W_e. Since $\gamma_i(AW_i; y) < \delta_\beta(W_i; y)$, $\delta(W_i; y)$ is also undefined. By the stage at which β defines $\Delta(W_i; y)$, we choose

$$\delta_\beta(W_i; y) > \gamma_i(AW_i; d^\alpha_\beta).$$

Thus we can always keep

$$\gamma_i(AW_i; d^\alpha_\beta) < \delta_\beta(W_i; p_t(x)),$$

until x_α is realized. When it happens, we first put d^α_β in D and keep the restraint on A. By the reasoning before, we can satisfy P without injuring N provided they are on true path.

Let us summarize the selection of $\delta_\beta(W_i; y)$ as follows:
Whenever $\gamma_i(AW_i; y) > x_\tau$, where τ is a node extending $\beta^\wedge \langle 0\rangle$, then we must choose

$$\delta_\beta(W_i; y) > \gamma_i(AW_i; d^\tau_\beta),$$

where d^τ_β is a flip point for x_τ.

Purpose of links. We now consider the coordination of strategies. Suppose that we have a positive requirement P and two negative requirements N_1, N_2, which are assigned to α, β_1, and β_2 respectively with $\beta_1{}^\wedge \langle 0\rangle \subseteq \beta_2{}^\wedge \langle 0\rangle \subseteq \alpha$. Now P wants to put a number x in A, and N_1 stops it, so P flips a point d_1 to force N_1 to change. The worry is that by the time N_1 allows x to enter A, N_2 could define more Δ_2 such that N_2 stops x entering A. Thus we need to delay N_2's action by setting a link from β_1 to α. As long as the link exists, N_2 is bypassed so that N_2 does not define more Δ_2. By the time N_1 allows x to enter A, we travel through the link to α directly, at this moment both N_1 and N_2 will not stop x entering A.

Construction

Stage 0: Set $A_0 = \emptyset$ and for all α on T, set $\Delta_\alpha = \emptyset$.

Stage $s+1$: We will specify an *accessible string* σ_s of length less than or equal to $s+1$ by induction, and do construction along σ_s. We break the stage into at most $s+1$ substages. Consider an $i < s+1$.

Substage i: Given an accessible string α of length i.

Suppose $i = 2e+1$. Then a version of P_e is assigned to α. If there is an x in A_s such that $\Phi_e(x) = 0$, i.e., P_e is satisfied, then let $\alpha^\wedge \langle 1\rangle$ be accessible. Go to substage $i+1$. Otherwise, i.e., P_e is not satisfied yet,

Case 1. α has not selected its parameters. Then for each β in $E(\alpha)$, α picks d^α_β such that $d^\alpha_\beta \notin D$, and not picked by other nodes and if $\beta' \subset \beta$ then $d^\alpha_{\beta'} < d^\alpha_\beta$. If for all $\beta \in E_\alpha$, $\gamma_{|\beta|}(d^\alpha_\beta)$ is defined and α has no witness x_α yet, then α also picks x_α greater than $\gamma_{|\beta|}(d^\alpha_\beta)$ for all $\beta \in E_\alpha$, and all restraint set by nodes to the left of below α. Let $\alpha^\wedge \langle 0\rangle$ be accessible. Go to substage $i+1$.

Case 2. α has a witness x_α and $\Phi_e(x_\alpha) \neq 0[s]$. Then let $\alpha^\wedge \langle 0\rangle$ be accessible, go to substage $i+1$.

Case 3. α has a witness x_α and $\Phi_e(x_\alpha) = 0[s]$. Then

(1) if there is no $\beta \in E(\alpha)$ which stops x_α entering A. Then put x_α in A; if there is a $\beta \in E(\alpha)$ which stops x_α entering A, then let β_0 be the longest such node. Put $d^\alpha_{\beta_0}$ in D, set a link from β_0 to α, denoted by (β_0, α).

(2) Initialize all nodes to the right of or extending α. A node τ gets *initialized* if all parameters are cancelled when $|\tau|$ is odd, and the functionals it is building is started over when $|\tau|$ is even; Also any links involving τ are cancelled. Go to stage $s+2$.

Suppose that $i = 2e$. Then a version of N_e is assigned to α. We say that s is an α-expansionary stage, if

$$l(e, s) > \max\{l(e, t) : t < s \wedge \alpha \text{ is accessible at stage } t\}.$$

If s is an α-expansionary stage, then

Case 1. There is no link starting from α. Then if $\gamma(AW_e; y)$ has been moved and is larger than some x_τ for some τ extending $\alpha^\wedge \langle 0 \rangle$, then add d_α^τ to the list E_y^α; define $\Delta(W_e; y) = \Gamma(AW_e; y)$ for all $y < l(\alpha, s)$ with use

$$\delta(W_e; y) = \max(\{\gamma(AW_e; d) : d \in E_y^\alpha\} \cup \{\gamma(AW_e; y)\}).$$

Set restraint on A up to $\max\{\gamma(AW_e; y) : y \leq l(\alpha, s)\}$. Let $\alpha^\wedge \langle 0 \rangle$ be accessible. Go to substage $i + 1$.

Case 2. There is a link from α to some $\tau \supseteq \alpha$. Then let τ be accessible. Cancel the link (α, τ). Go to substage $|\tau|$.

If s is not an α-expansionary stage, then let $\alpha^\wedge \langle 1 \rangle$ be accessible, go to substage $i + 1$.

Before go to stage $s + 2$, initialize all nodes to the left or extending σ_s.

End of construction

We now verify that the construction works.

Let the *true path* Λ be the leftmost path which is accessible unboundedly often. In fact, under $I\Sigma_2$, for any σ on the true path, σ is accessible $\mathcal{M}$-many times, i.e., for any $k \in \mathcal{M}$, there is a stage s_k such that σ is accessible for the k-th time at stage s_k. First we show that Λ exists. This is the place where we make crucial use of $I\Sigma_2$.

Lemma 4. *For any $e \in \mathcal{M}$, there is a unique α on T of length e such that*

(1) *for any s there is a stage $t > s$ such that $\alpha \subset \sigma_t$;*

(2) *for any β to the left of α there is a stage s such that for any $t > s$ $\beta \not\subset \sigma_t$.*

Proof. Fix e in $\mathcal{M}$, and consider the set of strings

$$\{\tau : |\tau| \leq e \wedge (\forall s)(\exists t > s)(\tau \subseteq \sigma_t)\},$$

which is a nonempty bounded Π_2 set. By $I\Pi_2$, which is equivalent to $I\Sigma_2$, it is $\mathcal{M}$-finite. Hence there exists an element, which is to the left or above every other element. Call it α. By definition, (1) is satisfied. To show (2), consider the set

$$X = \{\beta : |\beta| \leq e \wedge \beta <_L \alpha\}$$

which is $\mathcal{M}$-finite. By the definition of α, we have

$$(\forall \beta \in X)(\exists s)(\forall t > s)\beta \not\subseteq \sigma_t.$$

By $B\Sigma_2$ there is a stage t_0 such that for all $t > t_0$, $\beta \not\subseteq \sigma_t$.

It remains to show that $|\alpha| = e$. We prove by contradiction. Suppose that $|\alpha| = i < e$. It means that whenever α is accessible, there is a link which has endpoint α. By the claim after Lemma 5, if α is on true path, then there is a stage after which no link can have endpoint α. (Note: in the proof of Lemma 5, the assumptions of α are: it is accessible unboundedly often and there is a stage after which it is never initialized. Thus there is no cycle.) That establishes the lemma. □

Lemma 5. *If α is on the true path, then for any s, there is a $t > s$ such that at stage t, α is accessible.*

Proof. Without loss of generality, let us assume α never gets initialized after stage s. Let $t_0 > s$ be a stage at which $\alpha \subseteq \sigma_{t_0}$. If α is not covered by any link then we are done. Thus let us assume that there is a link (β, τ) such that

$$\beta^\wedge \langle 0 \rangle \subseteq \alpha \subset \tau.$$

Notice that (β, τ) can only transfer into another link (β', τ) for some $\beta' \subset \beta$. We only need to argue that this transferring process must end at some stage. Let X be the collection of all such β''s. To be more precise, X is set of strings $\beta' \subseteq \alpha$ such that there exists $t > t_0$, there is a link (β', τ) at stage t and for any s between t_0 and t, there is a link (η, τ) such that

$$\beta' \subseteq \eta \subset \alpha \subset \tau.$$

Claim. For all β' in X, there is a stage s', the link from β' to τ is cancelled.

Proof of Claim. Suppose that link is set up at stage s_0. As α is on the true path, there is a stage $s' > s_0$, at which $\alpha \subseteq \sigma_{s'}$. Then at stage s', the link must be traveled. Otherwise, there would be a link from some $\eta \subset \beta'$ to some $\tau' \supset \alpha$, which is impossible. This establishes the claim.

As X is an $\mathcal{M}$-finite set, by $B\Sigma_1$, there is a stage at which there is no link from any $\beta \subseteq \alpha$ to τ. Let s_1 be the least such stage. As there are neither overlapping nor nested links, at the stage the last link to τ is cancelled, there are no other links covering α. Let t be the least stage larger than s_1 at which $\alpha \subset \sigma_t$, then α is accessible at stage t. □

Notice that in the proof of Lemma 5, we actually proved the following claim, which is used in Lemma 4 and Lemma 6.

Claim. Suppose that α is on true path and $\tau \supseteq \alpha$, α never gets initialized after s and there is a link from some $\beta \subset \alpha$ to τ at stage s. Then there is a stage $t > s$ at which no link exists from $\eta \subset \alpha$ to τ and τ is accessible. Thus the positive requirement assigned on τ is satisfied.

Finally we argue that along the true path Λ, every requirement is satisfied.

Lemma 6. *For any $e \in \mathcal{M}$, the requirement P_e is satisfied.*

Proof. Let α be a node on the true path Λ such that $|\alpha| = 2e + 1$. Fix a stage s_0 such that after which the node α is never get initialized.

Since α is visited unbounded many often, its parameters x_α and d^α_β for all $\beta \in E(\alpha)$ are defined eventually.

Case 1. $\Phi_e(x_\alpha) \neq 0$. Then P_e is satisfied.

Case 2. Suppose that $\Phi_e(x_\alpha)\downarrow = 0$ at some stage s_1 at which α is accessible. Then if no $\beta \in E(\alpha)$ which stops x_α at stage s_1, then by construction, x_α is in A, so P_e is satisfied. Suppose there is some $\beta \in E(\alpha)$ which stops x_α entering A. Let β_0 be the longest such β. There is a link from β_0 to α. By the claim after Lemma 5, P_e is satisfied. □

Lemma 7. *For any $e \in \mathcal{M}$, the requirement N_e is satisfied.*

Proof. Let α be a node on the true path Λ such that $|\alpha| = 2e$. Fix a stage s_0 such that after s_0 the requirement N_α is never get initialized.

If $\alpha^\wedge \langle 1 \rangle$ is on the true path, then $\Gamma(AW_e) \neq K$, thus N_e is satisfied.

If $\alpha^\wedge \langle 0 \rangle$ is on the true path, then we show that $\Delta_\alpha = K$ provided $\Gamma_e(AW_e)$ is total and equal to K.

Fix a y in $\mathcal{M}$. Let s_1 be the stage such that for all $t > s_1$, $\gamma_e(AW_e; y)[s_1] = \gamma_e(AW_e; y)[t]$. Then the set

$$E^\alpha_y = \{d^\tau_\alpha : \alpha \subseteq \tau \wedge x_\tau < \gamma_e(AW_e; y)[s_1]\},$$

is an $\mathcal{M}$-finite set since $\gamma_e(AW_e; y)[s_1]$ is fixed. By $B\Sigma_2$, there is a stage s_2, such that for all $t > s_2$, $\gamma_e(AW_e, d)[s_2] = \gamma_e(AW_e; d)[t]$ for all $d \in E^\alpha_y$. By the choice of δ, after s_2, $\delta_\alpha(W_e; y)$ never moves. Let s_3 be the stage such that $W_e \restriction \delta_\alpha(W_e; y) = W_{e,s_3} \restriction \delta_\alpha(W_e; y)$, and $\Delta_\alpha(W_e; y)$ is defined. We argue that for all $t > s_3$,

$$\Delta_\alpha(W_e; y) = \Gamma_\alpha(AW_e; y) = K(y)[t].$$

It suffices to show that there is no $x_\tau \notin A$, $x_\tau < \gamma_\alpha(AW_e; y)$ and x_τ enters A after stage s_3.

Case 1. τ is strictly to the right of α or $\tau \supseteq \alpha^\wedge \langle 1 \rangle$, then by initialization or by the setting of restraint, x_τ are either cancelled forever or is larger than $\gamma_\alpha(AW_e; y)$.

Case 2. τ extends $\alpha^\wedge \langle 0 \rangle$. Then if x_τ enters A, it must not be stopped by α, i.e., $x_\tau \geq \gamma_\alpha(AW_e; y)$.

In any case, we have

$$\Delta_\alpha(W_e; y) = \Gamma_\alpha(AW_e; y) = K(y),$$

which establishes the lemma and the theorem. □

References

[1] C. T. Chong and K. J. Mourad, Σ_n Definable sets without Σ_n induction, Trans. Amer. Math. Soc. 334 (1992) 349–363.

[2] C. T. Chong, L. Qian, T. Slaman and Y. Yang, Σ_2 induction and infinite injury priority arguments, part III: prompt sets, minimal pairs and the meet operator, to appear.

[3] C. T. Chong and Y. Yang, Σ_2 induction and infinite injury priority arguments, Part I: Maximal sets and the jump operator, J. Symbolic Logic 63 (1998), 797–814.

[4] C. T. Chong and Y. Yang, Σ_2 induction and infinite injury priority arguments, Part II: Tame Σ_2 coding and the jump operator, Ann. Pure Appl. Logic 87 (1997) 103–116.

[5] C. T. Chong and Y. Yang, Recursion theory in weak fragments of Peano arithmetic; A study of cuts, in: Proc. Sixth Asian Logic Conference, Beijing 1996, World Scientific, to appear; `http://www.wspc.com/books/mathematics/3755.htm`.

[6] M. J. Groszek and M. E. Mytilinaios, Σ_2-induction and the construction of a high degree, in: Recursion Theory Week, Proc. Oberwolfach 1989 (eds. K. Ambos-Spies et al.), Lecture Notes in Math. 1432, Springer-Verlag, Berlin, Heidelberg, New York, Tokyo, 1990, 205–223.

[7] M. J. Groszek and T. A. Slaman, Foundations of the priority method I: Finite and infinite injury, preprint.

[8] M. J. Groszek and T. A. Slaman, On Turing reducibility, preprint.

[9] S. Lempp and M. Lerman, A general framework for priority arguments, Bull. Symbolic Logic 1 (1995), 189–201.

[10] J. B. Paris and L. A. S. Kirby, Σ_n-collection schemas in arithmetic, in: Logic Colloquium '77, North Holland, Amsterdam, 1978, 199–209.

[11] D. P. Miller, High recursively enumerable degrees and the anti-cupping property, in: Logic Year 1979–80: University of Connecticut, (eds. M. Lerman, J. H. Schmerl and R. I. Soare), Lecture Notes in Math. 859, Springer-Verlag, Berlin, Heidelberg, New York, Tokyo, 1981, 230–245.

[12] K. J. Mourad, Ph.D thesis, University of Chicago, 1988.

[13] M. E. Mytilinaios, Finite injury and Σ_1-induction, J. Symbolic Logic 54 (1989), 38–49.

[14] M. E. Mytilinaios and T. A. Slaman, Σ_2-collection and the infinite injury priority method, J. Symbolic Logic, 53 (1988), 212–221.

[15] M. J. Groszek, M. E. Mytilinaios and T. A. Slaman, The Sacks density theorem and Σ_2-bounding, J. Symbolic Logic 61 (1996), 450–467.

[16] R. I. Soare, Recursively Enumerable Sets and Degrees, Perspect. in Math. Logic, Omega Series, Springer-Verlag, Berlin, Heidelberg, New York, Tokyo, 1987.

[17] Y. Yang, Iterated trees and fragments of arithmetic, Arch. Math. Logic 30 (1995), 97–112.

Open problems

Part I. d.r.e. and n-r.e. degrees

1. (Hinman) Let $\rho(m, n)$ be the least p such that for some properly m-r.e. set C, every set n-r.e. in and above C is of p-r.e. degree. We know that $\rho(1, 1) = 2$, $\rho(2, 1) \geq 3$, and $\rho(1, 2) \geq 4$. Are these optimal estimates? Similarly for larger m and n.

   (Arslanov) If, in the definition of $\rho(m, n)$, we require in addition that the sets which are n-r.e. in and above C must be below $\mathbf{0}'$, then by Arslanov, Lempp, Shore we obtain $\rho(1, 1) = 1$. It would be interesting to ask whether the estimates $\rho(2, 1) \geq 3$ and $\rho(1, 2) \geq 4$ remain true with this additional condition.

2. Downey's Conjecture: The partial orders of the n-r.e. degrees for any $n \geq 2$ are elementarily equivalent.

   a. Are they even isomorphic?

   b. True for $n = 2, 3$?

   c. Partial question (Lempp): Are $(D_2, R, <)$ and $(D_3, R, <)$ elementarily equivalent, where D_n denotes the n-r.e. degrees and R a unary predicate for the r.e. degrees?

3. Are all lattices embeddable into the d.r.e. degrees? Even preserving 0 and 1? What lattices are strongly embeddable (à la superminimal pair)?

4. Characterize the finite final segments of the d.r.e. degrees. Do they include all finite distributive lattices? (This would constitute another proof of the undecidability of the first-order theory.)

5. Characterize the possible extensions of poset embeddings into the d.r.e. degrees. (This may be closely connected to the final segments question.)

6. Are the r.e. degrees definable in the d.r.e. degrees? Does bi- interpretability hold?

7 (Arslanov) Let REA(x, y) be the predicate "x is r.e. in and above y" defined on the Turing degrees. Is the predicate REA(x, y) definable in each (or some) D_n, $n \geq 2$? (If REA(x, y) is definable in D_n and D_m for some distinct $n, m \geq 2$, then the structures D_n and D_m are not elementarily equivalent.)

Part II. e-degrees

1. (Cooper) Are the total e-degrees definable in the e-degrees?

2. (Rogers) Do the total e-degrees form an invariant class in the e-degrees?

3. (Cooper) Is the jump definable in the e-degrees?

4. (Cooper) Are there rigid upper cones (i.e., upper cones in which any automorphism must be the identity) in the e-degrees?

5. (Cooper) Determine the complexity of the first-order theory of the Σ_2^0 e-degrees.

6. (Knight) Let T be a theory and $\mathcal{S}$ a Scott set appropriate for T (i.e., $T \cap \Sigma_n \in \mathcal{S}$ for all n). Are the following equivalent for all n and X: (1) $X \leq_T \mathcal{A}^{(n)}$ for all $\mathcal{A} \models T$ representing $\mathcal{S}$, and (2) $\chi_X \leq_e \Gamma$ for some Σ_n-type Γ realized in these models?

7. (Nies) Consider the interrelation between set properties and and e-degrees, such as in the known result: $X \in \Sigma_2^0$ $\emptyset'$-hypersimple implies the e-degree of X is not cuppable.

8. (Cooper) If $\mathbf{a}$ is a Σ_2^0 e-degree, does $\mathbf{a}' = \mathbf{0}_e''$ imply that $\mathbf{a}$ contains a set A which is Σ_2-high (in the sense of Cooper [1990])?

   Solution (R. Shore and A. Sorbi). For every Σ_2^0 e-degree $\mathbf{a}$, $\mathbf{a}' = \mathbf{0}_e''$ if and only if $\mathbf{a}$ contains a Σ_2-high set. See the paper *Jumps of Σ_2^0-high e-degrees and properly Σ_2^0 e-degrees*, this volume.

9. (Cooper) Characterize the lattices embeddable into the Σ_2^0 e-degrees.

   Remark (S. Lempp and A. Sorbi). S. Lempp and A. Sorbi have shown that a wide variety of finite lattices, including M_3, N_5 and S_8, are embeddable into the Σ_2^0 e-degrees preserving 0 and 1, and conjecture that all finite lattices are embeddable preserving 0 and 1.

10. (Cooper) Characterize the minimal upper bounds in the e-degrees and the Σ_2^0 e-degrees.

11. (Cooper) Characterize the order types of initial segments of the Σ_2^0 e-degrees.

    Remark (S. Lempp and A. Sorbi). The following characterization, due to S. Lempp and A. Sorbi, of the one-point extensions of embeddings of finite bounded partial orders into the Σ_2^0 enumeration degrees may be useful for this problem: if P, Q are finite bounded partial orders, $P \subset Q$, and $|Q-P| = 1$, say, $x \in Q-P$, then any embedding of P into the Σ_2^0-enumeration degrees can be extended to an embedding of Q into the Σ_2^0-enumeration degrees if and only if none of the following three conditions holds: (1) (in $U(P)$) $\exists i \in P - B(x)\,(i \leq \bigvee B(x))$; (2) (in $L(P)$) $\exists i \in P - A(x)\,(i \geq \bigwedge A(x))$; (3) $\exists i_0 \in I(x)\,(B(x) < i_0$ and

$i_0 \not\leq A(x)$); where $U(P)$ is a finite upper semilattice such that $U(P)$ extends P as a bounded partial order, and every partial order embedding $i : P \to U$, U any finite upper semilattice, can be extended to a partial order embedding $j : U(P) \to U$; $L(P)$ is the dual of $U(P)$; and $A(x)$, $B(x)$, $I(x)$ is a partition of P, with $A(x) = \{a \in P \mid a > x\}$, $B(x) = \{b \in P \mid b < x\}$.

12. (Cooper) Characterize the jumps of 1-generic e-degrees below $\mathbf{0}'_e$.
13. (Cooper) Are the e-degrees of the 1-generic sets below $\mathbf{0}'_e$ closed downward?
14. (Cooper) Are the e-degrees of the cohesive/hyperhyperimmune sets closed upward?
15. (Cooper) Are all e-degrees below $\mathbf{0}'_e$ introreducible?

    Solution (Jockusch). The solution uses Selman's theorem that B is e-reducible to A iff

    $$(\forall X)[A \text{ r.e. in } X \text{ implies } B \text{ r.e. in } X].$$

    We first observe that if A is introreducible and B is an infinite subset of A then $c_A \leq_e B$ (where c_A is the characteristic function of A). This is seen as follows: To apply Selman's theorem, assume that B is r.e. in X. Let C be an infinite subset of B which is recursive in X. Then $A \leq_T C \leq_T X$ (by introreducibility), so A and its complement are each r.e. in X. By Selman's theorem, A and its complement are each e-reducible to B, so $c_A \leq_e B$. Now setting B equal to A in the above observation, one gets that c_A is e-equivalent to A for every (infinite) introreducible set A, and hence that every introreducible e-degree is total. It thus follows that there are non-introreducible e-degrees below $0'$. Since every total function is e-equivalent to the set of its initial segments, and the latter set is introreducible, it follows that in fact the property of containing an introreducible set characterizes the total e-degrees.

16. (Knight) Determine the structure of the total e-degrees in the e-degrees. (It is known that the total e-degrees generate the e-degrees under infima.)
17. (Cooper) Determine the structure of subclasses (such as n-r.e. e-degrees, etc.) of the Σ^0_2 e-degrees.

Part III. Recursive model theory

1. (Knight) Characterize the abelian p-groups with computable copies.
2. (Knight) For every abelian p-group G, must there be a computable tree T such that $G = G(T)$?

3. (Knight) Does every low$_n$ Boolean algebra have a computable copy? (Known to be true for $n \leq 2$.)

4. Are the following equivalent for every recursive limit ordinal λ and every Boolean algebra $\mathcal{A}$: (1) $X \leq_T \mathcal{B}^{(\lambda+n)}$ for all $\mathcal{B} \cong \mathcal{A}$, and (2) $X \leq_T \mathcal{B}^{\lambda} \oplus \emptyset^{(\lambda+n)}$ for all $\mathcal{B} \cong \mathcal{A}$?

5. (Knight) Is there a nonstandard model $\mathcal{A}$ of Peano arithmetic for which $\{\mathcal{B} \mid \mathcal{B} \cong \mathcal{A}\}$ has a minimal element? (It is known that there can be no least element.)

6. (Knight) Is there a Scott set $\mathcal{S}$ such that $\{\deg(R) \mid R \text{ is an enumeration of } \mathcal{S}\}$ has a minimal element?

7. (Knight) Is there an enumeration R of the arithmetical sets such that $\deg R$ is minimal over $\{\emptyset^{(n)} \mid n \in \omega\}$?

8. (Knight) Is there a completion T of Peano arithmetic such that the degree of T is minimal over $\{\deg T_n \mid n \in \omega\}$ where T_n is the Σ_n-fragment of T?

9. (Goncharov) Characterize all theories T such that any recursive model of T is decidable. Same for n-decidable. (This is easy if T is model complete.)

10. (Goncharov) Characterize all theories T such that any decidable model of T has a decidable homogeneous/saturated extension.

11. (Goncharov) Let T be an arithmetical Ehrenfeucht theory with only arithmetical types. Must every model of T be arithmetical?

12. (Goncharov) Characterize all models $\mathcal{M}$ with $\dim_A \mathcal{M} = 1$. With effectively infinitely many constructivizations (up to equivalence). With $\dim_p \mathcal{M} \neq \dim_U \mathcal{M}$.

13. (Goncharov) Is there a model $\mathcal{M}$ with $\dim_A \mathcal{M} = 1$ but $\dim_A(\mathcal{M}, \overline{a}) = \infty$ for some tuple $\overline{a}$ in M?

14. (Goncharov) Let $\mathcal{M}$ be an autostable model (i.e., $\mathcal{M}$ has only one recursive presentation up to recursive isomorphism) of an uncountably categorical theory T. Must every elementary substructure of $\mathcal{M}$ be autostable?

15. (Goncharov) Let T be an uncountably categorical, arithmetical theory for which each Σ_{n+1}-fragment is uniformly Σ_n. Must T have a recursive model?

16. (Goncharov) Describe all recursive homogeneous Boolean algebras in terms of the Morozov invariant. (Known for decidable homogeneous Boolean algebras.)

17. (Goncharov) Describe all algebraic subsets $P \subseteq M^n$ such that $\dim_A(\mathcal{M}, P) = 1$.

18. (Goncharov) Describe all models $\mathcal{M}$ such that there is some algebraic subset $P \subseteq M^n$ with $\dim_A(\mathcal{M}, P) = 1$.

19. (Goncharov) Let $\mathcal{K}$ be the class of all groups of recursive automorphisms of recursive models. Is $\mathcal{K}$ closed under free group product?

20. (Lerman) Characterize all recursive linear orders which admit a nontrivial recursive self-embedding in all recursive presentations. (This is conjectured to be the set of recursive linear orders with densely ordered recursive subset.)

Part IV. Global questions about degree structures and the lattice of computably enumerable sets

1. Are the following fragments of first-order theories decidable: Π_2-theory of the r.e. Turing degrees (in particular the question of which finite lattices are embeddable into the r.e. Turing degrees) (Slaman) , Π_3-theories of the r.e. tt-degrees and of the r.e. btt-degrees (Nies) Π_3-theory of the lattice of r.e. sets (Soare) ?
2. (Nies) Are the theories of the following degree structures m-equivalent to true arithmetic: r.e. 1-degrees, r.e. btt-degrees, r.e. Q-degrees, Σ_2^0-e-degrees?
3. (Odifreddi) Are the following degree structures non-elementarily equivalent or at least non-isomorphic: r.e. tt-degrees, tt-degrees below $\mathbf{0}'$, and wtt-degrees below $\mathbf{0}'$? Same for all complexity-theoretic structures, e.g., for the recursive PTIME-m-degrees and recursive PTIME-exp-degrees.
4. (Nies) Are the r.e. Turing degrees and the r.e. m-degrees prime models of their theories? (For the former, try to show that the r.e. Turing with finitely many parameters are bi-interpretable with $(\mathbf{N}, +, \cdot)$.
5. (Soare) Are there definable degrees $\neq \mathbf{0}, \mathbf{0}'$ in the r.e. Turing degrees or the Δ_2^0 Turing degrees? (Partial question due to Li Angsheng: Are there r.e. degrees $\mathbf{a} < \mathbf{b} < \mathbf{c}$ and a first-order formula $\varphi(x)$ without parameters such that $\mathbf{b}$ is the unique solution to φ in the interval $(\mathbf{a}, \mathbf{c})$.
6. (Nies) Are the low, or the high, degrees definable in the Δ_2^0 Turing degrees?
7. (Jockusch) Is the Π_2 theory of the Turing degrees decidable in the language of upper semilattices with 0 and Turing jump?
8. (Soare) Which upward closed jump classes (i.e., classes of the form High_n or $\overline{\mathrm{Low}_n}$) are invariant in the lattice of c.e. degrees (i.e., are the image of an invariant class of c.e. sets under the degree mapping)?
9. (Soare) Given any noncomputable incomplete c.e. sets A and C, must there be an automorphism taking A to a set $\Phi(A) \not\geq_T C$?

List of talks

One hour plenary talks

S. S. Goncharov, *Computable enumerations*

A. Nies, *Coding in* $\mathbf{R}_m$ *and* $\mathcal{E}$

A. A. Razborov, *Lower bounds on the complexity for algebraic proof systems*

S. Safra, *Probabilistically checkable proofs and applications to the hardness of approximation problems*

R. A. Shore, *Effectiveness, complexity and reverse mathematics*

Recursion theory section talks

M. M. Arslanov, *Upper density of* Δ^0_2 *e-degrees*

I. V. Ashaev, *A splitting theorem in generalized computability*

S. L. Bereznyuk, *Computable numberings of unusual objects*

D. Cenzer, *Complexity theory and model theory*

P. Cholak, *Automorphisms of computably enumerable sets*

R. Coles, *Initial segments of computable linear orders*

A. A. Efremov, *Upper isolated d-r.e. degrees*

M. Groszek, *Effective aspects of Prikry's problem*

V. Harizanov, *Uncountably many isomorphic copies of a computable relation*

P. Hinman, *Jump traces with large gaps*

Sh. T. Ishmuhametov, *On a new construction of minimal Turing degrees*

C. Jockusch, *Generalized cohesiveness*

N. G. Khisamiev, *Constructive abelian groups*

J. Knight, *True approximations and models of arithmetic*

K. Korovin, *Compositions of permutations and algorithmic reducibilities*

M. Korovina, *New approach to computability on reals*

O. V. Kudinov, *Some remarks on strongly hypersimple sets*

G. LaForte, *Completing pseudojump operators*

S. Lempp, *Constructive models of uncountably categorical theories*

M. Lerman, *Proving theorems by using a framework for priority arguments*

E. V. Lipacheva, *Partial recursive automorphisms*

M. G. Peretjat'kin, *Finitely axiomatizable theories*

S. Yu. Podzorov, *Arithmetical hierarchy and recursive Boolean algebras*

M. Pour-El, *Computability in analysis: survey & open problems*

G. Sacks, *Remarks on Vaught's conjecture*

V. L. Selivanov, *On Recursively enumerable structure*

A. Shlapentokh, *Defining integrality at Infinitely many primes*

R. I. Soare, *Automorphisms and poems*

B. Ja. Solon, *The e-reducibility of sets and the problem of the property of being nontotal of e-degrees*

Y. Sui, *On the Jockusch conjecture*

Y. Yang, Σ_2*-induction and minimal pairs*

List of contributors

I. V. Ashaev
ashaev@univer.omsk.su

Omsk State University
Mira pr., 55-a
Omsk, 644077, Russia

D. Cenzer
cenzer@math.ufl.edu

Department of Mathematics
358 Little Hall
University of Florida
Gainesville, FL 32611, U.S.A.

R. G. Downey
rod.downey@mcs.vuw.ac.nz

School of Mathematical and
Computing Sciences
Box 600
Victoria University of Wellington
Wellington, New Zealand

V. Harizanov
val@math.gwu.edu

Department of Mathematics
2201 G Street
The George Washington University
Washington D.C. 20052, U.S.A.

P. Hinman
pgh@math.lsa.umich.edu

Department of Mathematics
3834 East Hall
University of Michigan
Ann Arbor, MI 48109-1109, U.S.A.

Sh. T. Ishmuhametov
ishm@mmf.univ.simbirsk.su

Ul'janovsk State University
Tolsltoj str., 42
Ul'janovsk, 432700, Russia

K. V. Korovin
KOSTYA@ssc.nsu.ru

Institute of Mathematics
Russian Academy of Science
Siberian Branch
University Prosp., 4
Novosibirsk, 630090, Russia

M. V. Korovina
RITA@ssc.nsu.ru
Institute of Mathematics
Russian Academy of Science
Siberian Branch
University Prosp., 4
Novosibirsk, 630090, Russia

O. V. Kudinov
kud@math.nsc.ru
Institute of Mathematics
Russian Academy of Science
Siberian Branch
University Prosp., 4
Novosibirsk, 630090, Russia

S. Lempp
lempp@math.wisc.edu
Department of Mathematics
University of Wisconsin
480 Lincoln Drive
Madison, WI 53706-1388, U.S.A.

A. S. Morozov
morozov@math.nsc.ru
Institute of Mathematics
Russian Academy of Science
Siberian Branch
University Prosp., 4
Novosibirsk, 630090, Russia

J. B. Remmel
jremmel@ucsd.edu
Department of Mathematics
University of California at San Diego
La Jolla, CA 92093, U.S.A.

A. Shlapentokh
shlapentokh@math.ecu.edu
Department of Mathematics,
East Caroline University,
Greenville, NC 27858

R. A. Shore
shore@math.cornell.edu
Department of Mathematics
White Hall
Cornell University
Ithaca, NY 14853, U.S.A.

B. Ja. Solon
solon@icti.ivanovo.su
Novosel'skaja str., 8 B, apt. 9
Ivanovo, 153038, Russia

V. Solovyev
Valery.Solovyev@tatincom.ru
Department of Computer Sciences and Cybernetics
Kazan State University
Kremljevskaja, 18
Kazan, 420008, Russia

A. Sorbi
sorbi@unisi.it
Dipartimento Di Matematica,
via del Capitano 15,
53100 Siena, Italy

Y. Yang
matyangy@math.nus.edu.sg
Department of Mathematics
National University of Singapore
10 Kent Ridge Crescent
Singapore 119260

Bei Fragen zur Produktsicherheit wenden Sie sich bitte an:
If you have any questions regarding product safety,
please contact:

Walter de Gruyter GmbH
Genthiner Straße 13
10785 Berlin
productsafety@degruyterbrill.com